Descent from Glory

John Adams in 1823. This copy of the portrait by Gilbert Stuart was painted by Jane Stuart, the artist's daughter, and hangs in the Old House. The couch where John sat while Stuart worked still can be seen there. John's constant talking with the artist did not prevent Stuart from capturing the old President's character. *Courtesy of the National Park Service, Adams National Historic Site.*

Descent from Glory

❧ ❧

Four Generations of the John Adams Family

Paul C. Nagel

OXFORD UNIVERSITY PRESS
Oxford   New York   Toronto   Melbourne

OXFORD UNIVERSITY PRESS
Oxford   London   Glasgow
New York   Toronto   Melbourne   Auckland
Delhi   Bombay   Calcutta   Madras   Karachi
Kuala Lumpur   Singapore   Hong Kong   Tokyo
Nairobi   Dar es Salaam   Cape Town

and associate companies in
Beirut   Berlin   Ibadan   Mexico City   Nicosia

Library of Congress Cataloging in Publication Data

Nagel, Paul C.
Descent from glory.

Bibliography: p.
Includes index.
1. Adams family.   I. Title.
CS71.A2 1983   929'.2'0973   82-6505
ISBN 0-19-503172-5   AACR2
ISBN 0-19-503445-7 (pbk.)

Printing (last digit):   9 8 7 6 5 4 3 2

Printed in the United States of America

for
Wilhelmina Sellers Harris

Acknowledgments

In this book I have described how various Adamses—both notable and forgotten—lived with each other as part of the astonishing family created when John Adams and Abigail Smith were married in 1764. That union began a story which was not completed until 1927 with the death of Brooks Adams, who left behind a collection of family letters and diaries extending back to the time of his famous great-grandparents. These sources, many never studied before, open a door to the private world of a family whose achievements in politics, diplomacy, and literature are unmatched in American history. Since every page of what I have written springs from this enormous body of manuscripts, I have not included footnotes, for their length and complexity would overpower the story. However, my annotated early drafts of this book will be available in the library of the Virginia Historical Society in Richmond.

Help from many quarters made this book possible. Not only did my wife, Joan Peterson Nagel, cheerfully endure the surrender of weekends and vacations, but she became nearly as familiar with the Adamses as I did. She prepared the genealogical table included here. My first thanks must go to her as partner and dearest friend—to use the terms relished by John and Abigail Adams. Members of the Adams family have been kind and encouraging. Particularly am I grateful to John Adams Abbott, M.D., his wife Diana Abbott, and to Thomas Boylston Adams and Dorothy Quincy Beckwith Nelson.

When I began my research in 1974, the Massachusetts Historical Society's staff helped me learn about sources beyond the 608 reels of

Adams Papers, which include no documents dated after 1889. Consequently, to examine the family's later correspondence and to use other materials concerning the Adamses, I spent many pleasant months in the Society's reading room on Boylston Street in Boston. If I have overlooked material there, it is my fault entirely. The Massachusetts Historical Society has permitted me to quote from the microfilm edition of the Adams Papers, from the Adams Papers—Fourth Generation, and from other important documents. I am most grateful to the Society for this and for the many personal courtesies shown to me by such generous individuals as Stephen T. Riley, who was director when I first appeared in Boston, and his successor, Louis L. Tucker; John D. Cushing, Librarian; Winifred V. Collins, who made suggestions and listened so charmingly; and Aimée F. Bligh, Patrick R. Flynn, Robert V. Sparks, and Ross F. Urquhart.

No less sizable is my debt to the editors of the Adams Papers, whose office is upstairs at the Massachusetts Historical Society. I regret that Lyman H. Butterfield, who was Editor-in-Chief when I set out to learn about the Adamses, did not live to read this book. Like that of everyone who recently has studied John Adams and his descendants, my work was strengthened by Mr. Butterfield's wisdom and friendship. His successor, Robert J. Taylor, carries on this tradition admirably. I am thankful to him in countless ways, as I am to his indefatigable colleague Marc Friedlaender, now Adjunct Editor after many years of working with all four generations of the family. More than once, Mr. Friedlaender in his cheerful and discerning way rescued me from discouragement and bewilderment. This book would not have been completed to my satisfaction except for the unfailing help of Celeste Walker, Assistant Editor of the Adams Papers, who pushed many obstacles out of my path. She has my gratitude and admiration.

There are many other debts I have to scholars and libraries. Important among these must be my thanks to the editors of the Henry Adams Papers at the University of Virginia, J. C. Levenson, Charles Vandersee, and Viola H. Winner. At the Boston Public Library the director, Philip J. McNiff, made matters easy for me. In Cambridge, I studied materials particularly in Harvard's Houghton Library, where William H. Bond and his colleagues saw that I lacked for nothing.

South from Boston and Cambridge is Quincy, the most important scene in the Adams story. There, the National Park Service staff at the Adams National Historic Site has been unfailingly considerate. I am especially grateful to Pat Sheehan and Marianne Peak. I shall say more about Mrs. Wilhelmina Harris, the Superintendent at the site, in a moment. Many citizens of Quincy were cordial and helpful, especially Dorothy and Walter Wrigley, Owen Della Lucca, and H. Hobart Holly. I thank them and other friends in the community.

Three institutions have supported my project. The University of Missouri granted me a sabbatical leave in 1974–75. Money for travel and a microfilm reader came from the Research Council of the Graduate School, University of Missouri–Columbia. At the University of Georgia I owe a very special obligation to William J. Payne, Dean of Franklin College of Arts and Sciences, and to Robert C. Anderson, Vice President for Research. Their encouragement and kindnesses carried me over some difficult moments. Most recently, the Trustees of the Virginia Historical Society have disclosed a breadth of view astonishing even for Virginians by their enthusiasm for this study of Massachusetts people.

Reading my scrawl and typing this manuscript were burdens carried gracefully by two associates. Linda Green of the University of Georgia worked tirelessly not only in typing and retyping, but in helping me clarify my thoughts. Here at the Virginia Historical Society, Carol Wicker has joined me in the ordeal of revision with great skill and kindness. My good friend over many years, J. Rodney Kellar of Minneapolis, read the manuscript in an early state, greatly to my advantage. The book owes much to advice from Sheldon Meyer and Stephanie Golden of Oxford University Press and from Sallie F. Reynolds. Their gentle urgings have made a profound difference in what was finally written.

The last obligation I must mention is my largest. It is to Wilhelmina Sellers Harris, the Superintendent of the Adams National Historic Site, and once secretary to Mr. and Mrs. Brooks Adams. Without Mrs. Harris' knowledge, support, and hospitality, this book would be far less than it is. The Adams family has no greater friend than she, nor have I. To her I gratefully dedicate this book.

Richmond, Virginia   P. C. N.
April 1982

Contents

Illustrations

Descent from Glory

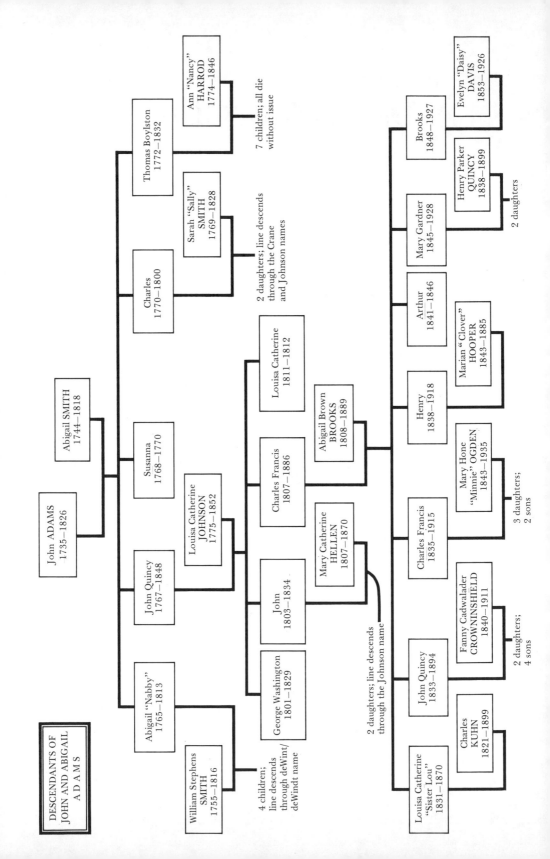

DESCENDANTS OF JOHN AND ABIGAIL ADAMS

John ADAMS 1735–1826
Abigail SMITH 1744–1818

Abigail "Nabby" 1765–1813
William Stephens SMITH 1755–1816
4 children; line descends through deWint/deWindt name

John Quincy 1767–1848
Louisa Catherine JOHNSON 1775–1852

Susanna 1768–1770

Charles 1770–1800
Sarah "Sally" SMITH 1769–1828
2 daughters; line descends through the Crane and Johnson names

Thomas Boylston 1772–1832
Ann "Nancy" HARROD 1774–1846
7 children; all die without issue

George Washington 1801–1829

John 1803–1834
Mary Catherine HELLEN 1807–1870
2 daughters; line descends through the Johnson name

Charles Francis 1807–1886
Abigail Brown BROOKS 1808–1889

Louisa Catherine 1811–1812

Louisa Catherine "Sister Lou" 1831–1870
Charles KUHN 1821–1899

John Quincy 1833–1894
Fanny Cadwalader CROWNINSHIELD 1840–1911
2 daughters; 4 sons

Charles Francis 1835–1915
Mary Hone "Minnie" OGDEN 1843–1935
3 daughters; 2 sons

Henry 1838–1918
Marian "Clover" HOOPER 1843–1885

Arthur 1841–1846

Mary Gardner 1845–1928
Henry Parker QUINCY 1838–1899
2 daughters

Brooks 1848–1927
Evelyn "Daisy" DAVIS 1853–1926

"The history of my family is not a pleasant one..."

Those who enter the private world of the Adams family must realize that no one was more fascinated by the story of John Adams and his descendants than they themselves. And so they produced an unrivaled body of introspection in the form of letters, diaries, autobiography, and biography, a storehouse of one family's thinking about itself. We of a later age can thereby appreciate the life and soul of this family which has been unmatched in the United States for public service and intellectual achievement. It is a moving, sobering saga—this personal story of John and Abigail Adams and those who came after them. No one ever described it better than their grandson, Charles Francis Adams, who said: "The history of my family is not a pleasant one to remember. It is one of great triumphs in the world but of deep groans within, one of extraordinary brilliancy and deep corroding mortification."

Charles Francis' statement is meaningless to observers who do not know intimate details of the Adams circle and who measure the family from America's perspective on worldly success. To such a superficial view, the only complaint John Adams' family might have had was at the defeat two Adams presidents met in seeking a second term. Except for that, the public achievement of four generations was one which no other family in the United States approached. As Charles Francis wrote, it is outwardly a story "of extraordinary brilliancy." John Adams, for instance, helped to shape America's independence through his service in the Continental Congress. He became one of America's first diplomats, during the period 1778–88, he served as the new nation's first minister to the Netherlands and then to Great Britain. He helped write the peace treaty with England which confirmed the success of the American Revolution.

Thereafter, John's public life was in the rising federal democracy of which he was the first vice president, serving two terms from 1789–97. He then had one term as the nation's second president between 1797 and 1801. His courage, forthrightness, and abrasiveness created dissent in his Federalist faction and narrowly denied him reelection, so that he retired at age sixty-six to live another quarter century on the farm he loved in Quincy, Massachusetts.

Despite his defeat in the election of 1800, John's career seems astounding. There were new satisfactions when his son John Quincy Adams carried forward the family's reputation in statesmanship. This second Adams was especially distinguished in diplomacy, being in turn America's minister to the Netherlands, Prussia, Russia, and Great Britain. He was also the architect of peace terms which ended the War of 1812. He served as Secretary of State from 1817 to 1825 in President James Monroe's cabinet. John Quincy's accomplishments in the State Department were as extensive as his term as president between 1825 and 1829 was often frustrating. Like his father, the second Adams president was fearless in standing for causes which the American democracy hesitated to endorse. President John Adams' determination to keep the Republic out of war had been as costly to his political success as was President John Quincy Adams' effort to strengthen America through such means as a national university and an interstate highway system.

Shortly after his defeat for reelection in 1828, John Quincy resumed public life as a representative in Congress where he fearlessly battled slavery until his death in 1848. During this time, a third generation of Adamses entered public service as John Quincy's son Charles Francis gave himself to historical writing and to opposing slavery. Eventually, he became congressman from his father's old district and then, from 1861 to 1868, he was the third Adams to serve America as minister to Great Britain. In this role, Charles Francis' achievement was memorable, for by his success in keeping England from siding openly with the South in the Civil War, he earned a reputation as the American statesman whose contribution to preserving the Union was second only to Abraham Lincoln's.

After returning to the United States, Charles Francis Adams retired, resisting public demand that he run for president. He appeared for one more international triumph when he represented his country in the Geneva Arbitrations during 1871–72. By this time, Charles Francis was more interested in watching his children carry forward in their turn the amazing record of John Adams' family. Each of his sons became nationally prominent, although in ways quite different from their forebears—and from each other. Of the four brothers, the weakest was the second John Quincy Adams who tried for a time to be a maverick politician in Massachusetts, running for vice president in 1868 on a minor party ticket. The second Charles Francis Adams became an advocate of railroad reform and eventually president of the Union Pacific Railroad. Afterwards, he published books and essays on historical and controversial topics. The two younger brothers, Henry and Brooks Adams, were distinguished as authors. Henry remains to many critics quite possibly America's finest writer of history. Both he and Brooks allowed their fascination by what they foresaw as the eventual decay of modern society to play a major part in their books. Brooks especially is noted for his warnings about the impending collapse of Western institutions.

This public attainment by one family is awesome, but to be fully appreciated, it needs to be placed beside the distressing story of the Adamses' private difficulties. Looking at this hidden side of his family's record, Charles Francis recalled how often the family's hopes had been frustrated. John Adams' brother Elihu died of dysentery in 1776 at age thirty-four; another brother, Peter Boylston Adams, became an ineffectual farmer who moved in the background of the family's life. He died in 1823 at age eighty-five. Then, among John's own sons, two had tragic lives of failure due to alcoholism, leaving only John Quincy to succeed. The same pattern appeared in the third generation, for of the three males born to John Quincy Adams, two were catastrophes for the family because of intemperance. Again, one child, Charles Francis, thrived. Charles Francis Adams, however, lived to see the grim pattern relent among his children. His four sons were all successes by some measure. However, with this fourth generation the public story of an extraordinary family closed. The unexceptional members of the fifth generation gave no sign of the potential and interest of their forebears, causing Henry and Brooks to speculate that perhaps the concentration of talent into only one member of each earlier generation had accounted for the family's quality. Had this peculiar vigor been forever diluted by its dissemination among the numerous remarkable children of Charles Francis Adams? His sons believed it had and seemed thankful that the painful greatness laid upon the Adamses was at an end.

Over the years, the Adams name itself led some members of the family to seek prominence. Others tried to be ignored. However, no Adams, success or failure, made a comfortable accommodation to life. Thus the lives of the eminent figures are no more revealing of the burdens of the family than the lives of the less renowned: John Quincy's sister, Nabby, and his two brothers, Charles and Thomas; the two brothers of Charles Francis, George and John; and in the fourth generation two sisters, Louisa and Mary. There are also the women who joined the family by marriage, notably Louisa Catherine Johnson, wife of John Quincy; Abigail Brown Brooks, wife of Charles Francis; Clover Hooper, wife of Henry; and Daisy Davis, wife of Brooks. The Adams story is as much about these individuals as it is about those who have a place in history.

The story is also shaped by a trait which marked the family and gave it much of its distinction. Near the end of the family saga, in 1914, the historian John T. Morse told Senator Henry Cabot Lodge that he had the clue to understanding the Adamses. "Being an Adams," Morse said, required that each "make some startling statement, it is a family propensity . . . it is the family way." Morse was referring to the Adamses' determination to be realistic about the weakness within mankind and society. Family members made this realism their outlook for nearly 150 years, allowing it to shape their careers in politics and literature. It gave them the "startling" quality Morse recognized. The result was that while Adamses earned the attention and acclaim of the public, they rarely hesitated to scold that public and to urge it to turn in another direction.

So it had been when John Adams as president stunned a war-bent America by insisting that peace with France must be maintained and by dismissing members of his Cabinet who opposed him. Constituents repudiated John Quincy Adams as a United States Senator when he refused to follow his Federalist Party's policies and instead supported the program of President Thomas Jefferson. Much of Boston was indignant when Charles Francis Adams demanded that the city recognize the unclean bond between Massachusetts industry and Southern slave labor. It was the same even in the final generation when Brooks Adams alarmed the few who read his books by his advocacy of a totalitarian society. Such courage, stubbornness, and candor made the Adamses often appear outrageously independent and brought them repudiation or, what was worse, neglect from the nation, a plight complicated by the abrasive traits for which most Adamses were famous. Rarely was any family member said to have much personal charm.

When John Quincy Adams first assailed the Southern slaveocracy as a congressman, he stood alone, threatened with censure and unseating by his outraged colleagues. As the North grew belatedly to respect and follow the old man's exhortations, the outcome seemed once more to illustrate where the Adamses stood in the course of American history. The family provided individuals of great talent who served their country well, but whose viewpoints were usually ahead of or above those prevailing in their day. Adamses wanted to help, to lead, to inspire their fellow citizens. However, a nation which increasingly claimed that individuals were sure to thrive in a free, competitive society could not long be patient with reminders that the optimism of the eighteenth and nineteenth centuries was not to be taken seriously. Inevitably, the Adamses were often repudiated or ignored, leaving them feeling misunderstood and unappreciated, but not surprised.

The mark then, of the Adamses' greatness, and what divided them from their peers, was this capacity to see with painful clarity the shortcomings within themselves and those about them. John Adams had matured in a time when assurances about human nature and individual rights were growing in America, thanks to the so-called Enlightenment and Age of Reason prevailing in Europe. John, however, always remained skeptical about such ideas, his misgiving arising from the religious teachings of his youth, when his father had begged him to become a clergyman. Much of what John learned came from the old Puritan and Calvinist messages about sin-ridden man's helplessness before God. This sobering view also contends that men should strive to be good stewards, should seek to labor for lofty causes despite the knowledge that mortals are capable ultimately of only evil and folly. While this paradox troubled and even mystified John Adams and his descendants, it led them into seeking public office and into writing books and essays to admonish both themselves and their fellow citizens.

All four generations of Adamses wanted to provide leadership and criticism that might move others to a more cautious and thoughtful engagement in democracy. Both John and John Quincy Adams tried to help advance the American nation, yet both were often sorely tempted to turn away, suspecting that such effort was doomed by the nature of mankind. While Charles Francis Adams approached this dilemma more cautiously, he too had the same anxiety. Like his father and grandfather, he managed to draw solace and a measure of understanding from elements in the Judaic-Christian scriptures and traditions. The final "great" generation, that of Henry and Brooks Adams, existed without the comfort of religious commitment. For the justice of God, Henry and Brooks substituted a scientific fatalism which stressed the same limits and ultimate failure of humanity that had fascinated their forebears. To these last prominent members of the great family, pessimism about human nature in a free society seemed as justified as it had to their ancestors.

By the time of the third generation, Charles Francis Adams recognized how his family's dissent caused it to be misunderstood and poorly appreciated. To help correct this, he published family letters, diary excerpts, and sketches which disclosed some pleasing aspects in the personal life of his forebears. Charles Francis resolved to do this in the 1830s after he began reading the letters that John Adams had written to his wife Abigail a half-century before. He recognized at once the "high-toned honesty" which had characterized his grandfather, and also that "gentler tones of affection are constantly to be found." It was in these kindly moods of John Adams "that the public understands him very little," said his grandson who now appreciated the importance of preserving the heaps of family letters and diaries. These not only displayed the real "spirit of action" among his relatives, but presented the Adams story in a much more intimate and appealing light.

As he read the manuscripts left by earlier generations, Charles Francis decided that historians actually revealed little about the persons of whom they wrote. The ordinary historical search, he complained, did not take the reader beyond the outer design: "We look for the workings of the heart when those of the head alone are presented to us." He proposed an astonishing change in biography, knowing that this might improve history's view of his father and grandfather. Authors should leave aside "the reasoning of the intellect," he suggested, and push on to find "the confidential whisper to a friend, never meant to reach the ear of the multitude, the secret wishes, not to be blazoned forth to catch the applause, the fluctuations between fear and hope, that most betray the springs of action." All these, for Charles Francis, were the genuine "guides to character," although he knew that these insights came from sources which usually vanished. Only "the coarser elements" ordinarily survived as the biographer's data, a loss which Charles Francis deplored. Yet when he had the opportunity, his delicacy kept him from including in his editions of family papers many of the most personal documents and passages. He left it for a later time to recreate the family's inward life, knowing that this story would eventually be needed for the world to understand the great but painful career of the Adamses.

For this reason, the chapters which follow tell of the Adamses at home, where the family derived its personality from moments of failure, sorrow, and frustration as much as from times of triumph. Both the personal and the public careers of the Adamses must be understood if we are to appreciate adequately those family attainments talked about in history books. This dawned upon one of America's most distinguished historians, Samuel Flagg Bemis, after he devoted much of his career to studying the public life of John Quincy Adams. When he put down his pen after completing the final volume of his biography, Bemis conceded that the work was unfinished since he said it still remained for someone to "probe Adams' inner life and character."

Our opportunity to know the Adamses as Bemis recommended comes because they recorded their life and thoughts so fully; because their manuscripts escaped destruction by carelessness and accident; and because later generations of the family decided in 1956 to give these Adams papers to scholarship and to the public. In these documents, the souls of Adams men and women linger much as when they loved and labored together within the family. The manuscripts contain a story never fully told until now. We must try to draw from these papers the moving history of those "deep groans within" and the "deep corroding mortification" which Charles Francis Adams saw so clearly but could not divulge. Such is the purpose of this book.

❧ 1 ❧

Beginnings

For several generations before John and Abigail's story began, there was nothing significant about the Adams family. Henry Adams came to America and settled in Massachusetts Bay, perhaps as early as 1632, evidently from Somersetshire in England where his father had been a John Adams. Little is known of the family in Great Britain, nor is there much to say about them after Henry established himself as a maltster and farmer in Braintree, a few miles south of Boston.

Henry had several sons. One of those accompanying his father from England was Joseph, whose son Joseph in 1691 had a son named John. This John Adams was the father of that John Adams with whom the family's great career began.

The elder John Adams, eventually known as Deacon John to distinguish him from his famous son, led a rural existence hardly different from that of his forebears except for an ardent involvement in the local parish and a determination that his son John, born in 1735, should go to Harvard College and become a clergyman. The Deacon's wife and John's mother, Susanna, was a lively person who went on to marry again after her husband's death, also surviving this second mate by many years. She was descended from the Boylstons, one of the colony's most vigorous and successful families, thus providing the only notable alliance in the Adams family's early story, except for Deacon John's mother, Hannah Bass, whose grandparents were John Alden and Priscilla Mullins.

How such a forceful and talented person as John Adams awakened from this background was a cause for wonderment to his introspective descendants, as it was to the second president himself. Later Adamses claimed that Boylston blood, Harvard inspiration, and John's choice of a wife, Abigail Smith, with her Quincy ancestry, were the reasons why their line departed from the pious, placid, barely literate existence of Henry Adams' immediate descendants. As the transitional figure, John Adams always pined for the rural and emotional simplicity which had once characterized his family. This yearning never left him throughout his long and utterly different life with Abigail. Her presence and aid he never failed to acknowledge as essential to his happiness and success, and with her he launched the Adamses upon a memorable career.

As a struggling young man, John battled to reach personal independence with the same ardor which would make him famous in other causes. By 1761, when his father died, he was comparatively poor, having only the hope of succeeding as a lawyer to sustain him. His father bequeathed him a small farm with a cottage where the young man lived and practiced law but also worked at agriculture. In those days nearly everyone out of necessity had something to do with farming, but for John it was more than a matter of sustenance. He always was happiest, in youth, middle life, and old age, when he could work in the fields, sustained by the presence at home of a loving wife and the books which also never lost their appeal. So, a farm, a cottage, and a small library were John's comforts even as he waited to be able to support a wife and family.

In the years before 1764, the young lawyer-farmer struggled against what he called his two foes, "absolute Idleness, or what is worse, gallenting the Girls," trying meanwhile to remember the words of his legal mentor, Jeremiah Gridley, that "an early marriage will obstruct your improvement, and in the next place twill involve you in expence." After nearly marrying Hannah Quincy, a young lady of his village whom John had found charming, he was even more wary of any marital bond which "might have depressed me to absolute Poverty and obscurity, to the End of my Life."

Nevertheless, his thoughts were never far from sexual matters. He attempted a poem which compared the merits of matrimony and celibacy. He contemplated the practice of divorce, especially when Hannah Quincy was mistreated by the man who married her. There was even an essay written to imaginary "Nieces," a device his great-grandson Henry Adams later carried to near-perfection, in which John put down his views on female character and deportment. He was particularly concerned about the tendency he found among women not to wash

sufficiently often. Were he to have daughters, he wrote, "I am determined to throw them into a great Kettle and Boil till they are clean."
But he was also convinced that they must learn to think and to enter
intelligent discourse. Finally, John said, producing children of honorable character was "the Highest Pinacle of Glory to which a Woman
can in Modesty aspire."

These bachelor meditations also persuaded John that couples intending matrimony ought to know each other well before the ceremony. He
accepted the New England custom of bundling, which allowed courting
couples to be left to themselves, wrapped together before the fire on a
winter's night with a board between them representing chastity. John
struggled to maintain a proper, if precarious, intimacy with women,
balancing his strongly sensual desires with his equally powerful yearning to master both self and profession. In old age, when he looked back
upon these early years, John marveled that he had not come to grief
by yielding to temptation. "My children may be assured that no illegitimate Brother or Sister exists or ever existed." Certainly one reason for
this triumph was that after surrendering Hannah Quincy to another
man, John fell so much in love with Abigail Smith that he—and she—
were willing to endure a waiting period of several years before marriage was practical.

From our perspective of more than two centuries, John Adams and
Abigail Smith seem an unlikely couple. He was nearly ten years her
senior, a bookish attorney in the village of Braintree. She lived in a
nearby hamlet, Weymouth, another seaside community south of Boston, where her father, the Reverend William Smith, was the town's pastor. The second of the Reverend Smith's three daughters, Abigail became as famous as her remarkable suitor. Her many strengths emerged
in a life combining both unusual opportunity and vicissitudes, beginning with encouragement from her father and from John that she stand
and grow upon her talents. In fact, her great-grandson Charles later
claimed that "Abigail Adams was the bigger man of the two; and that
to her, more than any other person, the high qualities shown by those
who came after them were due." Other family members disagreed with
him, but no one ever quarreled with the belief that Abigail was "a
woman of extraordinary mind and character." These were gifts which
kept John beside this rare bloom from the moment he discovered her
in the shade of Weymouth's parsonage.

Abigail had flowered under the care of her father, who saw that her
interest in books and writing was encouraged in his study. She was a

precocious seventeen-year-old when John met her, quick-witted and direct, as her letters to him before their marriage disclose. More aggressive and able than her two sisters and brother, Abigail had much charm and even beauty—enough to please John who, however, especially enjoyed her vivacity and her lively mind. Her father whose antecedents were humble, beheld the same promise of attainment in the young attorney which his daughter detected. Abigail's mother, however, who sprang from a distinguished lineage, hoped for a time that Abigail, who shared her pride in this origin, would choose to marry into a more notable family than the Adamses. Elizabeth Quincy Smith was descended from among the successful early settlers of the region, her father being Colonel John Quincy, a leader in Massachusetts Bay who presided over the handsome Mount Wollaston farm which had been established in Braintree by Edmund Quincy, one of the founders of the colony. Nevertheless, Abigail's interest in John Adams was eventually encouraged by both her parents.

Their courtship had all the ideal qualities: ardent yearning, deepening affection, and increasing compatibility. Even at this early stage in their relationship they shared a characteristic that became an important trait of those descended from them: an intense need for self-examination matched by a critical awareness of people about them. Arching over all this was love's charm. The Adams family was not begun from stilted restraint. John liked to talk of the two or three million kisses he had given Abigail—even two years before they were married. He called her "Miss Adorable" and teased her about learning to "conquer your Appetites and Passions," although he readily conceded he dreamed of her charms. On the other hand, while struggling to maintain the coy restraint her role required, Abigail managed clearly to tell John that she cared for him "with the tenderest affection." At night, she said, "I no sooner close my Eyes than some invisible Being bears me to you." She spoke of her own struggle between inclination and duty, assuring him that her affection equaled his. Both John and Abigail possessed so much sexual yearning, at the time called "excessive sensibility," that they actually became ill from anxiety and anticipation as the years of courtship were ending.

Still, there was blunt talk between them. Abigail advised her suitor that he had a forbidding presence which inhibited her as it did others: "I feel a greater restraint in your Company, than in that of almost any other person on Earth." She relieved this a bit by telling John that "for Saucyness" no one could match him, not even herself. As if to prove this charge, in a famous letter John sent Abigail a catalog of her faults, including directions on how she might improve her posture and gait: the habit of "sitting with the Leggs accross" would ruin her figure, he admonished. Abigail replied sweetly, chiding him with the arch suggestion that "a gentleman has no business to concern himself about the Leggs of a Lady."

These exchanges were closed beautifully by John as the time of their marriage approached. With remarkable prescience he told Abigail that "you who have always softened and warmed my Heart" would somehow polish and refine him. He predicted that through her he might even be enabled to "banish all the unsocial and ill natured Particles in my Composition" and reach a "happy Temper." Much of John's yearning for Abigail during the many separations which shadowed and shaped their marriage came from his knowing that she brought him a balance which he lost when they were apart.

Abigail's father performed the marriage ceremony for the pair on a late October day in 1764, after which the newlyweds arrived at the Braintree home John had eagerly readied for his bride. The eighty-year-old cottage inherited from his father was the family center during the first twenty years of their marriage. It was an ordinary house, containing nothing very remarkable in its several rooms below stairs and the fewer cubicles above, and was situated along a main road which led south to Plymouth. Not far to the north was the village center of Braintree. Near this cottage was the house where John was born and where his widowed mother still lived, and just over the rise called Penn's Hill lay the Atlantic Ocean.

In this locale, John and Abigail soon were well established, despite temporary moves into Boston in search of a better law practice. Over the years John steadily added to his possessions, acquiring in 1774 the cottage where his mother had lived and later so much adjacent acreage that by 1803 the Penn's Hill property contained 108 acres. Some of this land had once belonged to his earliest ancestors, making it especially dear to John, who cherished property where his forebears had lived in the sturdy, quiet manner he himself yearned for. John was usually impatient whenever necessity took him away from Penn's Hill.

The John Adams known well only to his wife, children, and grandchildren believed that to understand people, families, and societies one must first acknowledge the evil and folly of human nature, thus dismissing the optimism so popular with many eighteenth-century intellectuals. His private writings especially show that in his own family John tried to root out all foolish illusions. Among his contemporaries, he heard new talk about progress and human perfectibility, to which he would retort good-naturedly that most men sought "Peace, Nourishment, Copulation, and Society," until taverns, avarice, and partisanship misled them. Ensuing generations of Adamses received much instruction from John's letters about the disquieting reality of life, in which he accepted with grace and humor his own full share of the frailty imposed upon all human beings. No descendant quite matched him in this, though many tried. While John might occasionally rail and rebel against the world and his place in it, usually his family found him to be one who knew what God had made him and was reconciled to it.

John Adams and his progeny never could entirely resist entering into a society they doubted, but they did so with abilities they did not entirely trust. Ten years after their marriage, John wrote to Abigail, "You see how foolish I am. I cannot avoid exposing myself, before all these high Folk—my Feelings will at times overcome my Modesty and Reserve—my Prudence, Policy and Discretion." He found in his nature "a Feebleness and a Languor," and considered himself a very ordinary person whom events had brought forward, roused from "Sloth, Sleep, and littleness into Rage." He also conceded how sensitive he was, especially to any criticism but his own, and that his excessive response uniformly betrayed how weak he was. This paradoxical blend of toughness and vulnerability accompanied later Adamses into the world of affairs. Consciousness of this irony never saved John or those Adamses after him from despair and depression, especially when the world displayed little understanding and no gratitude.

The cost to John Adams of such a personality was severe. He frequently battled the melancholy that later haunted other family members and which occasionally incapacitated him. "I muse, I mope, I ruminate—I am often in Reveries and Brown Studies." At times he found his anxieties to be "unutterable," and once he suffered so from frustration and doubt that he collapsed into a deep withdrawal for several days, never really knowing how he managed to come out of it. He felt, however, that much of his difficult nature came from his compulsive determination to avoid self-deceit and personal betrayal.

Chiding himself for having high ambitions for his children, John recalled the lengths men went to secure fortune and title for their descendants. With human reason so frail and the lure of power and wealth so great, John was not surprised "that Men find Ways to persuade themselves to believe any Absurdity, to submit to any Prostitution, rather than forego their Wishes and Desires, their Reason becomes at last an eloquent Advocate on the Side of their Passions and [they] bring themselves to believe that black is white, that Vice is Virtue, that Folly is Wisdom and Eternity a Moment." Long before he experienced the full measure of sorrow and disappointment which both family and public life brought him, John Adams liked to stress what a "Mass of Corruption human Nature has been in general since the Fall of Adam."

With this view, the simplicities of family life, a village setting, and rural pursuits had ever increasing charm for John, who never drew away from the earthiness he believed lay at the core of all humans. The roughest side of physical labor delighted him, just as his sensual character did not alarm him. His writings are sprinkled with evidence of his tolerance for the sexual antics about him, although records of excess troubled him, as did the case of a local pastor who was famed for learning, piety, and gravity, but who, it was found, led the village in debauchery, as a pregnant servant could attest. This was outdone, however, by a seventy-seven-year-old deacon whom John cited as "the most salacious, rampant Stallion in the Universe," and before whom even young men were reportedly unsafe. Usually, however, John saw the happier side of sexuality.

John considered creation's finest ornaments to be farms, orchards, and the contentment of the family who dwelt in such a setting—especially when that environment included the area called the Blue Hills, whose pleasant vista made up the western view from the farm house where he and his bride settled. Work out of doors—hard labor—always refreshed him, and when his experience in the world of affairs depressed him, thoughts of simpler matters were his solace. "Let me have my Farm, Family, and Goose Quil, and all the Honours and Offices this World can bestow, may go to those who deserve them better, and desire them more. I covet them not." His longing for "rural and domestic scenes" remained constant until his retirement from public life. The largest and better part of him was honest in declaring preference for "the Delights of a Garden to the Dominion of a World." Amid the glitter of Europe, he called out: "My dear blue Hills, ye are the most sublime object in my Imagination."

John concluded his long political career with these yearnings intact. They were part of his remarkable charm, which was considerable despite a legend to the contrary. He was special in his family because he was the only Adams who looked at the world's opportunities and said with some sincerity: "I had rather chop wood, dig Ditches, and make a fence upon my poor little farm." Many years later, John's perceptive grandson, Charles Francis, acknowledged that the old gentleman's career was shaped by his zeal "to return to the only spot in which he really took delight, his home and his farm."

Even so, John's own mixture of ambition, dutifulness, and curiosity took him about as far from the simple ways of Braintree as anyone with his prospects could have imagined at the time of his marriage in 1764. Yet public life kept touching him at his most vulnerable points, seeming to bring him only more evidence about the helplessness of man to subdue selfishness and passion. It all astonished John Adams perhaps more than anyone. He liked to claim his story was unbelievable.

Within a few years after he and Abigail were wed, ability and hard work had made John one of the leading attorneys in Massachusetts. Soon he was a principal spokesman for colonial rights as Massachusetts and the other colonies began the exciting march to independence. In 1774, John went to Philadelphia as a delegate to the Continental Congress where he strenuously advocated freedom for America. Two years later, he joined Jefferson in preparing the Declaration of Independence. Soon thereafter, he drafted the Constitution for the Commonwealth of Massachusetts and then he began ten years of foreign diplomacy for the new American nation, serving in France, the Netherlands, and England. One of his achievements was coaxing financial aid from the Dutch, which proved crucial to the success of the colonial cause. He was also a forceful spokesman during the sessions in Paris which produced the Peace Treaty of 1783 establishing American nationhood. Soon afterwards, he became the United States' first Minister to Great Britain.

From this post John returned to Massachusetts in 1788. Convinced that his colleague Benjamin Franklin had been a deceiving rascal and that the Continental Congress had been filled with petty men, John was more than ever persuaded that anyone who believed in the primacy of decency, honesty, and reason was woefully deluded. Even so, he soon set out again from Braintree, this time to accept what he claimed was his duty to take federal office in a new nation. Finally, he returned to his farm and books after two terms as vice president of the United States, from 1789 to 1797, and then a term as president. He lost a bid for reelection to Thomas Jefferson in 1800. Thereafter, in a quarter-century of life as an old man amid his crops, cows, and books, John Adams contemplated these encounters with men and events, hoping to unravel the mysteries in human nature, liberty, and governance. His reverie often took him back to the time when he was creating a family and struggling to choose between it and the world of affairs.

Actually, with his children growing about him, young John Adams was wretched as a lawyer, despite his success. He groaned as he reminded himself that "law and not Poetry, is to be the Business of my Life." He had not the means simply to be the observer and commentator upon human affairs which he sometimes fancied he would have liked. Thus, three times before 1774 the cost of raising a family obliged him to move his brood from the country to Boston, where he chuckled at "his Squeamish Wife" for insisting they keep the bedroom shutters closed. Boston represented for John what he hoped to avoid in life—ambition, rivalry, petty business, and politics, whereas Braintree offered nature, retirement, and contemplation, especially beguiling as these surroundings were "the first seat of our Ancestors."

Thus the young husband and attorney found himself in a quandary even before he was summoned to his political career. To sustain his family, to buy more cherished land, and to acquire the books which he craved, John exchanged quiet toil in his fields, day-dreaming, and desultory reading for drudgery before the bar. The switch at least allowed him to blame his family's material needs for what was for him a painful shortcoming: the failure to master books and writing to his satisfaction.

Ambivalence tormented John as his interest veered from law and public life to books and study to farm and labor. While he liked to claim his legal practice was "the greatest Business in the Province," when asked what it had brought him he replied "Fatigue, Vexation, Labour and Anxiety." Despite all the advantages he had encountered, he was poor because, as he acknowledged, he was imprudent. "I have spent an Estate in Books."

In these early years, Abigail watched John shift from delight in his family and his farm to undisguised chagrin as other men passed him in life. He could acknowledge no one as his superior in ability. At these times, his despondency usually led him to insist afresh that wisdom meant being simply "Foreman upon my own Farm and the Schoolmaster to my own Children." He could not convince himself entirely, however, and soon he was drawn back to thinking of public affairs and dreaming of what he might contribute. At such times he would urge Abigail to preserve his letters so that "our Posterity" might learn about "these times of Perplexity, Danger and Distress."

In 1774, the call from the Continental Congress actually rescued John from the misery of trying to find fulfillment within the family circle, in farming, and by intellectual labor. Divided within himself, he could claim plausibly that it had required a public summons to take him from what he insisted he most preferred, his hearth and farm. Duty authorized him to desert the family and to make a name in politics, or so he said. It is, of course, impossible to know how genuinely mixed John Adams' feelings were when the excitement of that era took him from his family and books. However, his recollection of departing in 1774 for Philadelphia was revealing as he looked back in 1801 after his defeat for reelection. The old president stood confirmed in all his youthful pessimism, and was more than ever determined to dodge responsibility for exchanging his wife, his children, and his gardens, meadows, and books for the hollowness of public existence. With almost breathtaking irony, John said that Abigail and all his relatives were the cause of his decision to enter politics. The fact that all these persons seemed so positive about the need for courageous action was what forced him forward, he claimed.

The wife of eleven years, whom John left in charge of their children and affairs, was as complicated a personality as he. Abigail Smith Adams is no easier to understand today. Her role in the career of the Adams family was much more significant than the picture history has given us of her courage during the Revolution, her brilliance as a letter-writer, and her advocacy that women's talents be recognized. While these achievements are all notable, they have only limited usefulness to those who seek to know Abigail's personality and her endeavors as wife, mother, and grandmother.

To begin with, Abigail was in many ways a sharp contrast to John. Managing affairs and people had become her passion during her first ten years of marriage. She also took great pride in her triumphs in handling the family's business. While she sympathized with John's contentment in domesticity and rural felicity, she thought that this should not hinder the family's social progress. Proud of her ancestors, which included such famous names in the colony as Quincy, Boylston, Hoar, Shepard, Norton, Tyng, and Mason, Abigail never forgot that there were classes beneath hers. Yet she was very anxious when she compared her inelegant qualities as a colonial country wife with the fine ladies of Paris and London, warning John repeatedly that she would embarrass him when she set foot in Europe. To the end of her life she fought fiercely for the advancement of her family and nursed a more ardent hatred of Adams foes than did any male in the family. Her religious outlook never moved beyond the uncomplicated Christianity her father had expounded. In other ways, too, Abigail was very different from John. Never a serious student, she often claimed that preoccupation with books and ideas threw a person into depression. She preached, "All that is necessary for Man to know in order to be happy is easily obtained and the rest like the forbidden fruit serve only to encrease his misery."

Just as she did not share John's love for study, she usually did not join in his good-humored acceptance of sensuality. Despite the affections she displayed as a wife, she became increasingly prudish when she thought of the passion in others. Perhaps having herself glimpsed what intense passion a woman could encounter, she distrusted members of her sex and disliked the idea of women being drawn into sexual pleasure. Women, said Abigail, "are formed to experience more exquisite sensations" than are men. "We suffer and enjoy in a higher degree." She was invariably embarrassed and indignant when a widow remarried, especially when the new husband promised to be a virile companion. Widows who accepted younger men particularly drew her wrath, and she was disgusted when she thought that the display of a woman's bosom and thighs had become fashionable in America.

Abigail was distressed when her younger sister Eliza at age twenty-seven consented to marry a pastor from Haverhill, John Shaw, with whom the bride had earlier broken off a romance. What troubled Abigail was that anyone should let a yearning for matrimony lead to an unpromising union. "The spirit of Barter and exchange predominate so much here that people dispose of their own Bodies," said Abigail, although she conceded in her sister's behalf that "Men are a very scarce article to be sure." Soon Mrs. Adams was advising Mrs. Shaw about avoiding "the increasing way." She warned that neither she nor her two sisters had "nursing constitutions—twice my life was nearly sacrificed to it."

During their more than fifty years of marriage, John and Abigail Adams appeared to trade roles in the family. He became increasingly calm, tolerant, and unassuming, while she seemed more volatile, caustic, and proud. John grew lovable as time passed, while Abigail seemed to intimidate even her children. Eventually, doubts about human nature became more unrelenting in Abigail than in John. Before their marriage she had hesitated to accept his critical assessment of mankind. "Sometimes, you know, I think you too severe," she said, adding, "You do not make quite so many allowances as Human Nature requires." But she conceded that she was inexperienced, while John's being a lawyer led him "to a near inspection of Mankind, and to see the corruptions of the Heart," where he found things to be "desperately wicked and deceitful." However, a decade of married life converted Abigail to John's pessimism, and in 1774 she cried out how dangerous a creature man was and expressed her distaste for a world ruled by selfishness.

By 1775, John was off in that world amid exciting times, serving in Philadelphia at the Continental Congress and then in foreign diplomacy. For ten years he and Abigail were forced to be mostly apart, until she joined him in Europe in 1784, long after John had helped write the Declaration of Independence and make a new nation possible. When John entered public service, Abigail was thirty years old, a loving and fruitful wife whose frail proportions and always uncertain health belied her stern will and energetic courage. There was much beyond her hearth and nursery in which Abigail had become interested—farm management, political issues, local gossip; at times even a bit of philosophy and literature appealed to her. She was an extraordinary person whom John left behind in a difficult role.

He handed her their family, home, land, and meager funds to manage and demanded in addition that she supply him with consoling and encouraging messages. In one way, certainly, John Adams was different from his descendants. None of them accepted a wife as an equal comrade. In John's case, however, affection, respect, and necessity made him hail Abigail as his ally and tell her, "I must intreat you, my dear Partner in all the Joys and Sorrows, Prosperity and Adversity of my Life, to take a Part with me in the Struggle." Said he: "rouse your Whole Attention to the Family."

When John issued this challenge, distance and travel were enormous and frightening considerations. Beginning with John and Abigail's experience, long journeys and mournful separations became a shaping element in the lives of Adamses for several generations. This heightened for the family an appreciation of the physical and spiritual dangers in an unpredictable world. John and Abigail's first parting, in 1774, was alleviated somewhat since over the four years John could return from politics in Philadelphia to visit the Braintreee farmhouse. Also, he wrote her often, sharing with her the pleasing and thwarting experiences that public life brought him.

With his candor, John was controversial from the very start in the Continental Congress, for his unmistakable ability and his stout advocacy of colonial rights soon brought him into the center of excitement. In all of this Abigail supported him with enthusiastic letters and took proper interest when John published his famous essays signed "Novanglus" defending the American cause and when he proposed George Washington as commander-in-chief of the forces to defend that cause. John also led in establishing an American navy, while back in New England, he had the satisfaction of being appointed chief justice of Massachusetts, a post in which he was never able to serve.

As Abigail's concern for this larger world grew through John's sudden prominence, he kept adding to the exciting topics his letters shared with her. This was especially the case in the extraordinary year of 1776, as he wrote and argued for the creation of state constitutions to replace the colonial framework, while his colleagues in Congress recognized his abilities by placing him at the head of its Board of War and Ordnance. June and July brought Abigail descriptions of John's labors with Thomas Jefferson to draft a declaration of American independence and of how he made the principal speech in favor of this step. Before he returned home for a visit in the autumn of 1776, he had even begun charting the new nation's entrance into international diplomacy.

While John labored at his new tasks, Abigail worked to cheer and inform him. Her letters would surely encourage a loving husband to hasten home. She talked of yearning for him, of how tenderly her affection for him had grown through the years, and that it was now pent up and threatening to "break forth and flo through my pen." When John did not respond with sufficient warmth to suit her, Abigail clamored: "I want some sentimental Effusions of the Heart." "Lovers and Husbands" who were the "cold phlegmatick" type were not for her, she insisted. "I thank Heaven I am not so constituted myself and so connected." Whenever she thought of their being together, it brought overwhelming scenes to her mind—moments when "with the purest affection I have held you to my Bosom till my Whole Soul had dissolved in Tenderness." She confessed to John that at such memories she could only sit with her paper before her, "my pen fallen from my Hand." She delighted to recall that he was as susceptible to passion as she, a quality she coyly wanted him to understand would still be acceptable in a husband who now had the "stern virtue of a senator and a Patriot."

Abigail learned something about herself from the separation, telling John, "My pen is always freer than my tongue. I have wrote many things to you that I suppose I never could have talk'd." Usually her letters were a captivating blend of openly pledged affection—"I never close my Eyes at night till I have been to P[hiladelphi]a, and my first visit in the morning is there"—and of mundane concerns. She undertook to clean a house they owned in Boston which was in disrepair—"it had a cart load of Dirt in it," since one room had been used for chickens, another for sea coal, and a third for salt. "You can conceive How it look'd."

At home in Braintree, Abigail achieved her own heroism. In a famous 1775 scene revered in family annals, she took her eldest son to the top of Penn's Hill to watch the Battle of Bunker Hill across Boston Harbor. Then she stood fast during the ever-present rumors that British troops were about to overrun Braintree. She managed brilliantly amid the privations which made everyone in the community uncomfortable: even the diseases afflicting her family and neighbors seemed more severe, especially the epidemics of cholera, in one of which Abigail's mother died. She worried also when politics in the Continental Congress did not seem to go John's way, and Abigail used the advantage of a distant perspective to advise him.

Amid the details of heart and home, there was a time or two when Abigail thought it well to remind John that, beyond love and hard work, a woman must be regarded as a person sharing the rights of man, one of the issues then being debated by the Congress in Philadelphia. To this came John's rejoinder that male superiority had never been more than an absurd theory. "You know We are the subjects, We have only the Name of Masters. . . ." As for Abigail, he proclaimed her "my best, dearest, worthyest wisest friend in this World." Abigail, in turn, said that simply seeing his garments hanging in a closet brought her exquisite pleasure and pain. He was "dear to me beyond the power of words to describe," and when, one night, she dreamed he had returned to treat her coldly, her "heart acked half an hour after."

John was charmed by Abigail's epistles, which grew more lively and perceptive. "I really think that your Letters are much better worth preserving than mine," he told her. What could have consoled John more than Abigail's noble sentiment: "Here I can serve my partner, my family, and myself, and in joy the Satisfaction of you serving your Country"? Her management of family affairs pleased him as much as her writing. He was certain that neighbors would see how his farm was better conducted when he was absent. "I think you shine as a Stateswoman, of late as well as a Farmeress," he told Abigail, to which she responded that being busy kept her mind from the sorrow of his absence. When he seemed to falter at going back to Philadelphia after short visits with her, his children, his books, and his cherished farmland, Abigail pushed him away. He must return to duty while she remained behind with her tasks. This encouragement made John talk of how no man could hope to be illustrious without some woman—wife, mother, sister—being largely responsible.

John returned to Philadelphia but late in 1777 he was again in Braintree, this time to prepare for a much more wrenching parting from Abigail. He had been elected by Congress along with Benjamin Franklin and Arthur Lee as joint commissioners to France. As this separation approached, John recalled being overwhelmed by an inexpressible tenderness toward his family. After the flurry of talk and calculating caused by Abigail's determination that she and their children should accompany the new diplomat, it became clear that he must, after all, go without her, and Abigail reluctantly withdrew in the face of the danger and expense. What ensued was probably the most difficult year, 1778–79, in their marriage, for the Atlantic barrier broke Abigail's spirit and perspective.

John was so distracted by the intrigue of foreign diplomacy and the problems presented by his colleagues, particularly the unorthodox Franklin, that his European letters to Abigail grew curt, and were made more so by the danger of being intercepted by the enemy. At first, Abigail maintained a brave front, carrying on as mother, head of household, and correspondent until pain and suspicion overcame her. Initially she managed to write the correct sort of letter—too proper for Abigail, since her words had an unaccustomed triteness when she told John: "My Soul is unambitious of pomp or power. Beneath my Humble roof, Bless'd with the Society and tenderest affection of my dear partner, I have enjoyed as much felicity, and as exquisite happiness as falls to the share of mortals." It was enough for her, she said, "to glory in my Sacrifice and derive pleasure from my intimate connexion with one who is esteemed worthy of the important trust devoted upon him."

By early autumn 1778, however, Abigail's letters became vessels of indignation, suffering, and doubt. The few, terse words received from John made her regret letting him go alone. "I view myself in a situation by no means to be envied," she said, her earlier courage gone. Recrimination began—had John "changed Hearts with some frozen Laplander," she asked, adding that she would refrain from telling him how much his apparent indifference was costing her. With her suspicions of women in general, it was easy for Abigail to imagine what female wiles were doing to her faraway husband; she later developed the same fears for her sons and grandsons when it was their turn to be vulnerable. It alarmed Abigail that John seemed to have time to write about the talents of Parisian ladies. Evidently European life pleased him, she said, adding sarcastically the surmise that he would probably linger there happy in the view that it was his duty.

His responses failed to reassure her, and Abigail soon grew bitter, observing that it was apparent her role in his life was merely "to attend you whenever you think your own comfort, pleasure or satisfaction can be promoted by it." But despite all her hurt and anxiety, she never allowed her affection to be extinguished, and her lonely cries should have stirred the granite hills near the Adams farmhouse. Writing from their familiar second-floor bedchamber, she told John of how she had recently been overcome with loneliness when a young lady sang mournful Scotch ballads. Especially touching for her were the lines:

> His very foot has Musick in 't
> As he comes up the stairs.

Receiving Abigail's reproach and accusations, John broke under his load of frustrating work and personal guilt. He asked how she could so misjudge him. How could he neglect or forget "all that is dear to me in this World"? Why did she doubt him after all this time? "Can Professions of Esteem be wanting from me to you? Can Protestations of affection be necessary?" He was even blunter than she. "The very idea of this sickens me. Am I not wretched Enough in this Banishment without this." As her relentless tide of accusing letters swept in upon him, John begged her to stop. "Your wounds are too deep." "Let me alone, and have my own Way." John was bruised beyond words by Abigail's seeming lack of trust and her nagging style. It bothered him particularly to discover that he now had a wife who appeared oblivious to his burdens and sacrifice. He suggested that she cease mooning and begin to recognize her own age and situation.

Shortly before John returned to America on a brief visit in 1779, Abigail awoke to what she had been doing. She pleaded with John to realize her tone had been due entirely to anxiety over him and urged him to bury her complaints forever. They should forget this unhappy episode, she asserted. Certainly Abigail seemed wholly altered when John soon sailed again for Europe in November 1779, to be parted from her until August 1784. In their few months together in 1779,

John and Abigail had talked frankly not only about how each was paying a terrible price by this enforced separation but also about what
John's service to the fledgling nation might eventually mean to them.

Although the years following seemed interminable to her, the closest
Abigail came again to breaking down was once, late at night in 1782
on the eighteenth anniversary of their wedding, when she scribbled to
John of how depressed she was, and begged that he tell her more of
his love—"you write so wise, so like a minister of state." He had the
title and renown, she said, while she was left sitting in their cottage as
the price of the prestige he won. She asked him at least to put affection
into the mail—"it is little attentions and assiduities that sweeten the bitter draught and smooth the rugged road."

More and more, however, her letters were cheerful and full of support. She talked teasingly of confidence in her husband's ability to stand
impervious to Paris beauties. She applauded constancy, but at one point
she may have become undiplomatic when she reminded her sensitive
mate that she considered him beyond the age to succumb to the temptations of the fabled princesses who peopled Abigail's idea of Europe.
John perhaps forgot such barbs when Abigail assured him: "I know I
have a right to your whole heart because my own never knew another
lord."

It was John who next time succumbed to adversity and the strain of
being parted. His most difficult years came after he moved in 1780
from Paris to the Netherlands, where he was mostly alone, trying to
negotiate a treaty of amity and commerce with the Dutch. While succeeding admirably in October 1782, John was much bruised by the uncertainty of his larger role in Europe and by the political bickering at
home in Congress. His natural vulnerability made him fancy mistreat-
ment and plots against him on both sides of the Atlantic. As a consequence, after 1780 he suffered several emotional collapses. "I am in
earnest. I cannot be happy nor tolerable without you," said he, urging
Abigail to join him quickly.

He envied his good friends and neighbors, Cotton Tufts and Richard
Cranch, as they enjoyed the peacefulness of Braintree. "Why should
not my lot in Life be as easy as theirs?" To which he could answer
himself readily enough: "So it would have been if I had been as wise
as they and staid at home as they do." Abigail acknowledged that people did ask, rather annoyingly, how anyone who so professed to love
his hearth could be away so long, and would she have let him go, had
she known the cost? Her tactic was to avoid the first question and to
stress simply that, certainly, she would see him go again. John's needs,

however, had now become pathetic. When the war's end made her joining him a possibility, he called out from Europe, "I wonder if any body
but you would believe me sincere if I were to say how much I love you,
and wish to be with you never to be separated more?" He described his
solitude as "horrid."

Still, during 1782 and 1783 John managed one of his most important
contributions to United States history. He was a chief architect for the
treaty with Great Britain which closed the American Revolution in 1783.
Then he was assigned with Jefferson and Franklin to seek agreements
of amity and commerce for their new Republic with numerous European powers. At this time, he persuaded Abigail to come and be with
him. In 1784, with peace restored and the youngest children old enough
to leave, Abigail finally pried herself from the peaceable kingdom of
Braintree, stoutly maintaining it was not only her yearning but also her
duty to join "the best of Husbands and Friends." She dreaded going.
As a matron of forty, she had survived the ordeal of being without
John, brought her offspring and farm through the Revolution, and
had become a person of consequence in her family and her neighbor-
hood. Despite these remarkable experiences and the wisdom and courage she displayed throughout the war years, Abigail was timid about
straying from home. She was greatly in earnest when she warned her
children of that world beyond Penn's Hill in which evil ruled. Always
preferring to keep herself in surroundings no more than a day's travel
from home, Abigail agonized when duty or ambition drew her and
especially those dear to her into absences from the Braintree neighbor-
hood. When she returned after four years of Paris and London, Abigail announced: "I have seen enough of the world," conceding that her
stay had not been entirely happy. What she yearned for was "Domestick happiness and Rural felicity." Seeing life beyond Braintree had
only confirmed this desire.

Both John and Abigail craved the comparative security of hearth and
family, a wish masking the apprehensions they shared concerning human nature and society at large. While this outlook made their separations more distressing, its greater importance for the family's story
was in how it influenced the rearing of their four children. As John
wrote to Abigail in 1774, when he started for Philadelphia and his first
long absence, their most "ardent anxiety" must be molding the mind
and manner of each child. However, loving a child and yet chastening
it for maturity evidently became more of a problem for them than for
many parents.

John and Abigail established their family without delay. A daughter was born nine months after the wedding in 1764 and given her mother's name, but forever remained "Nabby." Two years later, on 11 July 1767, a son arrived just as his great-grandfather was dying, so the infant was called John Quincy Adams. At the close of 1768 another baby girl appeared, christened Susanna, after John's mother. This child died early in 1770, but a boy, Charles, was born shortly after. Two years later came Thomas Boylston Adams, named for the renowned ancestor of both parents. A baby girl was stillborn in July 1777, an event which stirred Abigail to send John in Philadelphia the simple Christian assurance so meaningful to her all her days: Life was uncertain, she said, and thus "we ought patiently to submit to the dispensation of Heaven." The loss of this baby brought the parents closer together. John wrote that never had news moved him so deeply.

For John and Abigail, the most strenuous part of marriage was to prepare their children for worthy lives, a project which soon had the youngsters struggling with apprehension, compulsive behavior, rebelliousness, withdrawal, and depression. They were afraid to leave a loving and attentive home and face the specter of human nature and the grim temptations which were pictured as waiting for them just beyond the hearth. Virtue was the word John used most often in discussing the goal of a parent and child. Only after moral excellence had been implanted should the youngster's desire to excel be encouraged, he said. In John and Abigail's minds, virtue began with the child's devotion to values and to precepts for living which were as clear as was the certainty that the world facing the youngster was corrupt and dangerous. In eighteenth-century family life, righteousness started with youthful devotion to parents who were the preceptors of virtue. Enormous meaning was thus conveyed when Abigail regularly informed her absent husband, "All our Little ones send duty." Indeed, the solemnity of duty fulfilled or duty avoided would be the sweet joy and the terrible burden for John and Abigail's four children.

There was nothing remote about evil for the Adams family. Even while her brood was small, Abigail pondered the sins of her brother, William, whose sad career darkened the life of his father and all the family, which included William's hapless wife, Catherine Salmon Smith, and their six children. This brother's irresponsibility, drunkenness, and deceit were vivid testimony of the world's savagery. Despite the most earnest parental efforts, as William's case seemed to prove, vice and viciousness could and would take early root and, as Abigail put it, "tho often crop'd, will spring again."

With these apprehensions, one wonders which duty was greater in the Adams household, that of being a parent or of being a child. The youngsters were earnestly reminded that their "good behaviour," then and throughout life, meant everything to their parents. "God grant I may not be disappointed," was John's frequent prayer, while Abigail shrank from sending the children to public school, lest they be confronted by corrupting examples. She wanted her three sons and daughter to "chill with horror at the sound of an oath, and blush with indignation at an obscene expression." John and Abigail shared the obsession that each child must have virtue so writ upon the soul that neither "time nor custom" could totally eradicate it.

There was another hazard. The Adams youngsters had an absent statesman for a father. The distance drove John frantic. "My poor children," he moaned, fearing that without him they would never have "Truth, Sobriety, Industry . . . perpetually inculcated upon them." His alternative was to admonish Abigail. The youngsters must be "kept pure." She should "take care that they don't go astray. Cultivate their Minds, inspire their little Hearts, raise their Wishes. Fix their Attention upon great and glorious Objects, sort out every little Thing, weed out every Meanness." In short, their children were to revere nothing but "Religion, Morality, and Liberty." So demanding was John that, at one point, Abigail had to defend herself. "Our little ones whom you so often recommend to my care and instruction shall not be deficient in virtue or probity if the precepts of a Mother have their desired effect." But she feared this might not be enough, and John was reminded that his family needed "the example of a Father constantly before them."

Abigail, of course, earnestly played her part—and John's too—as well as she could. The children heard all the moralisms of that time, as Abigail spoke of virtue, industry, decency, grace, honesty, usefulness, and especially God when she sought to form a child's mind. As soon as a youngster could read, John sent letters to each one bolstering Abigail's efforts. His daughter was scarcely nine years old when he sought to enlist her aid in persuading her brothers to be "good and useful Men—that so they may be Blessings to their Parents, and to Mankind."

On his second voyage to France in November 1779, John had taken the eldest two sons, John Quincy and Charles, leaving Abigail sorely alarmed that these children must go to sinful Europe. She comforted herself by writing them embellishments upon the family dogma. She particularly stressed that the heart's "moral Sentiments" were more important than the "Furniture of the head." Abigail seemed genuinely afraid that moral ruin would come upon her children and consequently upon herself as a failed parent. She begged the ten-year-old John Quincy: "Adhere to those religious Sentiments and principles which were early instilled into your mind." He must never forget his accountability to God. She put the point in a manner no child of young John Quincy's sensitivity could ignore. "I had much rather you should have found your Grave in the ocean you have crossed, or any untimely death crop you in your Infant years than see you an immoral profligate or a Graceless child." The great enemy "vice," she said, was on the verge of victory. Had it not carried off Abigail's brother William? "You must keep a strict guard upon yourself," she continuously warned the children, "or the odious monster [vice] will soon lose its terror, by becoming familiar to you." Thus Abigail made certain the children learned of their heavy responsibilities. The greatest of these was knowing that triumph over evil meant they would "do Honour to your Country, and render your Parents supremely happy, particularly your ever affectionate Mother."

John and Abigail's campaign of virtue extended to such matters as slovenly dress, time wasted, sloppy penmanship. The children were urged to keep a diary or journal and advised on the priceless value of exercise. All these admonitions, and they seemed endless, were usually capped by such general injunctions as "Every Thing in Life should be done with Reflection, and Judgment, even the most insignificant Amusements," or "above all things, preserve your Innocence, and a pure Conscience. Your morals are of more importance, both to yourself and the World than all Languages and Sciences. The least stain upon your Character will do more harm to your Happiness than all Accomplishments will do it good." When she admonished John Quincy about the vicious world around him, Abigail put to him with great effect the question which had become central to the Adams family conscience: "What is it that affectionate parents require of their children; for all their care, anxiety and toil on their account?" The answer: "Only that they be wise and virtuous, benevolent and kind." Thus, said Abigail, her children must remember that their parents "would not direct you, but to promote your happiness."

Readying children for entrance into a sinful world obliged John and Abigail to offer them a bewildering mixture of affectionate support and cruel distrust, striking even for an age when Christian virtues were still stressed. Perhaps this tendency grew out of the separations they endured, their peculiar sensitivity to the crisis of the time, and the alarming disappointments they had seen among their relatives. Even as they adored their youngsters, John and Abigail hastened to impose upon them knowledge of their shameful humanity and to remind them that children could not be trusted. A child's only hope, according to John and Abigail, was to reach adulthood so imbued with the righteous instincts its parents had nurtured that it could achieve a hairbreadth escape from life's temptations.

The tale of the Adams family would surely have proven simpler and perhaps more cheerful had John and Abigail been able to exhort their youngsters to look no further for satisfactions than home, farm, library, and church. However, since John yearned to lift his life into a larger and thus more dangerous sphere, he and Abigail, whose ambitions equaled his, could hardly avoid demanding worldly success from their offspring. There is no sign that they saw any disparity in the gospel and example they put before their family. The age they lived in was becoming accustomed to mixing the ancient belief in the power of evil with the new talk about progress and perfection.

While John as well as Abigail asserted: "Of all happiness domestick is the sweetest," he sent out, by his actions, confused and troubling signals to his children. The latter were warned of a world dominated by sin and folly, yet their parents seemed to find that world irresistible. Baptized at home in the waters of self-doubt, the Adams children were told that to be less than excellent in affairs both great and small meant they were ultimately betraying their family. It was natural, therefore, that John and Abigail's two oldest sons resisted having to leave the comparative safety of home. John Quincy and Charles were very unhappy when their father took them from their mother. Twenty years later, Charles's death from intemperance confirmed the most depressing features of Abigail's teachings about the inevitable presence of sin. Unfortunately, Charles' story was not the only tragedy to unfold in the family history. While John Quincy Adams escaped the streak of weakness which was evidently the cost to the family for its achievements, his triumph, solitary in his generation, was attained at the price of much personal unhappiness.

2

Growing Up

A youthful John Quincy Adams put the plight of his generation admirably when he mused: "What is the lot we have to expect in the world? I look forward with terror; and by so much the more, as the total exemption from any great evil hitherto leads me to fear that the greatest are laid up in store for me." Once their father took up duties in the Continental Congress, the simple life at the foot of Penn's Hill became a tiny island of safety to the four Adams children surrounded by the sinful universe their parents pictured. To be sure, there was overflowing affection between the parents and children, to judge by the way family records abound with exchanges of devotion. Far away and lonely at the Continental Congress in Philadelphia, John was consoled by memories of walking with his family among the orchards and fields at home. "Charles on one Hand and Tom on the other. . . . Nabby on your Right Hand and John upon my left." Amid the dangers of fomenting revolution as the price of national liberation, John took heart as he thought of what Abigail called their "flock of little ones." For each child there seemed an identical share of love and hope. John told Abigail in 1776 that the youngsters had "capacities equal to any Thing. There is a Vigour in the Understanding, and a Spirit and Fire in the Temper of every one of them which is capable of ascending the Heights of Art, Science, Trade, War, or Politicks."

Nevertheless, amid the hopes proclaimed, the hugs and kisses exchanged, and the encouragement liberally dispensed, the specter of a cruel and unpredictable future was never far distant. Thus, Abigail was obliged to dampen John's enthusiasm about each child by reminding him that their friends, the Warrens, had a "disordered son" whose name that family would not speak. "God grant that we may never mourn a similar Situation." Even now, she said, "Heartake" overtook her as she watched their own boys.

Because John Quincy twice left her to go abroad with his father, Abigail's record of concern for this eldest son is particularly eloquent. She attended to every aspect, from his appearance to his soul. She scolded him often for his unkempt dress, a fault he later never allowed his wife to mention. When John Quincy was in Holland after John had been named minister there, Abigail hoped that "the universal neatness and Cleanliness of the people where you reside, will cure you of all your slovenly tricks, and that you will learn from them industry, economy and frugality."

It was, of course, the threat of sin which drew most of the mother's concern. "If you could once feel how grateful to the Heart of a parent the good conduct of a child is, you would never be the occasion of exciting any other sensations in the Bosom of your ever affectionate Mother." At that moment she was especially fearful that, since John Quincy had "a constitution feelingly alive" and that "your passions are strong and impetuous," he would be driven into debauchery. He must strive anew for self-control, she told him. Abigail dreaded even mirth. "I never knew a man of great talents much given to laughter," said she, adding, "My own ideas of pleasure consist in tranquillity."

The impact of such exhortation was everywhere evident in John Quincy's youthful letters. To his younger brothers he wrote about the terrors of being exposed to "vice and folly" and that he hoped "never [to] be tempted by them." He liked to preach to the smaller boys: "We are sent into this world for Some end. It is our duty to discover by Close study what that end is and when we once discover it to pursue it with unconquerable perseverance." Such messages were composed under John Adams' watchful eye, since the children's letters were never mailed until a parent had approved them. John Quincy was likely to see his father nod happily when the eldest son urged his brothers back in America to join him in employing "those hours which are often spent in frivolous amusements, in gaining knowledge which will make us useful to our fellow men when we grow up."

Such themes common in the earliest letters of John Quincy Adams never lost their earnestness when he repeated them to his own children and grandchildren. At age six he was deploring how little progress he made in his reading because he spent "too much of my time in play."

There was, he confessed, "a great deal of room for me to grow better." He was so often tempted with frivolity, he said, that "Mamma has a troublesome task to keep me Steady, and I own I am ashamed of myself." One can imagine how these confessions warmed the father in distant Philadelphia, who was often scourging himself in a similar fashion; so did the son's repetition of his parent's outlook when he reiterated a "determination of growing better." The father and son therefore got on well in the years after 1777 when they were in Europe together, which in itself earned the lad high praise. John told Abigail that their son was "as promising and manly a youth as is in the world."

Being companion to his father at age fourteen, John Quincy had the opportunity, first in Paris and then in Holland, to attend European schools and to spend his free time watching diplomatic life and profiting from the company of his father. Encouraged to strive after deliberateness in the use of time, an injunction John laid heavily upon himself, the lad seemed to learn more from the world around his parent than from books. Then came an opportunity in 1781 whose consequences were enormous. John Quincy left his studies in Holland to accompany Francis Dana to St. Petersburg—he to be secretary while the latter sought a treaty between the United States and Russia. Fourteen months later, when John Quincy had returned after additional travels in Scandinavia and Germany, his acquaintance with the world and his perspective on politics were unequalled among young Americans of his time. This experience, when added to what his parents hoped for him, indicates that at age fifteen, John Quincy expected much of himself. This expectation remained with him to his death.

He returned to his father's side as the negotiations for ending the American Revolution were under way. He found John urging Abigail to join them in Europe and to make sure she did not leave their daughter Nabby at home. The daughter was now old enough to have begun a disquieting love affair, so John commanded Abigail to sail for Europe and bring Miss Adams with her.

As the first of John and Abigail's children, Nabby had received an overflowing measure of the ardent but fearful attentions of her parents. She became a prisoner of the need to please, to obey, and to do right. In 1776, when she was no more than eleven years old, her father wrote to commend her "remarkable Modesty, Discretion, and Reserve." Let this be enhanced, he advised, and sent his instruction: "To be good, and to do good, is all We have to do." The girl tended increasingly to withdraw. Abigail reported that Nabby's lofty bearing, which John so admired, was being seen by some neighbors as "pride and haughtiness," although the mother was convinced it masked "a too great reserve."

When Nabby was seventeen, Abigail was pleased at how prudent and discreet her daughter was—much beyond her years. Yet there were maternal complaints that Nabby was "rather too silent," a deficiency in Abigail's eyes that was heightened by the young lady's refusal to write letters. The mother began to wonder if Nabby had not too much fear of failure and scorn. Although the girl at last wrote her brother John Quincy in Europe, describing herself as "your Sister who is so sensible of her own unworthiness," Abigail wondered if this was not false humility, masking the sin of pride. Nabby redeemed herself somewhat when she finally obeyed her mother and sent John Quincy such messages as one begging him to become "an ornament to your parents, who watch with attention each improvement, and whose hearts would be wounded by a misconduct, and may it be our joint effort to study their happiness." Even in letters to cousin Betsy Cranch, daughter of Abigail's sister Mary, Nabby dutifully repeated the essentials of Adams teaching about an uncertain world and its dangers. She liked to write essays in praise of duty—"I never wish to stifle this voice," and she begged Betsy to criticize her, saying she knew she had often "done wrong." To Betsy, Nabby confessed how she yearned to be able to talk freely with others. Once, after being with friends, she reported "you never saw me quite so stupid before." Would Betsy take pity, Nabby asked, and "write me a conversation that I may make use of when I go into company"?

Nabby soon grew restless in the presence of her mother, for there always seemed to be more tension between Abigail and her daughter than between Nabby and her father. The girl began begging to be allowed to join John in Europe—Abigail reporting that the girl "thinks of nothing else but making a voyage to her pappa." When this had to be postponed, Nabby fell in love, hoping perhaps to become independent of Abigail's powerful presence and steady exhortations. Her suitor was an attractive young attorney, Royall Tyler, who came to Braintree to begin a settled life using the balance of his patrimony. Previously he had squandered a generous portion of his inheritance in what was then considered riotous living—he had even made a European excursion.

By the autumn of 1782 Tyler was carefully pressing for Nabby's affections, wooing the young lady's mother as much as the daughter. At first Nabby seemed cool and reserved toward him. Abigail was clearly thrilled by this prospective match and tried to break down Nabby's manner lest it discourage the young man. Their daughter was "a fine majestick girl who has as much dignity as a princess," Abigail reported to John, adding that "no air of levity ever accompanies either her words or actions." What the young lady needed, announced the anxious mother, was to be caught "by a tender passion." This would reduce her "natural reserve and soften her form and manners." Indeed, Nabby soon warmed, and the romance moved so rapidly that Abigail and her sister, Mary Cranch, at whose home Tyler boarded, were startled. Mary and Abigail agreed on a cooling-off period, so Nabby was sent to visit relatives in Boston and elsewhere.

At this time letters bearing news of the love affair reached John Adams in Europe. The poor father was thunderstruck. More than that, he was frantic at being several thousand miles from this domestic crisis. His daughter in love! John was outraged. "My child is too young for such thoughts," he said, apparently forgetting that Abigail had been considerably younger when he began his own suit. He was especially put off by Tyler's record of mild dissipation. During much of 1783, John wrote again and again. He said his daughter was not meant for a reformed rake: "My child is a Model." As for the cool haughtiness which had initially so alarmed Abigail, it was Nabby's greatest glory, said John. He preferred to have Nabby "silent" in company, "and I would have this observed as a Rule by the Mother as well as the Daughter." He demanded to know if Abigail had lost her prudence. "Your family as well as mine have had too much cause to rue the qualities" which even she acknowledged Tyler had once displayed—"if they were ever in him, they are not out yet," John insisted, and for months remained adamant. Nabby was to marry no one "who does not totally eradicate any taste for Gaiety and Expense," the penchants of a "Rascall." Once a person had succumbed to frivolity, no matter how penitent he might be, the luckless wretch was tainted forever. The only son-in-law he desired was one known for "Prudence, Talents and Labour." Let Nabby look for these traits, he commanded; let her "keep her Reserve."

When she received her father's command, Nabby's reply was somewhat odd. She had many blessings, she said, and if her dearest wish were granted, might that not be risky? "I should enjoy too much pleasure." She tried to convince herself that to be denied something ought to be a comfort. "It is best. It is right." Nevertheless, she wept in private while she tried writing her father in a vein to reveal that she shared his worship of obedience and duty. How grateful she was to him for his advice and wisdom, she said, adding: "I have suffered not a little mortification whenever I reflected that I have requested a favour of you that your heart and judgment did not readily assent to grant." Her letter was lush with vows of devotion to John's outlook and of her determination not to rely upon her own judgment again.

In this, however, Nabby failed, for by late 1783 she was more in love with Tyler than ever. Abigail, meanwhile, tried to assure John that their daughter still talked of valuing her father's approval. She would remain cautious in her love until there was real proof that Tyler did not deserve an Adams' affection. At this moment, however, Richard Cranch, Nabby's uncle, began praising Tyler in letters to John, messages strengthened by the fact that Nabby's suitor had purchased the handsome Vassall-Borland property in Braintree, one which Richard himself coveted. It was a step, Abigail also reported, that Tyler took so "that he may not be considered unworthy a connection in this family." All that was lacking now, she said, was John's consent.

Confronted by all this, John relented in January 1784, graciously asserting that the newlyweds could live in the house at Penn's Hill, while only Abigail need join him in Europe. Tyler had permission to use John's valuable law library, and John even foresaw that John Quincy might eventually study in Tyler's office. John acknowledged bending to the steady reports of Tyler's excellent character, talent, and diligence, and at last had the grace to add that in this decision Nabby's heart must make the choice. However, Nabby never had the chance, for John's letter of consent was delayed and did not reach Braintree until Abigail, having waited as long as she dared, had sailed for Europe. With her came Nabby, in need of soothing, said Abigail, because of her struggle between duty and desire.

In June 1784, Abigail and Nabby were joyfully reunited in London with John and John Quincy, and the four of them traveled together to France. They were soon settled in the house of the Comte de Renault in Auteuil, a village outside Paris near the Bois de Boulogne. Here they spent nearly a year while John worked with Jefferson and Franklin to negotiate commercial treaties between the new American nation and as many countries as possible. This proved to be one of the happiest intervals in the Adams story; John was much comforted with his family about him and Abigail discovered that she could after all compete with European women. Then, appointment as America's first minister to Great Britain summoned John to residence in London, where by June 1785 he and Abigail were living in Grosvenor Square, near Hyde Park.

Here they could walk together and talk about their family concerns. They sent John Quincy to America to study at Harvard, which created anxieties enough. But John and Abigail's immediate concern was Nabby, who had remained with them in Europe. Nabby's engagement was once again troublesome. For about a year her romance with Royall Tyler had seemed to go well, despite the distance separating them. Cordial letters flew betwen Tyler and the family, although Nabby's do not survive. The young man pledged patience, devotion, and hard work, while Abigail sent him the coziest details about family matters. However, in Braintree, Tyler evidently began behaving immaturely, although the story is unclear. He boasted of his link with the eminent Adams family, while at the same time he concealed messages to others which the Adamses had entrusted to his care. Astonished friends and relatives in Braintree were concerned about this as well as about the excessive sentimentality they saw in the young man. Tyler might have survived this episode had he not become careless about writing to Nabby, although he later claimed he sent letters on every available boat.

Abigail had done her best to instruct Tyler in the family's reverence for duty, urging him to have such comportment as would not betray those who had helped him. However, when the suitor's faults became apparent, Abigail turned away, insisting that she had always been dubious about him and that only John's belated consent had kept her silent. When Nabby's father heard of Tyler's curious performance and his daughter's doubts, he reverted to his earlier position and said that he would rather follow Nabby's coffin to its grave than to see her marry such a man. Abigail was careful to inform relatives, however, that Nabby had been left alone to struggle with the problem. The parents denied any effort to shape her decision. When she chose to end the engagement, there was family rejoicing on both sides of the Atlantic. Abigail announced of Nabby: "Being once free, I believe she will in future proceed with a caution purchased by experience."

Abigail was a poor prophet, for Nabby was soon engaged again, and this time she promptly married, in 1786. The union at first had Abigail's blessing. Events however quickly showed that poor Nabby had acquired a husband with just the sort of character deficiencies her father had warned against. Soon after the family had moved to London, they had found their circle enlarged by the presence of Colonel William Stephens Smith, who had been sent from America as secretary to John. Smith was a genial, rather dashing gentleman who had served honorably in the Revolution, both in battle and as aide to General

George Washington. He was ten years older than Nabby, and presented an opportunity, at least so Abigail thought, for Nabby to make a more desirable match than Tyler. The mother wrote excitedly to John Quincy, now back in Massachusetts, that his sister found this newcomer "a man of an independent spirit, high and strict sentiments of honour," and with much "gentleness" in his manner. Despite her disclaimers, the mother behaved as if her daughter, like most women except Abigail herself, could not be relied upon to act prudently during romance. She counseled Colonel Smith privately, for she told John Quincy that on such occasions Smith could hand her any letters meant only for her. Accordingly, John Quincy was urged to send to his sister's suitor those messages which he wanted only his mother to have, for Abigail felt she was frequently the best judge of what her busy husband ought to know about his children in America.

In this period, John Adams' views about his secretary are not known, but he could hardly have been entirely pleased. Smith extended considerably a holiday lark on the Continent, telling John, "I know you will make every allowance, particularly when you consider I have passed the period of ridiculous dissipation, and am now in pursuit of knowledge and improvement." He returned late in 1785 to find that Nabby had broken her engagement to Tyler. Quickly pressing his own suit, the Colonel made his case to Abigail rather than to John, insisting that she understood him best. He spoke fetchingly of his disadvantage, coming from a family without fame or fortune, "but such as I am, I ask *your* friendship and aspire to your Daughter's love." As for the customary statements from a prospective son-in-law about his professional and financial plans, Smith was quite vague. Nevertheless, Abigail was won completely, and Nabby cooperated. The pair was soon betrothed, with Abigail assuring one and all in Braintree of the Colonel's impeccable character. Nabby, she said, had entirely forgotten Royall Tyler. "You will say, is this not sudden," Abigail conceded to John Quincy, who was, in fact, astounded by this swift turn of events. The mother was very reassuring—Nabby's "pensive sedateness" was in command.

Near the time of the marriage Abigail may have hesitated, wondering, perhaps, if in fact she did not know what was behind Nabby's mask of reserve. Consequently, John and Abigail solemnly asked their daughter if she had cleansed her heart of all traces of her former love. She assured them she had. Even so, Abigail was troubled by the haste of the courtship—if by nothing more serious—and she tried to persuade an eager Smith to postpone the nuptials for a year. He refused,

and on the eve of the ceremony Abigail, highly disturbed, dreamed of
Royall Tyler. On 12 June 1786, the wedding took place in the family's
Grosvenor Square house, with the bride's mother wondering to John
Quincy if the event would be justifiable "in the sight both of God and
man." She renewed her campaign to persuade relatives in America of
her son-in-law's integrity and his charming manner, but now Abigail
seemed to be protesting too much.

The Colonel took his bride to a residence in nearby Wimpole Street,
from whence the pair came daily to dine with the Adamses. Nabby was
nearly as prompt as Abigail had been in producing a first child, a son
born in April 1787. Shortly after the baby's arrival, Colonel Smith went
to Portugal to deliver some papers, and was gone for many months.
During this time he wrote John Adams for the loan of 100 pounds.
Soon after his return Nabby went with her husband to New York, where
his family lived and where he was confident his charm and services
would be rewarded by a choice political appointment. The Colonel also
planned to travel about America and make a fortune in land speculation. Neither the political plum, however, nor quick riches came his
way. The Adams family would rue Smith's career, for it was one mostly
of disappointment and disgrace, as we shall see. He ran up large debts
from ostentatious living while growing increasingly desperate over his
failure to become an important man or to build a fortune. As soon as
he was back in America, the Colonel began leaving Nabby alone with
their children for long intervals. It made Nabby feel the more keenly
what she had expressed to John Quincy upon her departure for America, that her preference was not New York but the environs of Boston,
where her brothers and her other relatives were.

Meanwhile, in Massachusetts, John Quincy also had found himself
embroiled in difficulties. His first task on returning from Europe had
been to adjust to the demands of college life. When John had decided
to send his son to Harvard, he had written to the Reverend Joseph
Willard, president of the college, to smooth the path. He wrote that
after having allowed John Quincy to wander "in Europe for seven years"
and serve as a diplomatic secretary, he now faced "the necessity of
breeding him to some profession in which he may provide for himself,
and become a usefull Member of Society." John Quincy was so ad-
vanced, he said, especially in literature, that he could not simply enter
as a freshman. Harvard could have John Quincy Adams only if it saw
fit to place him in the third or fourth year. While the college unbent
sufficiently to admit young Adams free of tuition costs, it refused to do

so except at the standing which the faculty thought best. This was accepted by the family, and so John Quincy had returned home, armed with admonitions to be certain that neither the "vices" nor "fopperies" of Europe accompanied him.

He returned with a heavy heart—and an impressive wardrobe. His rural relatives gawked at his sixty-five pairs of stockings, four pairs of shoes, and many garments in his favorite colors of black and blue. They also marveled at this kinsman who, because of his recent advantages, had the reputation of a prodigy. John Quincy knew well enough that he faced great expectations, for his cousins as well as his adult relatives had kept him conscious of the extraordinary opportunity he possessed to fulfill family hopes. He wrote in his diary that it would be much more pleasing to remain with his parents in London, foreseeing clearly what was ahead in America: "to return to spend one or two years in the Pale of a College, subjected to all the rules which I have so long been freed from: then to plunge into the Long and tedious study of the law and afterwards not expect (however good an opinion I may have of myself) to bring myself into Notice, under three or four years more; if ever." All this John Quincy freely conceded was very discouraging, especially because of his fierce determination to cease depending upon his parents—to "get my own living."

For a time the joy of reunion with relatives and friends in Massachusetts was a diversion, especially when the welcome was spiced with admiration. He charmed his cousin, Betsy Cranch, by singing French songs to her and reading aloud. He even shared his diary with her, which made a vivid impression upon the young lady. "He is monstrously severe upon the follies of mankind," Betsy reported in her own journal. She was not the only family member who fretted over John Quincy's assured ways. Abigail heard from her sister, Eliza Shaw, with whom the prodigy lived in the winter of 1785–86 doing remedial work in Greek and Latin before entering Harvard, that the young man was too strongly opinionated. Even John Quincy's youngest brother Thomas noticed it, wondering why John always felt obliged to differ with others. From England Abigail ordered her eldest son to abandon this high regard for his own views, although within himself John Quincy sorely lacked the self-possession he struggled to manifest.

The young man was sharply critical of what he considered one of his major weaknesses, the inability to govern himself in the best use of time. Yet he found life in a rural village tiresome, a view he held all his life. He marveled at how his relatives seemed able to convert "a life of

Tranquility" into "a life of bliss." He was different, he felt. "Variety is my theme and Life to me is like a journey in which an unbounded plain looks dull and insipid." In sharp contrast to his father's view, John Quincy considered Penn's Hill to be a "dull" place. It "convinces me how grossly the whole herd of novel and romantic writers err in trumping up a Country life," he said. Man's best circumstance was one "which calls forth the exertion of faculties, and gives play to his passions."

By this time, however, his reveries on ambition, politics, and Greek and Latin literature were readily interrupted by thoughts of a romantic nature. If one subject almost completely engrossed John Quincy upon returning to America, it was women. He thought about what it would be like to marry, and he told cousin Betsy that he found girls irresistible. He had complained to her that "he falls in love one moment and is over the next," she reported. This preoccupation followed him when he was admitted to Harvard in the spring of 1786. There he led a discussion before Phi Beta Kappa in which he praised reason as the basis of marriage, and condemned a union of passion, in which mates soon became "satiated by enjoyment" leaving an "astonished couple [to] find themselves chain'd to eternal strife."

John Quincy disliked most of college life, experiencing difficulty in concentrating and often missing recitations. He wrote his father that he knew as much as his fellow students but that he and they had read different books. The tone of faculty members he found haughty, and thus "hard for me to submit to," although he was comforted here by family philosophy, noting that deferring to authority "may be of use to me, as it mortifys my Vanity, and if anything, in the world, can teach me humility, it will be to see myself subjected to the commands of a Person that I must despise."

"It is a most unhappy circumstance for a man to be very ambitious," he went on, "without those Qualities which are necessary to insure him Success in his attempts. Such is my Situation." He was at this time nineteen years old. Conceding that he wanted to grasp at every dream, he feared his ambition would make him envious and lead him into evil, the great menace he saw lying in wait when he left Harvard. "Soon, too soon, I shall be obliged to enter upon the stage of general Society on which I have already met with disgust, and which with satisfaction I quitted." While one part of him craved an exciting, diverting existence, another portion shrank from accepting responsibility for a role of his own. It disturbed him that his parents exaggerated his capacities. John

Quincy graduated from Harvard in July 1787. His parents were still in England and therefore missed his commencement oration, "The Importance and Necessity of Public Faith to the Well Being of a Nation." The address was drawn from the family's gospel, for the young speaker was already gloomy as he saw the decline of national morals exemplified in a recent example of human folly, Shays' Rebellion in Massachusetts.

After he had so expressed himself, hardly daring to think of what he might be able to do to lift America, John Quincy began one of the most difficult times of his life. Just as he had to endure collegiate experience, so he could not escape a season of tutelage in law, a profession into which his family assumed he must follow his father. Thus John Quincy planned to spend three years after graduation studying under Theophilus Parsons, a well-known Newburyport attorney. Anticipating this move, however, John Quincy felt overpowered by insecurity. "I am good for nothing and cannot even carry myself forward in the world," he said, while he peered into the future which the proud but apprehensive John and Abigail were urging upon him. Looking ahead, his assessment was worthy of his father: "Vanity! Vanity! all is vanity and weakness of spirit."

He grew more fearful of bringing sorrow and disappointment to the family, as so many in the younger generation seemed to be doing. "There is a great chance that I myself shall at some future period serve as an additional example of this truth." He became so depressed that Abigail's sister, Mary Cranch, marveled that a person who had been all over the world could be so cast down at the prospect of going to Newburyport. It was a place where, the young man confided to his aunt, "he cared for nobody and nobody cared for him." Often now John Quincy wept, and hearing of this, Abigail insisted upon a physical rather than emotional explanation. His moods must be due to an acid stomach, so he should consume limewater and milk. She sent him a recipe for the mixture. She also directed him to exercise, for it would "give vigor to the Animal functions."

Once he reached Newburyport, however, John Quincy found affairs better than he had anticipated. He made friends and attended parties, some lasting until 4:00 A.M., affairs which often distracted him from his studies. And there were the ladies, about whom the young Adams liked to muse in his diary. Such diversions, however, did little to relieve the fundamental problem. For John Quincy, the world was so demanding, so intimidating, that he feared to try it on his own. He was taught to scorn it, as his father did and at the same time encouraged to conquer it—also like his father. It was the old dilemma visited upon a new generation, keeping John Quincy sorrowing in his diary while he was unable to write to his parents. His recurring self-image, a "mere cypher" unemployed and without character, made him so depressed that he became frightened. His troubled sleep was interrupted by "the most extravagant dreams." At this time he could no longer think of the future—honor and reputation amounted to nothing, he said; they were worthless, delusive goals. By early 1788 he felt completely hopeless. "What am I to do," he cried, hoping "to god I shall not go on in this way." He feared now that society would pity him, a prospect which made him pray "oh! take me from this world before I curse the day of my birth."

Then, somehow, the tough, dogged spirit which was both bred and built into this child of John and Abigail pulled him from this despair. In 1788, John Quincy uttered a supplication which he repeated forty years later in 1828 when he lost his attempt at a second term as president. "[G]ive me resolution to pursue my duty with diligency and application, that if my fellow creatures should neglect and despise me, at least I may be conscious of not deserving their contempt." This exhortation sustained him through a long career in which mistrust of self and society vied with ambition born of a stern spirit of stewardship, the conflict fundamental to the family's sobering philosophy and already illustrated in the life of his father.

By this time, 1788, John and Abigail were returning to America, concerned about two younger sons who were now old enough to have problems comparable to John Quincy's. While he and Nabby had dominated their parents' concern, Charles and Thomas had been growing up in America, in the care of Abigail's sisters, Mary and Eliza. From the start Charles was known for two qualities, his charm and his very delicate temperament, neither, as events would prove, associated with Adamses who survived and flourished. Almost nothing can be learned about Charles from his own pen; only a few letters remain, and these were written during his brief career as an attorney before his death in 1800. They were bland documents obviously designed to avoid responding to questions raised by his parents. Much earlier, of course, Charles had had his share of attention and admonition in the Braintree farmhouse. When he was six, his father told him, "You are a thoughtful Child you know, always meditating upon some deep Thing or other." Praising the lad's "good Disposition," John kept urging him to improve himself, an achievement "which will be an inexpressible Satisfaction to your Mamma, as well as to me."

Indeed, Charles as a child seemed especially close to his mother. It was "Charly" who usually hovered nearest to her, eager to share the worries and happiness which "mar's" face might register. As Charles broke inconsolably into sobs on parting from his mother when John took him to Europe in 1779, perhaps not even Abigail could sense what the moment cost the lad of nine years. She worried over "my delicate Charles," and in spite of assurance from John that "Your delicate Charles is as hardy as a flynt," the boy soon drooped with homesickness. He eagerly endured an almost interminable voyage alone in 1781 to be back home with his mother, returning with orders from his father that he be put in school "and keep him steady.—He is a delightful child, but has too exquisite sensibility for Europe."

There is little doubt that Charles was a person whom others found appealing. As his mother said, "he possessed the faculty of fastening every body to him." She feared that he would be "spoilt by the fondness of caresses of his acquaintance." When Charles' need for and delight in being the "favorite" of others persisted, Abigail sent him a sermon drawn from family doctrine: "Praise," she told him, "is a dangerous Sweet unless properly tempered." She sensed that he was drifting toward arrogance and conceit. Lest his "Little Bosom swell with pride," she reminded the boy that all humanity was so imperfect that humility instead of pride should govern every heart. This was a hard lesson for Charles to learn, especially when his Aunt Eliza reported that the girls sought him out as partner at the dancing parties which were a staple of the Haverhill village life in which Charles grew up while Abigail was in Europe. When Charles entered Harvard (while John Quincy was also there), one wonders how much comfort Abigail derived from her sister Mary's assurances that Charles was so "amicable" that "he will be beloved wherever he sets his foot."

Although a good student, Charles was the first of several Adams men to fall into errant ways at Harvard. From the beginning the family viewed that institution with both affection and mistrust. Abigail put the matter nicely: "I look upon colledge life as a sort of ordeal. If they proceed unscorched it is in some measure a security to them against future temptations." Charles emerged with no such strength. He had led a campus rebellion; John Quincy had caught him rifling his private papers; he had begun to drink intemperately; and he had developed a marked aversion to the rebukes and advice of his family. As John Quincy

said, Charles "does not like to be censured." All this was observed by
the youngest Adams, Thomas, who confided to his Aunt Mary Cranch
(and she to Abigail), "that a young fellow must be very steady and very
resolute to pass thrō college without being often led into some scrape
and that if he had forty sons he should be afraid to send one of them
to have their education there."

The youngest of the family, Thomas Boylston Adams had been a
very shy lad, slow, clumsy, unsure of himself to a pathetic degree, and
often incapacitated by rheumatic pains. Depressing circumstances
seemed to trail him, and in time his response was a predictable melancholy. In 1775, when he was three, he had been gravely ill from cholera, and his Grandmother Smith had stayed with him, nursing him
back to health. In so doing, she had become sick herself and died at
the age of fifty-three. This story was, of course, impressed upon him
by the family, who reminded him of his responsibility for this loss.

Then, in 1777, as Abigail reported, while his brothers and sister were
gathered for letters from their father in Europe, "when she delivered
out to the others their Letters, he inquired for one, but none appearing
he stood in silent grief with the tears running down his face, nor could
he be pacified till I gave him one of mine.—Pappa does not love him
he says so well as he does Brothers."

When Abigail went to Europe, the twelve-year-old Thomas, whom
she now called "a Rogue [who] loves his birds and his Doves," was sent
to live with his Aunt and Uncle Shaw in Haverhill. Here he seemed to
find quiet understanding and affection, and was evidently encouraged
to make animals his companions. Thomas refused to write to his parents, even when John Quincy returned to entreat him—"he says he
knows not what to write, that he is well, and that I can as well do for
him." Aunt Eliza said "he is now innocently playful. I hope he will not
learn to do evil." She was thinking of Harvard which Thomas entered
in 1786. From the distance of England his parents overruled the protests of the Shaws and Cranches who argued that Thomas was too immature for college. Even John Quincy fretted that Thomas was too
young "to be left so much to himself."

Yet off to Harvard Thomas went, Abigail believing it was a prudent
step since she thought John would soon no longer have a government
salary, making a college education too costly for the family. She soothed
herself by sending letters to Thomas which sternly scolded him for the
laziness and rowdiness which she simply assumed were in her son's behavior. These finally made the innocent lad erupt with a cry of indig-

nation. Abigail replied contritely that she had been misunderstood and that his parents were pleased with reports of his diligence. So far, she said, the family had been lucky. Perhaps after all, it would yet be spared the blight of "undutifull and vicious children." For insurance, however, she closed the letter to Thomas with a characteristic sermon on virtue, admonishing him to aim for perfection. The evidence is too scant to know what Thomas thought of such goals, although he was, certainly, the youngster most docile in the presence of John and Abigail's teachings.

The problems of John Quincy, Charles, and Thomas were not the only challenges John and Abigail faced on their return to Massachusetts in 1788. They had to reestablish life for themselves while John anticipated being chosen vice president by the new Electoral College. While they were still in Europe, Abigail had arranged for them to move from the Penn's Hill house to a handsome house on a farm once owned by the Vassall and Borland family, and, briefly, by Royall Tyler, Nabby's luckless suitor. The land was located nearer the old north precinct church and its burying ground where so many Adamses rested.

Actually, the house which John and Abigail bought offered very little more room than the farm cottage on the other side of town, but it was much grander in appearance, having served as a country residence for people made well-to-do by the West Indies trade. This new residence, with its ninety-five acres of surrounding farm land, became known to generations of Adamses as the Old House—a name never far from the mark, as the 1732 building was always somewhat ramshackle. John Quincy went to look at the property shortly before his parents returned—"I was not perfectly pleased with it," he reported, though he hoped the family's presence might in time improve it. Of one thing he was certain: "I shall never make it the standing place of my residence."

John and Abigail were elated to be back in the familiar scenes of Braintree's north precinct. Within six years the area became a village on its own, named Quincy after one of Abigail's ancestors. Quincy remained the family seat for Adamses for four generations, and it seemed to John and Abigail the only safe and comfortable spot in the world. But John's enjoyment of it was soon disturbed. In April 1789, less than a year after returning to Braintree from London, he was in New York City presiding over the new Senate as the nation's first vice president. Thereafter, until 1801, John's home much of the time—and Abigail's when she could leave Massachusetts to be with him—was in New York first, then in Philadelphia when the federal capital removed there late in 1790, and finally in the new city of Washington in 1800 when he and Abigail were briefly the first occupants of the White House.

These scenes were far from John's fields and books in more ways than distance. The only circumstance which would have made John more unhappy than his entanglement with national politics and high office was not being elected vice president. It was the family paradox again—for an inner impulse compelled John to be active in a political scene he considered treacherous and unrewarding. Even as he took his place in the federal capital, the news that violence and delusion were features of the revolution in France only served to make him more cautious in his expectations of republican government, even in America. His conservative views and his unintentionally abrupt, lofty, even rude manners gave many observers the impression that John, the simple farmer from Braintree, was really an aristocrat who favored elitist rule. In fact, he agonized over the vulnerability of the United States he had done so much to create. Not even citizens of America were ready, he felt, to participate in democracy without a careful arrangement of political and social balances and restraints. The behavior of Congress under his scrutiny tended to confirm John's misgivings, as did the followers of Thomas Jefferson and James Madison, who claimed that the nation could trust citizens to be reasonable.

Public affairs soon had so discouraged John that he drew even more comfort from thoughts of private life spent in agriculture. It was home, farm, library, and especially his wife and children which John said brought solace for all his "tryalls." And when he was obliged annually to leave the Old House for the burdens of the vice presidency and presidency, his reaction was to moan: "Oh my sweet little farm, what would I not give to enjoy thee without interruption?" Just as distressing for him was the fact that often Abigail was detained in Massachusetts by illness or household problems. John was always lonely, and he constantly clamored for Abigail to join him. When she raised the question of who would care for the farm and house when she was away, John grew impatient. "I will not live in this state of separation," he would fume, telling his wife she could abandon the Old House "to any body or no body. I care nothing about it—But you, I must and will have."

While John might command Abigail to "leave the place to the mercy of the Winds," he scarcely meant it. For instance, he took sharp interest when Abigail, herself a splendid manager, tried to entice John's brother, Peter Adams, to handle the farm. Poor Peter claimed that it was too difficult satisfying John, who expected anyone working his land to produce "so much more than it was possible for him to do"; and besides, Peter said, he knew that John considered him "a knave and a fool."

So once more it was left to Abigail to handle their properties while her husband was away. She took full advantage of the opportunity. She directed the enlargement of the Old House, adding the parlor (the "long room") and, above it, the presidential study. Knowing her husband's preference, she created as a surprise for him a quiet place where John could assemble some of his books and work peacefully. Farm management, household direction, village affairs were concerns Abigail enjoyed fully. Demanding though these tasks were, however, she continued to take time to admonish and instruct her husband and children on their daily lives. She reminded John of the green gown he was to wear for sleeping, and that he must not forget to put a bearskin over his bedcovers. Those charged with the President's well-being when Abigail was in Quincy were told that John should soak his feet in hot water while taking her mixture of rhubarb and calomel for his habitual cough. She clucked over him charmingly from afar, telling him how lonely she was for him, how she disliked sleeping by herself, and assuring him, "I know you want your own bed and pillows, your hot coffe" and all the "little matters" to which she knew he was accustomed.

Abigail also gave special attention to her mother-in-law. After Deacon John Adams died in May 1761, John's mother had married John Hall five years later. In 1780 she was again a widow, and lived seventeen more years under the roof of her other son Peter, whose house was located about 300 yards up the road from John and Abigail's new home. Peter acquired such a place only because his wife, Mary Crosby Adams, had inherited the property. When John and Abigail tried to get the widow Hall to move in with them, a step they felt could improve the station of a vice president's mother, the old lady politely refused, claiming the comfort of habit. However, she walked to the Old House often for visits, even in deep snow, especially when John was away for political duty—assignments which thrilled his mother, who craved the latest news from Abigail and searched every newspaper she read for stories about her son. Abigail, in turn, comforted her, urged John and their children to send her little presents, and took her to the two church services which occupied nearly the entire Sabbath. When John shipped his mother a barrel of flour from Philadelphia and Abigail presented her with a warm winter gown, the durable Mrs. Hall, nearing ninety years of age, was elated. The old lady died soon after John's inaugural as president in 1797, leaving him deeply grieved. "My Mother's countenance and conversation was a source of enjoyment to me that is now dried up forever in Quincy," he told Abigail.

In some ways Abigail's solicitousness created problems for John. She talked a great deal in public about her statesman husband's contributions and sacrifice, thereby displaying the keen ambition she shared with John. While John rejoiced in her support, he begged her to be more restrained in conversation and behavior. "Let us hold our tongues," he urged. The Adamses, he insisted, must have a modest manner of living; Abigail was admonished to obliterate the family arms she liked to see on their carriage. Sometimes Abigail was chastened by this, insisting that the luster of public life meant nothing—she would gladly step down—and while she conceded it was nice having a carriage, she assured John: "I never placed my happiness in equipage." Still, fretful over the world's snares as she was, Abigail remained the person most ambitious for the family's political success. It was she who rejoiced or agonized as her men did well or poorly in public endeavor.

When she remained in Quincy, where she had nieces and six servants in attendance, Abigail relished being the focal point for the Adams family. Her zeal in urging John to place relatives on the public payroll often embarrassed and annoyed him, but he usually complied. Her sister Mary's husband, Richard Cranch, a studious fellow so lost in theology that he had difficulty providing for his family, was made postmaster in Quincy. Abigail sought as much patronage as possible, whether for a distant cousin or her hapless son-in-law, Colonel William Smith. This practice brought glee to John Adams' political foes, who were growing numerous, thanks to his outspoken ways. To her sisters, Abigail was always attentive. She worried especially over Eliza Shaw. Since Haverhill lay on the other side of Boston, Eliza could not be supervised as readily as Mary Cranch, who lived in Quincy.

It was, of course, upon their children, no matter that they were now adults, that Abigail and John lavished their overpowering concern, reluctant to release their offspring into the world all Adamses were taught to dread. Despite John's demands that his sons make haste to attain independence, he paid them an allowance he could ill-afford—about $100 each per quarter. This was literally a retainer, to keep them as lawyers and to prevent them from straying beyond the family orbit. Talk of other careers or moving west was sternly discouraged. John and Abigail mistrusted their children, for their philosophy obliged them to doubt all human nature trapped in the perilous struggle with an evil society. Abigail begged her busy politician husband not to forget to send fatherly admonitions. "You can do much service to your sons by your letters and advice." And when John might hesitate out of concern for the children's independence, a consideration that rarely hindered Abigail, she prodded him by saying: "You will not teach them what to think, but how to think, and they will know how to act."

Though Abigail was now middle-aged, she was always fearful that she might have to "blush" because of a child's conduct, for when "much is given, much shall be required." She assured John, "I know their virtues and I am not blind to their failings." John must realize, she said, that the most cherished thing he could bequeath his children was a "glorious reputation of integrity," for "your children's children will be incited to virtue by your example." To her sons Abigail threw down an unequivocal challenge—they must avoid every pitfall since the family was now "placed in a conspicuous view." For the members of the second generation, therefore, their father's attainment in a worrisome world made the task of pleasing their parents more difficult and confusing.

This handicap was evident in the lives of all three boys. When Charles left Harvard and joined his parents in New York in 1789 to begin a legal career, John Quincy wrote his brother "a very serious letter" and talked with him "in such manner as must I think lead him to be more cautious." Doubts raised by Charles' conduct in college led John and Abigail to call him to their side. Said John Quincy: "I am well convinced that if any thing can keep him within the limits of regularity, it will be his knowledge of my father's being near [him and the] fear of being discovered by him." Abigail talked freely of her grief over his worsening behavior and the barroom company he kept, while Aunt Mary urged Charles to redouble his guard against "every temptation to evil" and warned that, unless he chose better associates, he would "imbitter the declining years of his Parents."

When Vice President Adams and his lady were obliged to leave New York in November 1790 when the federal capital was moved to Philadelphia, they left Charles reluctantly, hoping to keep up their guardianship by mail. They were quickly disappointed. Charles' surviving replies are secretive about his work habits, filled with grandiose talk instead about politics and friends. Once he attempted to counsel the anxious John Quincy by advising his elder brother "to take the world a little more fair and easy." All this masked the truth that Charles was having great trouble dealing with life. In a moment of rare candor, he confessed to his mother that he worried over what would happen to him and his brothers if John Adams failed in politics. "What will be the path for his Children," he asked. "Will any encouragement remain to follow the road of public virtue?" Charles fled his problems by turning to drink. When John wrote to Charles hoping to "fix his attention" on virtues cherished by the family, Charles evidently destroyed the letters, perhaps because they expressed grief over stories of the son's misbehavior. The son vaguely urged that his father must "not too easily take up ideas prejudicial to me." He insisted that tales circulating about him were false—these were all things "which I attest Heaven are not true." His parents sorrowfully wished that they could believe him.

When Thomas, the youngest, graduated from Harvard in 1790, his search for identity also became a painful tale of struggle against parental domination. Summons from his parents to study law in Philadelphia immediately sent Thomas into a rheumatic attack. He permitted this to delay him in setting out for the federal capital where his father dreamed of being surrounded by three lawyer sons. Finally, Thomas arrived in mid-September 1790, after John Quincy had tried to intervene by urging his parents to recognize that "Tommy" lacked any inclination to be an attorney. He should be encouraged to be a merchant, said the eldest brother. Thomas himself told Aunt Eliza how he yearned for the quiet and affection of Haverhill, and when he talked of breaking away and establishing himself in the West, his father grew indignant. The young man must remain in Philadelphia where he was promised family support until his law practice flourished. Thomas did not relish this dependence.

Under such pressures, Thomas' modest store of self-confidence began slipping away. He informed his mother, "I never was much in love with myself and I feel less so now than ever." His father scolded him for spending too much time with Quaker families who had befriended the quiet, good natured young man. Not only did John fear time was thus wasted, but young ladies were present, and this son must not be tempted into an early marriage. The tension between Thomas and his parents had become a serious matter when, suddenly, he had the chance to escape by becoming his brother's secretary, for John Quincy was about to enter America's diplomatic corps. The elder brother's life had taken on a new dimension, to the great excitement of the family.

As John and Abigail's sons came to manhood, all feared in their separate ways lest they, as John Quincy put it, make their parents "lament as ineffectual the pains they have taken to render me worthy of them." And only one of them, John Quincy, was able to survive this. His career was marked with achievements, yet for a time his prospects seemed as uncertain as those of Thomas and Charles. When John and Abigail had first returned from England, John Quincy gave little promise of strength. His parents found him wishing "that he had been Bred a Farmer, a Merchant, or an any thing by which he could earn his Bread." In his diary he described life as mortifying, and talked of how he had achieved nothing, of how difficult it was to work, and of the shameful lure of brute pleasures. Eventually, he almost stopped writing this diary.

It was well into John's vice-presidential years before John Quincy began to find more peace within himself as a young lawyer. His parents insisted he must remain a bachelor, indefinitely, it seemed. He began to derive some satisfaction from creative success as a political essayist, so that the letters of self-reproach grew fewer, and there were even some triumphs before the bar and some new friends. On a lovely spring day in 1794 he told his father, "I find myself contented with my state as it is," conceding that the old burning impatience was slipping away. "I see very few things in this life beyond the wants of nature that I desire." He was especially gratified that his essays could give him a voice in political affairs without his being actually contaminated by them. John Quincy even enjoyed life in Boston, where he began to feel settled.

These modest achievements were exasperating to John who, for the moment, forgot his own youthful yearning for a quiet life. He promptly sent John Quincy an order which must have shaken the young man: "You come into life with advantages which will disgrace you if your success is mediocre. And if you do not rise to the head not only of your Profession, but of your Country, it will be owing to your own *Lasiness, Slovenliness* and Obstinacy." As it turned out, however, John Quincy had little time to ponder any defense, for news came that he was appointed minister to Holland, a cause of less jubilation for him than for his excited parents. He learned of his entry into public life through his father. At that time, 1794, and for years afterward, John Quincy was embarrassed by the thought that his participation in great affairs began because his famous parent arranged it. He resented being uprooted from the life he was pleased to be making for himself, and he was not wholly convinced by his father's protestations of innocence in the son's selection for duty in the Hague. Appreciating his modest credentials for the appointment and recognizing his father's zeal, he tried to evade the subject of just how he had received the position, saying that it was "of no service to indulge conjecture upon the subject." The elder Adams, however, somewhat guiltily, continued to beg John Quincy to believe that "he had never uttered a word upon which a wish on his part could be presumed that a public office," as John Quincy reported it, "should be conferred upon me."

There is some basis for his doubts; manuscripts show that Secretary of State Edmund Randolph and Vice President John Adams had been talking about John Quincy's selection before the nomination was sent to the Senate, although the father professed to have had no prior knowledge of this step. When his son's appointment was safely approved, John hastened home from Philadelphia so he could personally advise the fledgling diplomat. The younger man was sent off with the usual warnings and ordered to improve his appearance. He was also informed that he would now have a chance to "see with how little wisdom this world is governed," something Vice President Adams believed he was observing first-hand in the new Federal government.

John Adams had not been reelected vice president without a strong negative voice from the rising antifederalist faction, which voted for George Clinton of New York. This faction, led by Thomas Jefferson, who had resigned as secretary of state, was sympathetic with the egalitarian principles of the French Revolution and—even worse in Adams' view—with Thomas Paine's *Rights of Man*. All this confirmed for John Adams the conservatism he had been preaching for so long. Appropriately, it had been John Quincy's "Publicola" essays in the *Columbian Centinel*, in which he repudiated the outlook ascribed to Jefferson and Paine, which had first drawn him to the attention of President Washington. When John Quincy was chosen for a significant diplomatic post, the American republic seemed in grave danger to the Adams household.

With his new summons to duty, John Quincy began once more to write in the diary which he had been neglecting. The young man wrote freely of his hope that his appointment had been due to genuine national need, and not because of his father's maneuvering. This obsession that an Adams must be called to public life rather than seeking office gripped John Quincy as it had his father and would his descendants. Such a summons was considered the only acceptable reason an Adams might leave rural contemplation for a dangerous involvement in the world. Auspicious though his selection might appear, John Quincy grumbled nonetheless: "I wish I could have been consulted before it was irrevocably made. I rather wish it had not been made at all." But go he did.

In preparing for the long absence from home, however, the new Minister told his mother that Europe was the place where brother Thomas might discover a career of his own, beyond the well-intentioned but stultifying concern of his parents. He said it was essential to have a member of the family with him, especially as congenial a person as Thomas. Thomas accepted the role of secretary to his brother, and off they went, John Quincy to the start of a brilliant career and to find the woman of his life, and Thomas to discover that John Quincy's dominance was almost as overpowering as that of their parents. They departed for Holland, leaving John and Abigail with plenty of other worries. The Vice President and his lady also watched as their daughter Nabby became trapped in the wreckage of her marriage and as their unstable son Charles set out to add another unfortunate marriage to the family's story.

3

Courtships

John and Abigail's children passed through difficult courtships, and there were no happy marriages in the second generation. This distressed the elder Adamses, whose experience had been very different, despite the long separations John's duty necessitated. Her children's troubles seemed especially to darken Abigail's already grim view of human instincts, an outlook hardly relieved by the steady weakening of her health during middle age. On the other hand, John seemed more ardent in his letters to Abigail as hers became restrained. To his proud assertion that two decades of marriage had not diminished his yearning for her, Abigail replied that the years should subdue "the ardour of passion," which should be replaced by "friendship and affection" of a more solidly rooted character.

Seemingly confident of her own self-mastery, Abigail became increasingly indignant over the weakness of other women. In 1794 when her cousin, Mrs. John Hancock, now widowed, proposed to marry an "able bodied enough sea captain," Abigail was outraged. How frail was woman, she wrote to John, refusing to dignify this attachment as love. Did not the widow Hancock know that at her age, "the hey dey in the blood" was supposed to be "tame," at least sufficiently to allow prudent behavior? John could not let this pass without comment: Mrs. Hancock's second marriage was a cause for rejoicing, he said. Until now she had experienced "a peevish, fretful feeble child for an husband," so surely she might be congratulated on going "to the arms of a generous, cheerful, good-humored and able-bodied man. . . . I am not censorious. Not I."

Abigail's sister Eliza Shaw, who seemed made of less stern stuff than Abigail, also aroused her concern. Eliza had caused worries when, many years before, she had married the impecunious Reverend John Shaw against Abigail's advice. A warm, gentle person, Eliza proceeded to make her husband's parsonage in Haverhill a haven where the Adams boys gladly went to visit. Eliza had two children of her own, and then, after a ten-year interval, she again became pregnant in 1790 at age forty. Abigail seemed to find the event both alarming and embarrassing. "It is really a foolish Business to begin after so many years, a second crop," she complained, thereupon coming down with one of her frequent "Nervous Headaches" which incapacitated her for a week or more at a time. The baby, which was christened Abigail Adams Shaw after her aunt, proved to be the longest-lived and ablest of Eliza's children.

In 1794, Reverend John Shaw died suddenly after a day of especially vigorous preaching. Eliza was left destitute, but revealed a fortitude which apparently astonished Abigail, who was sometimes uncomfortable when her sisters, who usually deferred to her, showed strengths of their own. Eliza did not remain a widow long; within a year she married again, choosing another clergyman, one who had preached at her late husband's funeral. The Reverend Stephen Peabody operated a small academy in Atkinson, New Hampshire, barely across the state line from Haverhill. Here she cared not only for her husband's students, but for Abigail's grandchildren, who found her what she strove devoutly to be—an embodiment of Christian love.

Aunt Eliza had taken care of Nabby's two sons when the Adamses decided it was necessary to find a home for the boys safe from Colonel Smith's pernicious influence. Of their children's marriages, the one causing Abigail and John the most sorrow was Nabby's union with the Colonel, which, despite high hopes for it during the courtship, had proved to be a disaster. After arriving from England in the spring of 1788, Colonel Smith had settled Nabby and their baby son out on Long Island just beyond what is now Jamaica, New York, but was then a desolate area. She was left mostly alone, pregnant again, while her husband spent his time in New York City indolently awaiting a political appointment. Few of Nabby's letters survive, but these show a young wife whose remarkable fortitude crumbled gradually. Her lost letters can still be glimpsed through recapitulations appearing in other letters exchanged by the family.

Late in 1788, Nabby had a second son, causing Abigail to rush to the rescue. She found the old tension between mother and daughter renewed. Not only did Nabby resist her mother, but she demanded to know everything Abigail wrote to John and begged to be taken to see her father. Abigail admitted to John Quincy, who had advised her against intruding upon Nabby, that he had been right. Matters grew even more discouraging when Colonel Smith's expectations of a grand existence came to nothing. By the winter of 1789, Nabby and her two sons were able to escape the countryside by moving in with John when he rented a house in New York City after being elected vice president.

The Colonel's only contribution at the time seemed to be in begetting children, giving Nabby seemingly perpetual pregnancies—a third son was soon born but did not long survive. Then came a daughter. While this trend alarmed the Adamses, they fumed as much over their son-in-law's idleness. "A gentleman would not have been so long unemployed as he has been," Abigail said, recalling John's unavailing early efforts in London to interest Smith in making a living.

By now the Adamses conceded to each other that Nabby had married the sort of man John had warned against. The Colonel kept up his travels, first a jaunt to England and then excursions for land speculation in upstate New York and Detroit. In the course of these trips, Smith decided that he deserved to be named minister to Great Britain, and he asked his appalled father-in-law to argue his case before President Washington. As justification for his request the Colonel liked to cite "the honor of our connected families." While John turned a deaf ear to Smith's brash requests, Abigail continued to help him, growing annoyed when President Washington hesitated to give his former aide a government appointment. Meanwhile, amid disappointment and debt, Nabby became even more silent, only rarely complaining about her plight. She did describe her situation as "Mortifying," telling her father, bitterly, that her husband had not "been treated by those in Power as he was entitled to expect."

John Adams was less generous about Colonel Smith. He wrote to Abigail calling his son-in-law "an adventurer of an Husband," a person "tormented by his ambition." To make matters worse, young Charles Adams fell in love with Colonel Smith's sister, Sally. To John and Abigail, the Smith clan had at best meager qualities to bring to them, a judgment borne out when Peggy, another of the Colonel's sisters, suddenly married an impoverished Frenchman who claimed a title and promised to bring his friends great wealth in a hurry, a talent which endeared him at once to Nabby's in-laws. Said Abigail, with fine restraint: "I believe the grace of consideration has been but sparingly bestowed upon any member of the Family."

Soon the Frenchman, St. Hilaire, was proved to be a fraud, but not before his tale of a fortune had led Colonel Smith deeper into debt. Smith had even persuaded poor Charles Adams to invest some of John Quincy's money in St. Hilaire's foolishness. Unaware of this development, John and Abigail waited to see if this latest disaster might bring Nabby's husband to his senses. And indeed, he did decide to halt construction of his new mansion, which he had designed as a copy of President Washington's Mount Vernon.

The Colonel, however, could not long curb his determination to live extravagantly. He continued to exist on speculation and credit. Shame, cried John Adams, shame, to those who enjoyed "rolling in luxury upon the Property of others." As for his daughter, who lived in isolation while Smith cavorted, John's only lament was "Poor Nabby!" Busy as he was on the eve of his inauguration as president in 1797, John Adams thought painfully of his grandsons left to the example of their father. He wrote to Nabby imploring her to cultivate in her boys qualities which must have seemed bitterly familiar to her—"prudence, patience, justice, temperance, resolution, modesty, and self-cultivation." Let them have "command of their passions, the restraint of their appetites."

Nabby momentarily dropped her mask and confessed that she was nearly desperate, for the Colonel had taken her once again to rural New York where there was no school or clergyman and where she was often alone. Foreseeing nothing but humiliation for herself, Nabby prayed "that my Heart may be steeled against the misfortunes which seem to await us." However, by the time Abigail arrived to look in on her, Nabby had drawn into herself once more. Her mother reported: "I could not converse with her. I saw her Heart too full. Such is the folly and madness of speculation and extravagance." Nabby did consent to send her sons, little William and John, safely beyond their father's influence. They went to the loving guardianship of Aunt Eliza and her new husband in New Hampshire. Nabby herself, however, would not give in to her parents' pleas to leave Smith and come with her daughter to the family circle in Philadelphia where John had taken office as president. Again and again, just as she was preparing to do so, rumors reached her that her husband was coming home laden with wealth. When the Colonel failed to appear, a humiliated Nabby broke down and wrote to John Quincy. "I have had so many trialls and struggles in my mind to contend with that I only wonder that I have retained my senses."

Throughout the years of her father's presidency, Nabby appeared so distraught from watching the Colonel chase "vissions" and "Ideal

Schemes" that her parents often feared she might attempt suicide. She confided to her mother that existence was a burden and that she was "little short of distraction." Small wonder, since her husband, after all but abandoning her, was now striving to defend himself against stories that he had been pretending to be an agent of the president. With a great show of outrage Colonel Smith asked how it was "that the President should for a moment allow himself to believe *me* capable of so base and dishonorable a colouring?"

Still soft-hearted toward her son-in-law, Abigail sent him a note which he relished as the only sympathy drawn from the Adams family. It brought him to concede to Abigail that perhaps he did not know his own faults—how else to explain "the horrid state that I find myself in?" Self-scrutiny, however, was not Smith's strength. He preferred blaming all his troubles on the "false statements of wicked and designing men." He talked alternately of a campaign to redeem himself or of retiring in disgust from the world. He was never able to perceive why his "honest fame," as he put it, had not brought him public employment.

President Adams reluctantly sought to name Smith adjutant general in 1798, but the Senate rejected the appointment overwhelmingly, despite George Washington's endorsement. A desperate Smith kept begging his father-in-law for some army post, all hope of other preferment having vanished. He was supported by Abigail, who urged his case as she did that of other kinsmen in need of work. John's distress was apparent—"My relations have no scruple to put my feelings to the test," he complained, especially with Smith's repute so low that he assured Abigail that their son-in-law could not pass the Senate even if nominated as a lowly lieutenant. The Colonel's explanation to his father-in-law was that "the fault is in my stars, not in me." Before leaving office in 1801, the President gave way to urging and managed to have his son-in-law appointed surveyor of the District of New York and inspector of revenue, although John's view of the Colonel was unchanged: "All the actions of my life and all the conduct of my children have not yet disgraced me so much as this man," he said. "His pay will not feed his dogs, and his dogs must be fed if his children starve. What a folly!" No one in the family disagreed. Nabby, however, remained loyal, never actually repudiating her husband.

The administration of President Adams was so beseiged by political troubles that Nabby's pathetic state was almost more family trouble than John could bear. In 1797 and 1798 he sought to offset the threat of war with France, a conflict some of his increasingly divided federalist associates desired, while at the same time he was so worried by dangers to the Republic that he signed the Alien and Sedition Acts in June and July of 1798. These were days of torment for the President, who brooded alone while Abigail lay gravely ill at home in Quincy. The cup of public and personal woe seemed full, but there was still more disheartening news.

Nabby's disastrous marriage made John and Abigail worry about what romance would do to their sons, fears which had not been allayed when Charles married Sally Smith. They hoped, however, that John Quincy, in Europe, was safely beyond the threat of love. Surely, they thought, he would not choose a bride unseen and unapproved by the family?

Apprehensions about John Quincy's romantic tendencies had begun years earlier, in the days before he was named to the foreign service. Soon after he had completed his Harvard studies, John Quincy's eagerness for love had been the talk of the family. He himself had professed to see in himself a battle raging between emotion and judgment. Abigail, to help him in his internal struggles, campaigned to ally him with a distant cousin three years his senior, Nancy Quincy, a plot which the mildly amused young man called his mother's "darling project." To his relief, Nancy married someone else, while in 1790 John Quincy fell in love with Mary Frazier, with a passion so ardent that Abigail had to warn him that he lacked the resources to marry. Under family pressure, he gave up Miss Frazier, announcing that his capacity for affection was now ruined. "I am proof against everything," he said, half sadly, half in boast to his cousin Billy Cranch.

After this break John Quincy had worried about himself. He admitted feeling drawn to women not known for prudence and discretion, and wondered if, "alas, my taste is naturally depraved." He began to drink, often and heavily, for which he rebuked himself mercilessly after every "foolish adventure," every "lamentable mistake." When in 1792 he "closed the year with folly," he predicted that he was destined to be a "sacrifice at the shrine of Intemperate Enjoyment." "The subject is awful," he confessed, but what was worse, "there is a fatality against which I find it is vain to resist."

To help control his interest in women, John Quincy cultivated a haughty view of female capacity. He announced that the opposite sex was not intended to compete with men intellectually. "There is something in the very nature of mental abilities which seems to be unbecoming in a female." He sailed off to Holland and his new career, finding in Europe only further evidence that "it is vain to labour and toil against the prescriptions of Nature. Political subservicncy and domestic influence must be the lot of women." Nor did he, during the rest of his life, alter this opinion. Not even memories of his remarkable mother dissuaded him. After all, it was she who had compelled him to surrender his love for Mary Frazier, while Abigail's very different sister, the gentle Aunt Eliza, had been the person to whom John Quincy had turned for sympathy and found it.

John Adams had seen no reason to discourage his son's romance with Miss Frazier, but he learned of the situation too late, as he had in the case of Nabby and Royall Tyler. "I will not thwart him," he told Abigail. His only fear was that "my sons by early and indiscreet connections should embarrass themselves." If John was stirred by John Quincy's love affair, Abigail remained unmoved. Waving aside his talk of a broken heart, she told her son that Providence had kept him single for his foreign mission. Besides, early marriages were dangerous because they were often based on the charm and bloom of youth, and she warned: "Time will trim the lustre of the eye, and wither the bloom of the face." This allowed Abigail to repeat a favorite theme she often stressed to calm her bouncy husband. She instructed John Quincy to defer love until a time when he and his chosen could be intellectually attracted. This basis brought more permanent satisfaction, for then "the ardour of passion settles into the more lasting union of friendship." It was a theme her son had used in one of his student orations at Harvard—before he met Miss Frazier.

From Holland John Quincy tried to make Abigail understand what he had experienced in surrendering Mary Frazier. "I hope you will not think me romantic," he said, but the loss had been more profoundly wrenching than any found in novels. It had been a very close call, he admitted. He told Abigail he had almost disobeyed her decree to abandon his "ardent affection." It had been "voluntary violence," leaving such "wounds" that he predicted he would be forty-five before he recovered.

The young diplomat could not have been a poorer prophet, however, for his amorous desires promptly surfaced soon after he set foot in Europe. This new romance began at the residence of Joshua Johnson, United States consul in London. The Adamses had known his family from their days in Europe. Joshua Johnson had been abroad since 1771, first as an American mercantile representative in England, then in France where he had fled from London as a refugee with his family during the American Revolution, and finally in 1783 back in London where he was appointed the first American consul in 1790.

John Quincy and Thomas found the Johnsons to be a delightful family, accustomed to an elegant life in their house on London's Tower Hill. Joshua was from a large Maryland family. His brother Thomas was the first governor of that state, 1776–79. Joshua's wife, Catherine Nuth, was an Englishwoman whose genealogy is uncertain, although her father may have been a secretary for the India office. She was a tiny, beautiful, brilliant person whose husband openly adored her.

Thomas Adams could claim being first to recognize the great jewel among the eight children of Joshua and Catherine Johnson. This was the second child, Louisa Catherine, who had been born on 12 February 1775 in London. From 1778 to 1783 Louisa had lived at Nantes, in France, where she received a Roman Catholic education, although her family was Anglican. Her father seemed to consider her the most talented and mature of his children, and in a family where music was everyone's delight, Louisa had special gifts. As she herself recalled, she was a musical perfectionist and utterly delighted with her life in London after 1783. "I was so entirely happy I never looked or dreamt of any thing beyond the home," Louisa remembered. "Sense and talent I almost worshipped."

Perhaps it was the talent evident in the twenty-nine-year-old John Quincy Adams which made him pleasing to Louisa when he was back in London in November 1795. He had arrived on a trivial diplomatic errand of such slight requirements that he spent most of his time at the Johnson residence, while Thomas handled his brother's duties in Holland. At first the young diplomat's object was unclear; and the Johnson family, including Louisa, assumed that he was paying court to the eldest daughter, Nancy. John Quincy kept his views to himself, which tormented his mother, who heard alarming gossip from London that her son was once more courting. He preferred to be somewhat mysterious even in his diary, although not when he was upbraiding himself for fancied blunders and shortcomings during his daily visits with the Johnsons, which usually began before dinner and lasted until early morning. When there was a "remarkably pleasant" day at the Johnsons, John Quincy was uneasy. If a day were happy, "that indeed is the greatest objection against it, for what will be tomorrow?" He often wrote how at the Johnsons he was "sullen and silent the whole time." His attendance upon the Johnsons was certainly not all grim. There were plays, concerts, dances, games of whist, art galleries, and walks in the park, although John Quincy then usually found himself accompanied by several of the Johnson daughters and their mother.

Years afterward, Louisa described this courtship to her children, re-

membering how John Quincy made a good impression upon being introduced to the family by the celebrated American painter John Trumbull, who had come to England to work with Benjamin West. Trumbull
jollied the Johnsons about John Quincy, calling him a fine fellow and
capable of making a good husband. Despite the young Adams' appearance in his Dutch-style clothes, amusing to the refined ladies of Joshua
Johnson's salon, Louisa recalled that they all had liked John Quincy's
high spirits and stirring conversation. All, that is, but Louisa's father.
He suspected that the young diplomat had come to woo his eldest
daughter, which alarmed him, for Johnson's southern background
made him dislike Yankees, claiming they made poor matches. His wife,
however, was drawn to John Quincy, and so his suit proceeded for
three months, with everyone, including Nancy, believing she was the
prize. Louisa went about her life "quite unconcerned on the subject,"
and even teased "Nancy's beau." In turn, he claimed not to care for
some of the songs Louisa sang, although in his diary he confided that
she performed wonderfully well.

Three months into his long London stay, John Quincy at last disclosed his preference. In the midst of dinner one evening early in 1796,
with the kind of bumbling abruptness which so vexed him about himself, he handed Louisa some verses he had composed for her. There
followed an exciting scene—the very proper governess to the younger
girls snatched the paper from the blushing Louisa, and the entire table
rose in confusion. Since he often talked of wishing for nothing but a
poet's life, this side of John Quincy made him appealing to Louisa;
their happy sparring turned into love, "an affection that lasted probably much longer than would have done love at first sight," Louisa sur-
mised. She recalled that her sisters now suddenly were teasing her as
the new situation chilled "all the natural hilarity of my disposition."

The next three months had moments of both happiness and misery
for the couple, who soon were engaged. Louisa began discovering the
real person with whom she was to spend her life. Only in his spare
time was John Quincy a poet, and his other moods alarmed her. On
one occasion, not realizing his sensitivity about his appearance nor that
Abigail had often scolded him for slovenliness, Louisa urged him to
take pains to dress handsomely. He did his best, and when she compli-
mented him, it was too much. The young man "took fire," as Louisa
remembered, haughtily informing her that "*his* wife" must never inter-
fere with his dress. At this "high and lofty tone," Louisa, who herself
rarely lacked for temper, told John Quincy to forget their courtship

and walked away. They soon made up, but Louisa later wondered if, had she been more mature, she might then have seen that her husband's "unnecessary harshness and severity of character" would make life with him a perpetual trial to her own joyous and affection-craving spirit.

In his turn, John Quincy discovered that the Johnsons believed the marriage should take place before he returned to Holland, and that Louisa ought to follow as promptly as he prepared a place for her. His hesitation over this caused the family, especially Mrs. Johnson, considerable anxiety. "Madame again very grave," he reported to his diary. Finally Louisa's mother could wait no longer and summoned John Quincy for a talk, in which he gave "a full explanation of my views and intentions." Indeed, he wished to marry Louisa, but this had to await greater prosperity, a point, the young diplomat said, upon which "I *must* adhere." Although neither Louisa nor her family "was satisfied," John Quincy kept to the sort of lofty posture which later often outraged Louisa during their marriage. It was a mood which rang in his statement: "The right and the reason of the thing are however indisputably with me, and I shall accordingly persist."

Persist he did, causing distinct unpleasantness, for the Johnsons were mystified by his cool procrastination. "The usual asperities arose," he would report after an evening with the family, adding, "I must get away." Yet there were times when the couple had satisfying moments, and when John Quincy left for Holland on 27 May 1796, he said: "Took my leave of all the family with sentiments unusually painful." He was not entirely unhappy, however, about leaving a style of living in London which had become a constant reproach to his conscience. His diary had become the altar before which he made his confessions to the family gods of duty and denial. Since he had often lingered at the Johnsons' until after 2:00 AM, he had rarely arisen before 10:00, then required an hour to dress, thereafter reading or writing until 3:00 PM. Once his daily walk was accomplished, he returned to the Johnsons' to take up the social routine again. His comment "What a life!" was reproach rather than delight.

John Quincy and Louisa began a correspondence which helped unfold the relationship of these two remarkable personalities. His letters indicate not only how weary he was of the submissive posture his parents had forced upon him, but that he was consciously determined to dominate in his own circle. He donned a self-centered manner quite different from his father's more considerate style. Simultaneously,

Louisa's letters to her fiancé put in brilliant relief her qualities of understanding and patience which enabled her, at fearful personal cost,
to remain John Quincy's devoted mate, a wife whom, it must quickly
be said, he cared for deeply in his own exasperating way. In Holland
for a year, 1796–97, John Quincy undertook to enhance Louisa's personal qualities by mail, only occasionally mentioning that he missed her
or reminding her of his undying affection, or wishing he could hear
her play the harp. He did acknowledge that she was "one whom I love
dearer than my life," indeed, she was "the delight and pride of my
life." However, this affection was often hidden amid admonitions
grounded in Adams family doctrine and, as he happened to know,
wholly alien to Louisa's pleasurable world. He tried, nicely enough at
the outset, to keep Louisa mindful that in life with him, she might "find
it necessary to suppress some of the little attachments to splendor that
lurk at your heart, perhaps imperceptible to yourself."

One large attachment Louisa could not contain. She did not hesitate
to seek her betrothed's permission to join him in Holland for a visit, if
not for marriage. In reply, John Quincy talked loftily of how he must
have more money if they were to wed. "Let us my lovely friend rather
submit with cheerfulness to the laws of necessity," he said, retreating in
apparent shock from the impropriety he saw in her willingness to come
to him. She must rise above "childish weakness or idle lamentations,"
he admonished, proposing instead the solidly Adams decree of fortitude: "We should be indeed unfit for the course of life in prospect
before us if we indulged ourselves in dreams of finding all our way
strewed with flowers or its borders lined with decor."

To Louisa's shrewd hints that he was overly impressed by rank and
dignity, John Quincy sent another typical Adams message. "You think
me ambitious, and will therefore perhaps suspect the sincerity of this
declaration," he said before affirming: "I never had a wish to be placed
so high in the world at so early a period of my life." Gladly, he declared, would he return to private life. But she must understand the
distinction between ambition and duty, and know that it was the latter
which kept him from her. He wrote her that he had no desire to see
John Adams elected president. The higher his father went, the harder
must the son labor to prove himself worthy, he said, adding that his
burden was already large, and I "do not wish to see it increased."

Louisa's side of the correspondence flourished amid wit and candor,
and John Quincy was often in retreat before her eager, passionate af-
firmations. Prettily, she told him of her struggle to show the fortitude

John Adams in 1766. Pastel by
Benjamin Blyth. *Courtesy of the
Massachusetts Historical Society.*

Abigail Adams in 1766. Pastel by
Benjamin Blyth. *Courtesy of the
Massachusetts Historical Society.*

The birthplaces of John and John Quincy Adams beside Penn's Hill in Quincy. Cottage in the foreground is where John Quincy and his brothers and sister were born and reared. This photograph shows the houses in the 1880s as they appeared shortly before the family loaned them to the Quincy Historical Society. A century later they became the property of the American people by gift from the Adams descendants. *Courtesy of the National Park Service, Adams National Historic Site.*

The Old House as sketched in 1798 by E. Malcolm. This is how the place appeared when the Adams family purchased it ten years earlier. *Courtesy of the National Park Service, Adams National Historic Site.*

Louisa Catherine Johnson Adams soon after her arrival in America in 1801. Portrait probably by Edward Savage. *Courtesy of Mrs. Henry L. Mason.*

John Quincy Adams in 1796. Portrait by John Singleton Copley, a copy of which hangs in the Old House. The painting was done in London while John Quincy was courting Louisa Catherine Johnson. The artist's wife commissioned the portrait as a gift to Abigail Adams. *Courtesy of the National Park Service, Adams National Historic Site.*

Left, the second Abigail Adams, "Nabby," who was the wife of Col. William Stephens Smith. Portrait by Mather Brown was painted in 1785 and now hangs in the Old House. *Courtesy of the National Park Service, Adams National Historic Site.* *Right,* Thomas Boylston Adams in 1795. Miniature by Parker, whose first name is unknown. Thomas had it made in Holland to appease Abigail, who was clamoring for a likeness. The portrait presents Thomas at one of the few cheerful points in his unhappy life. *Courtesy of the Massachusetts Historical Society.*

The Quincy village center in 1822. Eliza Susan Quincy's watercolor shows the old burial yard where many Adamses repose. Beyond is the church structure which was soon replaced by the Adams Temple that survives today. To the left toward the bay was the Mount Wollaston farm so dear to four generations of Adamses. *Courtesy of the Massachusetts Historical Society.*

he preached. Alas, she said, "though I listened to the Teacher, I lost the lesson." Gently, she kept up the pressure for him to marry her: "there is not any thing on earth can afford me equal happiness to accompany you to whatever part of the globe the fates may destine you." She was frank about her sense of inferiority, "my conscious deficiency appears manifest and I already think I see you blush for my awkwardness," yet she wanted to believe that "you will forgive and encourage me by your kindness to mend."

Even as Louisa's warmth filled her letters, John Quincy insensitively inquired of her if her devotion and love could withstand a difficult life ahead with him. Whether he was projecting his own self-doubt, whether he was intimidated by a woman whose vibrant qualities were not those of the domesticated female he preferred in a wife, or whether John Quincy simply was so inhibited that he drew away from the forthright Louisa, can only be conjectured. However, by the close of 1796 his haughtiness had pushed Louisa into indignation. Did he not realize how much their separation devastated her? she asked. She was so miserable, so gloomy that her family called her "the Nun." She begged him to believe her avowals. She thought if he were there, she could contrive to persuade him "how much, how sincerely I love you."

Troubled by her lover's sudden and discouraging questions, Louisa asked bluntly, "why write to *me* in this stile? . . . I have never doubted *your* affection and I cannot conceive why you should suspect *mine*." When John Quincy seemed to become more obtuse and clumsy in the face of her loving insight, Louisa stated in some exasperation what she would repeat all her life, that "you seem to me to have very little knowledge of my disposition." She wrote, "alas, how are you changed," and summoned all her charity to say that she hoped he would never have to feel the "anguish" which his "late severity" had caused her.

Louisa's straightforward manner may have brought Abigail's style to John Quincy's mind. This could explain why he became so impatient and pretentious when he replied to her sparkling attempts to point out that he had a better side. One thing Louisa must understand, he said: she must never display temper toward him, for it repelled him. And then a threat: "I do most cordially wish, my amiable friend, that you may never have occasion to know whether I should possess a proper degree of spirit or not, in opposition to you." To this he added the cruelest touch yet, asserting that he had been blinded by "an irrational love" when he wooed her in London. It had left him temporarily unable to point out her share of humanity's imperfections! He did concede that, while he might not now seem as "tender and gentle" as he had in London, he still loved her. It was simply that the calm reason which had fled him then was now restored. Surely she knew that marriage must be a "mutual exhortation and encouragement of each other to every honourable project and every laudable employment." Therefore, he said, Louisa must remember that any plea from her that he "take time from myself, in order to give it to the world" was "little better than frivolous." Seeing that she was beaten, Louisa chose to reply very simply. "My whole life shall be devoted to render you happy."

In the meantime, John Quincy had been conscious of what marriage into the apparently wealthy Johnson family would mean to him in his search for independence. Mindful of Joshua Johnson's claim to vast American holdings in Georgia, John Quincy told his parents he wanted no more public employment and no legal career. "There is some probability that I may be induced to make a settlement in one of the Southern States," he confided to an alarmed Abigail who was still in the dark about his engagement. Knowing how eager Louisa's father was to see her married, John Quincy slyly awaited new propositions from Joshua Johnson which might assure the leisurely literary career young Adams coveted.

Even as he discussed his marriage to Louisa with the Johnsons, John Quincy toyed with Abigail, merely saying to her, "I begin to think very seriously of the duty incumbent upon all good citizens to have a family." To which the artful son added: "If you think this the language of a convert, perhaps you will enquire how he became so? —I am not yet prepared to answer that." Gradually, however, Abigail suspected a Johnson girl had ensnared her son. Groping for a target, she criticized Joshua Johnson's wife, a person wholly unlike herself. She warned John Quincy that every Johnson daughter might inherit from her mother "the taste for elegance which her manner is conspicuous for." Such daughters would also be handicapped by being "half-blood," the patriotic Abigail pointed out, meanwhile chiding John Quincy on how quickly his boasted immunity to women had been pierced by "youth and beauty." Her words to him were simple: "weigh well," she said. When John Quincy's parents finally had details of this latest romance, they were doubly troubled because it made the son talk of abandoning the career upon which they had launched him. Thus, they resented news about his impending marriage for more than ordinary reasons— Louisa was not fully American; she had been reared amid pampering and luxury; and the Johnson influence seemed to pull their talented son from public life and from New England.

John Adams took time from worrying about war with France to urge that his son be aware of the danger in a wife "of fine parts and accomplishments, educated to drawing, dancing and music." Thinking of his son's future in public life, the father stressed that a discreet and prudent mate was essential. Said John: "I give you a hint and you must take it"—the son should watch his purse if he married a Johnson. Abigail wrote of her fears that courtly life for Louisa would "be the final step" to unfit her for the discharge of "those domestick duties which cement the union of hearts and give it its sweetest pleasures." Both parents recalled what London life had done to corrupt Nabby's choice of a mate.

Parental displeasure was a painful experience for John Quincy, exacerbating his impatience with the sprightly Louisa. In addition, Abigail blamed Louisa for a misdemeanor of which the young lady was wholly innocent—the secrecy with which John Quincy teased his mother. As a parent who had difficulty in releasing a son to another woman, Abigail became more resentful and spiteful than she might have if John Quincy had earlier taken her into his confidence. This created a hostility between the two women which never quite disappeared.

At last, however, after her son had been betrothed for six months, Abigail grudgingly gave in, sensing she had angered him by her reluctance. Her hesitation had reflected only good intentions, she wrote him, stating again the fear that a young and inexperienced Louisa would come to the United States with "anti-American" tastes and sentiments. However, she now urged John Quincy to marry Louisa: "give my love to her and tell her I consider her already my daughter." With consent came a new warning, however: Louisa must know she married Abigail's son—John Quincy should tell his bride that she must follow him; he was not to settle in the South for her sake. Here Abigail came back to the most distressing feature of her son's plans. She believed that in choosing a girl alien to both New England and America John Quincy had made himself—and the Adams family—vulnerable to much embarrassment. There was even danger of decline, especially when he seemed in no mood to continue as a public figure or to resume the practice of law.

On this matter John Adams was as direct as Abigail: "I do not approve of your Project of quitting the Diplomatic career at present; much less of your Thoughts of settling in the Southern States." This stern family voice, along with the chidings about Louisa, understandably disconcerted John Quincy. His replies to his parents wavered between assertions of humble obedience to sullen reiteration of his determination for independence as he approached thirty. He spoke of how his parents were too ambitious for him, of how they did not understand him. On the other hand he wrote, "I never can find contentment or delight at a distance from my parents." He was bound to them "by more than ordinary ties." He sounded as if he were sternly reminding himself of what the Adams bond had to be. Even so, he stood his ground on one point, telling his father that a leisurely life pursuing literature was his goal and shrewdly pointed out the laudable purpose in this plan. He reminded John that as a literary man he would seek to raise America in Europe's eyes.

As he slowly prepared for marriage in the summer of 1797, John Quincy chose not to tell his parents that the great event was at hand. He was still rebellious, refusing to surrender his plan to settle in the South, an intent he assured Abigail was not simply at Louisa's behest. Nor did he change his distaste for returning to legal work. As for remaining in the diplomatic service, he wanted his parents to know that he did so only because George Washington wished it. He would, he said, never keep a public place under his father's auspices. Yet as John Quincy went to London and his wedding, he had been hit squarely by Abigail's parting shot: how thrilled she was that his career abroad had not sullied his father's image. "Persevere, my son, and be the ornament and glory of your country, and the solace and comfort of the declining years of your parents."

But now John Quincy faced another stratagem from home. Late in his administration, President Washington had requested that he leave The Hague and become minister to Portugal. The young diplomat seized upon this excuse to inform Joshua Johnson that marriage to Louisa must be postponed because he had no way to take a bride to Lisbon. However, Johnson was now even more impatient, for he was planning an immediate departure with his family to America, and Louisa's relatives were determined to see her wed before they left London. Consequently, Louisa's father immediately arranged that one of his commercial vessels would take the bridal pair to Portugal. This put John Quincy in a difficult spot, made more so when his prospective father-in-law disclosed that the Johnson wealth was an illusion and that he must abandon huge debts in England and make a fresh start in America where he hoped to regain his money through claims to a vast property in Georgia.

For a time, John Quincy encouraged Louisa to accompany her parents, asking that she submit to his elevated talk of denial and sacrifice. The young woman could not resist observing that there was another approach to life beyond the one he had learned from youth. "Life is short, and admits not of much real felicity," she told him, "therefore we ought not reject the good that [it] offers, by watching for evil." Surely, she said, they could together overcome the many hardships he thought might arise if she joined him now. Then she made one last thrust. With just the slightest touch of sarcasm, Louisa said that if anything inconvenienced him, she was "the last person on earth to desire it." And since he made it clear that marriage now would hamper him, "I have relinquished the pleasing hope of our meeting." Louisa had given up.

But Joshua Johnson was not so easily outdone. He blandly announced to John Quincy that the good ship *Mary* was waiting to become a honeymoon gondola. Caught between his pledge and his preference, John Quincy did little to disguise his displeasure, complaining to Joshua that he did not wish for such extravagance. He harshly advised Louisa once more to lay aside her foolish expectations of love and marriage. She "put too much gilding" on the prospect, he warned, and added, "you have promised yourself too much, and I regret already your disappointments." With wretched grace, John Quincy decided he had no choice but to head for London and marriage, arriving on 12 July 1797. He made one last show of reluctance by waiting to see Louisa until his second day in the city. It was a slight she never forgot.

John Quincy married Louisa on 26 July, with the Johnsons and Thomas in attendance. Afterwards the party went to visit a country seat, returned to London for an interminable dinner, and finally released the young couple to themselves near midnight. The fortnight which followed was filled with parties, leaving John Quincy grumbling at these "appendages to the ceremony, from which I should have been willingly relieved." Louisa recalled the time more happily, until disaster overtook them. Her father's weakened financial structure broke down. The circumstances are unclear, but Louisa's brother always blamed the collapse of their father's fortunes on John Quincy. The delayed marriage caused Johnson to reach America too late to salvage his property there, or so some family members thought. Others felt that providing a boat for John Quincy and Louisa had been the fatal step, the final indiscretion by which Johnson's debts caught up with him. As it proved, Johnson fled to America on 9 September with his career in ruins. He could not pay even the 500 pounds he had pledged to John Quincy as the start of Louisa's dowry. Worse, perhaps, for John Quincy was that his plans for writing under warm, southern skies had been dashed.

Her father's disgrace greatly affected the remainder of Louisa's life.

The dread memory reappeared often in her vast collection of letters and reminiscences, showing that the fall of the parent she worshipped left her vulnerable before a husband whom she feared as much as she loved. Louisa's dreamlike world had been destroyed, and her marriage lay forever in the shadow of these ruins. Most mortifying for her was the certainty that everyone, particularly the Adams clan, was convinced that she and her parents had contrived by deceit to ensnare John Quincy. Such were Louisa's thoughts as she sat trembling in a London hotel room listening to the knocks of creditors whom her parents had barely escaped. Louisa would always believe that entering the Adams family as she did blighted "every future prospect." The memory, she said, "has hung like an incubus upon my spirit."

John Quincy was tight-lipped over the Johnsons' sudden poverty, and Louisa knew he was deeply distressed. He allowed only one comment to escape later when, in sympathizing with his mother's disgust over the worthlessness of Colonel Smith's clan, he said, "I am too familiarly conversant with such things." Louisa, however, believed that the loss of a dowry and prospects of a literary career annoyed John Quincy less than the notoriety which might come to his family for being connected to the Johnsons. "I am perfectly conscious," recalled Louisa, "that to a mind like *his* the wound would never be healed." Not only did she claim afterwards that she was never able to talk openly with her husband about Joshua Johnson but that she had to watch him prejudice their children "to whom my connections have been objects of scorn." A quarter-century after her wedding, Louisa asserted: "Happy indeed would it have been for Mr. Adams if he had broken his engagement," and thus spared himself "a Wife altogether so unsuited to his own peculiar character."

Life for the newlyweds was further complicated by a change in plans. John Adams had become president earlier in 1797 and decided that he needed the talented eyes and ears of his son in a more significant spot in Europe than Portugal. He ordered him instead to Prussia as America's minister in Berlin. John Quincy and Louisa left London on 18 October 1797 for this new destination, where they spent nearly four mostly unpleasant years. While they had servants, usually six—a butler, a personal maid for Louisa, a footman, a housemaid, a coachman, and a cook—in comparison with others in the diplomatic community and with the life Louisa had known at home, they had to live with a certain austerity required by John Quincy's salary. Louisa herself sewed the bed-curtains for the unheated chamber in which she and John Quincy slept. Her husband managed often to be delighted with his bride and worked especially to convince the dubious Abigail that his mate was well chosen. "My wife is all that *your* heart can wish," John Quincy assured his mother, adding, with what was probably unintended ambiguity, "I am as happy as a virtuous, modest, discreet and amiable woman can make me."

As daughter-in-law of the president of the United States, Louisa found herself addressed in Berlin's court circle as "your Excellency" and "Princess Royal." Her beauty, talent as a musician, and grace as a dancer won many hearts, but her husband, uneasy with his rank in an alien world, was constantly apprehensive lest Louisa embarrass them amid the peculiarities of etiquette in Prussia. Louisa found John Quincy to be a mixture of concerned affection and overpowering severity, a combination which mirrored his rearing. Louisa's love, in turn, was blended with awe; she talked of her husband's self-discipline, his transcendence of foolish temptations, and his "unshaken purity and integrity beyond all praise." Of those early years she recalled, "My love and respect for him was so unbounded" that she remained conscience-stricken at the burden she brought their marriage. His "superior talents" made her "venerate" him, Louisa said, "although I often shrank from the severity of his opinions in passing judgment on others less gifted than himself."

Nevertheless, Louisa had the gift for knowing how terribly vulnerable and insecure John Quincy felt. She did all he would permit to support him, including candid replies when he sounded the refrain about intending to flee public life. She knew he would be wretched in private life, and she tried to tell him so. He had been reared in "flattering circumstances," Louisa said, and thus he had "insensibly acquired a taste for them." Louisa's powers of discernment were evident from the start, but her husband tried to keep them subdued. When the political and diplomatic issues in Berlin drew her attention, John Quincy was quick in his rebuke. Women had no business thinking politics, she was told, so that a crestfallen Louisa "sought no further." In short, living with John Quincy Adams brought nothing less than a cultural shock to Louisa, who never forgot how her father had been unfailingly deferential and attentive to her mother.

Fortunately, there was another and very different Adams at hand to comfort Louisa, John Quincy's brother Thomas, who had come to Berlin with the newlyweds. Louisa was forever grateful for Thomas' comfort in 1797–98. The young man was genuinely taken with his sister-in-law, describing to Abigail Louisa's appealing "softness of temper," her tender affection for John Quincy—she was "a most lovely woman," he said, worthy in every way of his brother. Both at home and in Berlin society, Thomas cheered Louisa while John Quincy earnestly played the diplomat's role everywhere but in his household. It was Thomas who kept near Louisa at parties and danced with her, which delighted Louisa, who soon became a ballroom sensation. When dancing, Louisa said, she felt all the tension and loneliness in her existence drain away.

Life for the youngest of John and Abigail's sons had not been pleasant in Holland and Prussia. Thomas' poor health continued to bedevil him, and he was ever conscious of his dependence upon his elder brother. His mother kept up a stream of letters full of stern advice, much of it designed to steer Thomas away from the lure of women. The long-suffering son finally sent a lively reply: "You enjoin upon me insensibility to female charms. Five and twenty years ago, you would perhaps have spared a young man so impracticable, not to say *unfeeling,* a task. Ask of my creator to remould my nature, and perhaps I shall yield obedience to it." But Thomas quickly assured his mother that, despite "this rhapsody," he still pledged her his "filial respect and affection."

Louisa's gratitude to Thomas was boundless, especially since he "made me the fashion—and I became a belle." It was he who "soothed me in my afflictions, corrected gently my utter want of self-confidence, flattered me judiciously, and by his unerring judgment, often prevented me from committing mistakes." Thomas left them in September 1798 to return to America, and Louisa was almost inconsolable. His first letter after his departure was to Louisa alone. "I absolutely kissed it," she wrote in reply. "Would to heaven we could have you back again." She wrote him that John Quincy was also disconsolate, unable to mention Thomas without weeping. "I never saw him so much affected by anything in my life."

In this first year, Louisa discovered other aspects of her husband's makeup, an emotional side he never would be able to contain: he was afflicted with nervous collapse every time his wife or children were ill. At the beginning of their marriage, Louisa experienced a series of unsuccessful pregnancies, during which her husband could only hover near her in alternate states of despair and hope. John Quincy filled his diary with gloomy forecasts as Louisa failed repeatedly to carry a pregnancy to term. He tried to find a lesson in this disappointment. "It is neither wise nor good to murmur at the ways of Providence," he wrote four months after their wedding, as Louisa lay gravely ill from a miscarriage. "I have been highly favoured: beyond my deserts and even beyond my wishes—Shall I receive good, and shall I not receive evil?— The mind at least submits, however the heart will rebel."

Even Thomas had been stirred by Louisa's difficulties, confiding to his journal, "poor little woman; how she suffers! Matrimony, these are thy fruits! Bitter Bitter." Abigail was more business-like about Louisa's miscarriages. "I regret that Mrs. Adams has got into a habit which I fear will injure her constitution," she told Louisa's fretful husband, who now viewed each episode as further assurance of a barren marriage, which he claimed was a "horrible" prospect. Yet in this recurring ordeal, he drew consolation "from the loveliness of temper and excellence of character of my wife."

Public life brought to John Quincy nearly as much unhappiness as Louisa's health did. He never was comfortable as a diplomat in Berlin where he was, in effect, working for his father and holding distinction, he feared, simply because he was an Adams. John Quincy took no pains to disguise his discomfort; he was pleased that his parent was president but wished that he, as his son, could be quietly at work amid books in private life. This made his father indignant. John was unsympathetic with his son's distress, calling his outlook "too refined" and "the worst founded opinion I ever knew you conceive." Although John considered his son at least as qualified as any man in America, John Quincy was still unconvinced. He wished to be recalled immediately, saying that his father's judgment was excellent except where his son was concerned. The distressed John Adams described their child to Abigail as a "very proud young man," but the President stood firm. John Quincy and Louisa remained in Berlin until the Adams administration ended in 1801. Then the pair, with a baby son born after nearly all hope had been abandoned, made their way to America, a land Louisa had never seen. There, they shared with both the Adams and Johnson families more sorrow and disappointment than romance and hope.

❧ 4 ❧

Disappointments

Thomas Adams returned to America late in 1798, three years before John Quincy and Louisa, arriving in time to share some of the most distressing events in his family's history. On the public side he found that his father's presidency was moving into troubled times. England and France were at war, and many American citizens sympathized with one or the other. For some, including many of Vice President Thomas Jefferson's followers, France represented Europe's hope for freedom and equality, while powerful individuals in John's Federalist Party saw England as nobly defending rule by those fit to govern.

This conflict of opinion in the United States created much agitation in the press, criticism of President Adams, and demands from many Federalists that the nation go to war against France, whose abuse of America's rights as a neutral seemed to justify retaliation. John wanted to avoid this step, believing that the Draconian measures necessary in wartime would destroy the delicate balance between state and federal governments so recently created. He also feared that war would upset the division of authority among the three branches of federal government, the check and balance system the President considered essential if the United States was to be spared tyranny by the few or mob rule. Such an arrangement must be allowed to mature, John contended, even at the cost of keeping peace with revolutionary France.

John's own party, however, pressed harder for war. Many Federalists looked to Alexander Hamilton, who had been powerful in Washington's administration, as their real leader. Even some of John's cabinet members were loyal to Hamilton, who believed that a war atmosphere would be the time to end the confusion in America over federal versus state rights. Hamilton wanted a strong central authority through which the nation could be led by the rich and well born.

When Thomas Adams arrived in Philadelphia, he found Congress swayed by the Hamilton faction's preference for hostilities with France and for silencing public criticism of the Federalists. John shared many of their misgivings, and he allowed the Alien and Sedition measures to become law, hoping these would restore harmony. But when members of his cabinet and Congress still demanded war against France, the President courageously stood his ground, securing agreement with the French which assured peace. Many historians now contend that President Adams' strategy was the step which preserved the federal Union while preventing the rise of an autocratic central government. However, the cost to him was high. He had dismissed those members of his administration whom he considered disloyal, thereby forcing an open break in his party and bringing public repudiation of him by Hamilton in 1800. The resulting split in Federalist ranks prevented John's reelection.

These public difficulties were grievous enough. Yet the President's children may have been an even more painful burden. John was elated to greet Thomas on his arrival in Philadelphia and immediately invited him to become his private secretary. The young man resolutely refused, however, announcing that he must find his own life—no easy task even then for a president's child. He decided to begin legal studies, which he detested, and admonished himself that "whatever pain it may cost me I must necessarily isolate myself from my family." Unfortunately, poor health, an aversion to hard work, and a fondness for strong drink all obliged Thomas after all to rely on his parents during much of 1799–1800.

In addition, John and Abigail were alarmed about John Quincy, an unwilling diplomat in Berlin who talked of abandoning public service altogether. Nabby's difficult marriage to Colonel William Smith was growing worse, and finally, Charles seemed clearly determined to ruin himself. John Adams was so distressed with his children at this time that he confided to Abigail he hoped there would be no more grandchildren. "I cannot bear the trouble of children at my age." Nabby and Charles alone, he said, "will bring down my grey hairs with sorrow to the grave."

Even Abigail slipped into depression, especially when the President wrote that remembering their "shiftless children" was "enough to distract me." She freely admitted that painful thoughts of their offspring were now always with her: "In silence I do reflect upon them daily." And yet, she wrote, trying to comfort her burdened husband, it was futile to look back. Had not they done their best as parents? She grieved especially over Nabby, who was innocent of wrongdoing but whose emotional plight probably explains why her skin was covered with a rash. Meanwhile, the Colonel begged for the humblest military appointment available. John replied that he would rather Smith went to the grave, so stripped was he of honor. Thinking of George Washington now retired to the quiet of Mount Vernon, John cried to Abigail, far away in Quincy: "Happy Washington! My children give me more pain than all my enemies."

Although Charles' relatives saw signs of stability in him when he was admitted to the bar in New York in 1792, and when he married Sally Smith in 1795, these were brief reprieves in an otherwise relentless deterioration. What particularly frustrated John and Abigail was that Charles usually avoided giving any personal information in his letters, and he carefully dodged their questions. Instead, most of his few surviving manuscripts are verbose reports on New York politics as Charles saw the scene—he fancied himself an ally of Alexander Hamilton and he was fascinated by "Baron" von Steuben, the Prussian officer who had aided Washington's army and remained in America. Charles occasionally lectured his father on the dangers in the French situation, but there was little else save an occasional indignant request that his parents not believe the malicious reports circulating concerning him. Meanwhile, John Quincy demanded to know the whereabouts of his $4000 and the interest on it. Charles made no reply. His wife Sally disclosed his tragic plight before Charles confessed any difficulties.

A broken-hearted Abigail said she knew what ailed her son—"the poor child is unhappy, I am sure. He is not at peace with himself," although her letters of exhortation must have given scant comfort to Charles. Late in 1798 Charles finally told Abigail how he had misused John Quincy's money, adding, "I have not enjoyed one moment's comfort for upwards of two years on this account. My sleep has been disturbed, and my waking hours embittered." He excused himself on the grounds that Colonel Smith could have coaxed the money from John Quincy himself. Abigail was not impressed: "Charles, you know, never had the power of resistance."

Soon Charles was beyond help, and to the close of his life he seemed to avoid his parents. He got little sympathy from his father. John Adams chose not to try to see his dying son, whom he described as a "madman possessed of the Devil." At least the biblical David's son Absalom had "some ambition and some enterprise," he said, adding: "Mine is a mere rake, buck, blood, and beast. . . . I renounce him." After this, John spoke of Charles no more, giving attention instead to Charles' wife and two daughters. Abigail, too, concentrated on this new generation, foreseeing a bleak future for her grandchildren. "When I behold misery and distress, disgrace and poverty, brought upon a Family by intemperance, my heart bleeds at every pore."

Loving attention, aid, and admonition—perhaps too much of these—had been lavished on Charles. Yet this evidently likeable young attorney turned a promising career into rubble in less than a decade. Reasons for this tragedy are not hard to find; the Adamses themselves hinted at a congenital weakness that made alcohol dangerous to them. Facing the stern expectations of his relatives, repeated as loudly to him as an adult as when he was small, Charles may have found drink the easiest method of endurance. Or, it may have been a means of defiance, just as he chose his wife and certain of his friends in the face of family opposition. The price of being an Adams was more than Charles seemed willing to pay, and his rebellion brought him swiftly to represent the very qualities his parents most feared. The head of the family must have felt he had no alternative but to renounce such degradation.

Abigail watched closely as Charles entered his final collapse in the summer of 1800. There was an awful fascination about the undoing, this descent into complete degeneracy; Abigail saw "the whole man so changed," swallowed as he was by "ruin and destruction." She could not turn away from her son as abruptly as John had. "All is lost, poor, poor, unhappy wretched man," Abigail said, and soon thereafter visited him for the last time by pausing briefly in New York on her way to John's side in Washington. Charles died three weeks later on 1 December 1800, abandoned by all but his wife and Nabby, who had taken her brother in when he had no other place to go in New York.

The doctor whom Abigail consulted about Charles had shrugged aside the case as just another hopeless instance in which a man had lost his better nature. To the Adamses, however, Charles was guilty of turning away from his responsibility, a subject on which the family had always had much to say. When news of Charles' death reached Abigail, she seemed relieved. He "cannot now add another pang to those which have pierced my Heart for several years past." After taking a strange pride in the fact that at least Charles had avoided the appearance of

an alcoholic ("He was bloated, but not red"), Abigail said the most generous thing possible under the circumstances—"He was no mans Enemy but his own." She believed that Charles' fate should teach the family humility so that the God who had permitted them to be so wounded
"may also mercifully heal us."

Little was heard about Charles after his death, for the family wanted
to forget him. Even the last gesture made to so many other unfortunate
relatives—burial in the family vault in Quincy—was denied him. His
father found it difficult ever to speak of Charles' tragedy, but he allowed one anguished cry to escape. "Oh! that I had died for him if that
would have relieved him from his faults as well as his disease." In contrast to John's hurt silence, Abigail extravagantly deplored the failure
of Charles to measure up. How sharper than a serpent's tooth it was,
she said, to have a "graceless child." Charles' fate was so mortifying that
his brother Thomas, expressing the family's sentiment, said: "Let silence reign forever over his tomb."

Charles' decline had come during one of the most difficult moments
in his father's career, when, as incumbent, he was defeated in the presidential election by Thomas Jefferson. Seeing himself at the moment a
failure and watching his children seemingly throw away their opportunities, John wrote Abigail an outburst of frustration and disappointment that must have stunned her. "Blind, thoughtless, stupid boys and
girl!" "What I cannot remedy I must endure," she replied, as she set
out to help him move to the new capital in Washington where they
briefly became the first occupants of what was later called the White
House. From there John arranged one of his most important contributions to American history, the appointment of John Marshall, of Virginia, as chief justice.

Eager to put politics behind him, John Adams left Washington early
on the morning of Thomas Jefferson's inauguration. He went alone,
for Abigail had preceded him. He was preoccupied with family sorrow
more than political disappointment, and felt his presence at his successor's swearing in was not important. He himself now put little stock in
ceremonies, remembering that no member of his family had been in
Philadelphia when he took the same oath in 1797. Instead, he pressed
on to his beloved Quincy where the prospect of working with his land
and books, peacefully beyond the arena of human folly, sustained him
immeasurably. The family marveled at his calm in the face of defeat.
John rarely mentioned his religious consolation for, as he liked to re-

mind Abigail, "I never write or talk upon divinity. I have had more than I could do of humanity." However, in the dark days surrounding 1800, he acknowledged that he found the Episcopal service especially comforting. It emphasized benevolence, submission, resignation, and reformation, which he felt to be his goals. "Alas, how weakly and imperfectly have I fulfilled the duties of my own religion," he said.

The retired statesman's remaining sons were gratifyingly loyal. John Quincy told his father that all "thinking and impartial" persons would see the splendor of President Adams' contributions: "You were the man, not of any party, but of the whole nation." Thomas wrote that John Adams departed for Quincy with "the veneration of *all* your countrymen." Indeed, he said, if this were "the only inheritance left to your family, they might esteem themselves rich in possessing it." Insisting that his father could depend upon the sober decisions of posterity, John Quincy urged that he write his memoirs. "Alas! Alas! Alas!" John responded good-naturedly, no one would believe his autobiography, and indeed he himself sometimes doubted that he could have had such a remarkable career. As for Abigail, she found her husband now possessed a tranquility she had never before seen in him. John referred to his electoral defeat simply: "My little bark has been oversett," leaving him to look forward at last to happiness. And for his children, he was certain that "my retirement" was a favor. "They will now have fair play," instead of the disadvantage of a parent-president. "If I were to go over my life again I would be a shoemaker rather than an American statesman," John told Thomas, and he insisted to John Quincy that it was exquisite pleasure knowing "that nothing remains between me and my grave but my plough."

On the other hand, Abigail took the enforced retreat to Quincy less stoically. She was disgusted, she told Thomas, that, despite the great service the Adams family had given America, no one would grant them their justice. "My spirits are sometimes ready to sink under my private troubles and public ingratitude," she said. While John was still in office, however, this indefatigable matriarch did not cease pressuring him up to the last moment to use his office for family advantage. Always eager to extend her maternal empire, Abigail had embraced the destitute Joshua Johnson and his family. Thanks to Abigail, who found Louisa's mother very appealing—she was evidently charming even in her helplessness, something the sturdy Abigail found irresistible—Joshua Johnson had been named superintendent of stamps. Abigail confided to Mrs. Johnson that she disagreed with George Washington's belief that relatives of the president should not be appointed to public posts. One of her final campaigns was for her sister's son, William Cranch, whom John appointed to the District of Columbia's judiciary.

Back in Massachusetts, Abigail concentrated upon governing her relatives, which she felt included, besides her own children, the families of her sisters, Mary Cranch who still lived in Quincy, and Eliza Peabody who lived in Atkinson, New Hampshire. In addition, there were John Quincy's pathetic in-laws. The Johnsons' plight grew steadily worse, for Joshua found his hopes for landed wealth in Georgia were baseless. Nor could he retain his postal appointment, and when he died in 1802, he left a penniless crew in Washington who importuned Abigail for advice in fiscal and marital matters. She took particular satisfaction in persuading Eliza Johnson, one of Louisa's sisters, to reject the hand of an impoverished clergyman. The realistic Abigail used her favorite maxim that when love inevitably cooled, such a union would bring little but sorrow; and she was pleased when her protégée married Senator John Pope of Kentucky. (Unfortunately the bride did not long survive the event.)

A far more serious concern for Abigail, of course, was the Johnson daughter who had married her son. The family had two difficult questions to answer when John Quincy appeared at home late in 1801. Would he be willing to undertake something in life that would bring new glory to their name? How would the young English girl, Louisa Catherine, perform as a wife for this eldest son who was so promising? John and Abigail believed that his recent defeat now darkened their son's future. As John Quincy himself put it, he entered Boston "to begin the world anew," stung by Abigail's blunt comment, "Your prospects, my dear son, are not very bright." She clucked sadly over the family's political fall because it meant private, remote stations for her sons, and especially for John Quincy, who possessed, as she told Thomas, "a capacity which has been cultivated and improved, may I say to you, beyond any other native American." Now her eldest son faced only a "humiliating prospect."

Much as Abigail fretted over John Quincy's situation, she became preoccupied with Louisa. The Adamses, Cranches, and other kinsmen who gathered to gawk at the English lady John Quincy had fetched home as a wife could hardly have suspected Louisa's eventual strength. It was not Abigail but Louisa who relished fishing, horseback riding, theatricals, musicals, and dances with her husband and sons. Louisa had already astonished her husband, a person completely unmanned in the presence of illness or disorder, with her tenderness and strength when he had fallen sick in Europe. Yet the Louisa who first appeared before the Adamses seemed unfit for life. No one could have foreseen that this frail creature would in crises ahead show a power and fortitude unmatched even by Abigail.

Louisa herself never forgot the circumstances of her arrival in the United States. The reunion with her own family in Washington was saddened by the poverty which had overtaken them. Her mother was loudly uncomfortable about missing the niceties she had enjoyed in England, while Joshua Johnson was depressed and ill. Louisa was left in Washington with her baby son George to share this unhappy scene while her husband rushed away to Massachusetts. He soon sent word that she should come to New England, traveling by herself. When Louisa rebelled at this, John Quincy reluctantly made the difficult trip back to accompany her and the child. They lingered in the capital long enough to dine with President Jefferson, an occasion to which the Johnson tribe was also invited and which all agreed was a gathering "of chilling frigidity." They made an overnight visit with Martha Washington at Mount Vernon which proved more cordial.

Then, on 3 November 1801, Louisa was almost physically pulled away from her family and her dying father, and she and John Quincy started toward Quincy. She was afflicted with a severe cough and the baby with acute diarrhea. An impatient John Quincy had to make stops in Philadelphia and New York so the patients could be treated. The renowned Dr. Benjamin Rush examined Louisa and announced, "She is under great apprehensions and still more depressed in her spirits than really ill." Finally, after enduring the lurching and bumping of a stagecoach journey, Louisa arrived at the Old House at noon on 25 November, more dead than alive. She was immediately presented to the family "for whose sake," as she put it, "I had been thus hurried on from the South to gratify their wish."

Except for the loving welcome John Adams showed the sensitive Louisa, the meeting was a disaster. "It was lucky for me that I was so much depressed and so ill," Louisa recalled, "or I certainly should have given mortal offence." As if to test what mettle might remain in her wretched daughter-in-law, Abigail took Louisa on social calls and to an auction of the furniture left by the much-loved uncle Norton Quincy. There they bought items for the Boston house to which Louisa and John Quincy presently moved. The arrival of Louisa generally confirmed Abigail's greatest fears. She assured herself that her son had chosen a wife who was frail, withdrawn, spoiled, and unqualified for her responsibilities. Efforts to be kind to the woebegone Louisa were numerous but seemed to underscore her incompetence and her alien nature, and she later recalled, "I became cold and reserved, and seldom spoke at all—which was deemed pride."

Nothing appeared to go well for her. Once he had moved his wife to Boston, for reasons of economy John Quincy dismissed the baby's nurse, brought all the way from Prussia, leaving little George inconsolable, his mother in tears, and John Quincy himself short of sleep. This unpleasantness was increased by calls from many new acquaintances who wanted to inspect Louisa and tell her about her husband's old romance with the charming Mary Frazier, after which that lady herself paid a visit. Although Mary had found a new suitor and would soon be married, Louisa pondered this additional evidence of what she fancied was her inferiority, and how by capturing John Quincy she might have denied him happiness with Miss Frazier. Her despair grew when her husband submitted to family decree, entering law practice and seeking local office in Boston politics. The young woman, once a petted daughter and then a favorite of the Queen of Prussia, found her life among the Adamses a cruel contrast to her past and to her dreams.

John Quincy, meanwhile, had his own broken hopes to contend with. Instead of a career devoted to literature, he was faced with law and politics. The first he found boring and degrading, while the second now meant participation in partisan affairs. After observing what party battles had done to President Adams, his son wrote to Abigail in 1801, "I will sooner turn scavenger and earn my living by cleaning away the filth of the streets, than plunge into this bottomless filth of faction." He felt none of the contentment he had experienced as a bachelor ten years before. In desperation he contemplated a last attempt to break free of the Adams influence, by calling upon Thomas, who was still in Philadelphia, to join him in starting a new life far from cities in general and Boston in particular. He proposed that they move to the western frontier so that they would not "wither away our best days . . . yawning off existence over the black letter." Thomas was elated and instantly made ready for a career in unsettled country, to seek "honest though homely independence." He was shrewd enough, however, to mistrust his brother's ultimate determination to sacrifice the luxuries of urban living. But, said Thomas, if John Quincy were really in earnest, "I am your man for a new country and manual labor. Head work is bad business, and I never was fond of it."

Thus, for a moment, the future of the Adams family was suspended between the dreams of John and Abigail and the impulse in the second generation to leave these dreams behind. Such a move might have been the making of Thomas, but John Quincy either lacked the courage to do it, or else he somehow knew that frontier life would not fit with his dandified ways and intellectual bent, and would not satisfy him any more, really, than law or politics. Then, too, the protests of his parents against a move so wasteful and distressing must have been influential. John Quincy talked it out with them during weekend stays in Quincy in the winter of 1801–2, visits they insisted on even if it meant he must leave Louisa and little George behind in Boston. To Thomas, Abigail wrote in great dismay, urging him to think clearly about what existence in western New York would entail. Life there, she said, "would soon render you discontented and unhappy." She bluntly predicted that Thomas, now nearly thirty, would never succeed in Philadelphia. If he went west he was sure to lose what polish he possessed and fall to the crude level of those around him. Abigail advised Thomas to come home to Quincy and settle down.

Thomas Boylston Adams evidently possessed most of the charm and comfortable qualities John Quincy so conspicuously lacked. Just as Louisa had been drawn to him in Berlin, so ladies and children generally took to him. He was a great tease, a fine dancer, and an appealing singer. His Aunt Eliza spoke of her delight in his "facetious, engaging manners," and he was likened in appearance and style to a happy English country gentleman, ruddy face and portly frame included. However, under this demeanor raged the same battle with self and world which gripped John Quincy. Thomas suffered far more grievously in the combat. He came in time wholly to lack confidence; little about himself pleased him, and his eventual dependence upon his parents and his brother only enlarged his self-scorn. He feared that much of it began at Harvard where he recalled being "a stupid, idle boy." Yet, he was actually bright enough, and some of the letters between the brothers show that Thomas could hold his own in discourse and that John Quincy respected him for it. It was Thomas' faltering nature which John Quincy mistrusted, making him chide and exhort the younger man incessantly, especially as the years passed and Thomas evidenced little sign of mastering himself.

For Thomas, Philadelphia had been, at first, a haven where he had many friends from the time he boarded with a Quaker family and among whom he was treated like a man and like an equal. They had shown great delight at his return from Prussia, rushing up to greet him, " 'Thomas, how dost thou do?' " as an amused President Adams mimicked them. Here Thomas undertook for a time the practice of law, while his father neither hid his misgivings nor stifled his advice, fearing as he did for his son away from the safety of Quincy. "I know his chastenings are meant as kindness, and his experience in life enables him to dictate lessons of prudence, of assiduity and economy to his children," Thomas complained in 1802 to his mother, but he wondered why his father had to ask if Thomas really thought he had the talent to succeed in Philadelphia.

In fact, both parents continued to urge Thomas, who was often ill with rheumatism and depression, to come home to the Old House. Despite the fact that he had joined with a Harvard chum briefly in editing the *Port-Folio,* a notable early Philadelphia literary journal, Thomas was never so successful that he could stand without financial allowance or "remittance" which John sent him. Slowly his letters to parents and brother grew callow and obsequious in tone, redoubling the efforts of relatives to bring him home. His father told him: "You will be a slave for life in Philadelphia," while even John Quincy urged him to "exchange the noise and bustle of a great city for rural retirement." After talking with his brother, Thomas surrendered, returning to a place in his parents' house, but not before John Quincy learned how this retreat bruised Thomas emotionally. Recognizing this, the elder brother sent his mother some blunt advice—she must not pressure Thomas when he arrived, for he would only be happy if "he is left entirely and in the most unqualified manner to his own choice and humour in his mode of life and his pursuits." Knowing that his brother was still determined to prove he could become independent, and recalling his own struggles against family pressure, John Quincy told Abigail to give Thomas advice only if he sought it.

At first she accepted this counsel, conceding that it must cost Thomas dearly "to bring his mind to quit a place where he hoped to have obtained an independence and resided for life." On 11 December 1803, Thomas reached Quincy and the unappealing shelter of the Old House. Abigail, of course, could not long stand by passively, so she set out to find Thomas a wife. Earlier she had warned him against premature marriage, and particularly against one of her pet grievances—women who dressed to "seduce the unwary, to create inflammatory passions, and to call forth loose affections by unfolding to every eye what the veil of modesty ought to shield." Thomas was urged not "to conceal any of your troubles or anxieties from your mother." She soon discovered to her astonishment that her youngest son had fallen secretly in love.

To quiet his mother's incessant questioning, Thomas acknowledged that he hoped someday to marry Ann Harrod, known as Nancy, of Haverhill. At once, Abigail went to work. She invited Nancy to the Old House so that she could look her over; she admonished Thomas anew that marriage must wait until he could support a wife; and she talked disquietingly of the less than satisfying marriages her other children had made. But Abigail was soon pleased with Nancy Harrod, perhaps because the latter's sturdiness and simple ways made her an admirable contrast to the frail and fastidious Louisa. When she learned Nancy and Thomas had been quietly courting for several years, Abigail announced she was eager to see this "worthy woman rewarded for her steady attachment." Although the elder Adamses conceded that Nancy would not have been their first choice, they knew that on her mother's side she could point to direct descent from the Treat family, which included a Connecticut governor and a clergyman.

So Thomas took up life in Quincy, struggling with depression—the "Blue Devils" he called it—while his parents fostered his love affair, all the while insisting that he establish himself independently as a lawyer before marrying. It seemed hopeless. Thomas confessed to John Quincy that the practice of law "haunts me like a spectre, for I some time start with terror from a profound reverie on writs of attachment." Although he had left by now his own law office for politics, John Quincy supplied Thomas with hollow encouragement—he must "make the best of it," "show an interest in local affairs," and be prepared to make personal sacrifices. Evidently Thomas dutifully sought to be active, although he remained depressed and unable to maintain himself and a bride.

Finally, in 1805, his father and brother helped Thomas be elected a representative to the state legislature, so that plans could be made for his marriage. John Quincy encouraged the matter further by writing that he would be pleased to hear of Thomas' "being provided with a warm bed-fellow." Under Abigail's direction, the couple made plans to move into the Old House after the wedding. Abigail wrote Nancy that if the younger woman agreed to this arrangement, "I shall do all in my power to render your situation agreeable to you and hope that you will feel that you have only exchanged one parents house for another." What the bride thought is unknown since Thomas carefully burned her letters.

On 16 May 1805, only ten days after his election to the Massachusetts legislature, Thomas and Nancy were finally united. The wedding took place in Haverhill. Inexplicably, no other Adamses were present. This may have been Thomas' finest moment, but it was brief. By bringing his bride to Quincy and his parents' home, his marriage served as a new token of his dependency. To make matters worse, his excursion into politics was a swift failure, no great surprise to his family, which already talked sadly of his inability to shoulder heavy burdens. Within a year, Thomas withdrew from the legislature, "owing to some untoward circumstance," as John Quincy put it. Thereafter he struggled ineffectually for independence, drifting toward a reliance upon alcohol. He produced a succession of grandchildren for the Old House, and bleakly told John Quincy, "I must submit to be a pensioner for a time to come."

While Thomas struggled and lost, John Quincy had scored only superficial gains. Although he might again appear to be a rising success, his emotional battles left him wounded almost as severely as Thomas was. After deciding not to move west, he was more restless than ever. He was expected to amount to something, but how was he to do this when ugly factionalism was the rule? "I would fain be the man of my whole Country," he told his diary. The answer, of course, lay in his spiritual legacy, which prompted him, like his father, to tackle the world's ills on his own stern terms. At least that was what John Quincy insisted he was doing when he muffled his scruples, joined the Federalist Party, and was elected to the United States Senate in 1803 after briefly serving in the Massachusetts legislature. In 1802 he had been narrowly defeated in his race for a congressional seat.

John Quincy's parents were dubious about his move into public life, seeing that he, no less proud and independent-minded than his father, was prone to speak scornfully of statesmanship when the Republic was in the hands of factions. Adamses had hoped that democracy in America could calmly solicit leadership from men because of their abilities. Instead, the rising party system seemed, at least to an Adams, to make governance rely on libel and propaganda in order to draw attention and votes. John's experience as president appeared to prove this, and he feared that the squalor of national politics would soon disenchant John Quincy, leaving him miserable. Yet it was soon evident that John Quincy had a greater ambition for and fascination with public life than his father had had. For one thing, he was bored by the rustic existence his father cherished. While he seemed offhand in reporting to Thomas about his entry into state politics—"a man may as well be busy about nothing for the public as for himself"—his tone became angry when he was defeated for a place in Congress by fifty-nine votes of what he termed "fair-weather" Federalist Party supporters. By the time he started out for Washington John Quincy confessed to his diary that he was consumed by politics.

The life of a senator in the federal capital proved no more satisfying, however, than the life of a Boston lawyer. Nor did he find the duty of public leadership any more rewarding than John Adams had. From the very start, as Louisa watched apprehensively, Senator Adams seemed determined to set himself at odds with the Federalist Party. The family suspected this party had earlier betrayed John Adams as president by not uniting behind him; but even this seemed hardly enough reason to justify John Quincy's voting and mingling with the faction most New Englanders detested, the party led by President Thomas Jefferson.

The Senator soon acquired the uneasy distinction of being the only New England Federalist to support the Louisiana Purchase, and he generally backed President Jefferson's efforts to defend America's rights as a neutral in the continuing war of France against England. Jefferson's measures were detrimental to New England's shipbuilding and trading interests, and Senator Adams' constituents were furious, especially when he attended a caucus of the Republican Party. Even John and Abigail were bewildered, and the latter sent a sharp reminder to her son that "No man in Congress is so delicately situated, when we take into view your connections."

Whether John Quincy realized it at the time, it was a judicious shift in one respect, for the Federal Party was dying. The Senator did not speak of this, however, but defended himself by talk of integrity, duty, and of conspiracies opposing him. Meanwhile his family affairs were not much better, for his strained financial situation and the limited choice of residences available in Washington left John Quincy and Louisa no alternative but to move in with the Johnsons, most of whom dwelt with Walter Hellen, Louisa's brother-in-law, who was prospering from speculation in tobacco. Besieged in his public role while cramped at night in upstairs quarters and surrounded by in-laws, John Quincy was miserable from the start. Louisa, at least, had her relatives to console her.

Soon, however, there was trouble here as well. In Quincy, Louisa had been cheered by the prospect of regular trips with her husband to Washington and thus to her family circle. To her dismay, however, this plan was threatened when John Quincy suddenly announced that she must stay either in Washington or Quincy, since he could not maintain two residences. This meant separation, no matter what she chose to do, since her husband would be in Washington for the Congressional season and otherwise living with his parents. At first Louisa decided to stay in Washington and her husband went back to Massachusetts alone in 1804. He departed hurt and angry.

That summer, the letters between the couple were bruising. Eventually, they were reconciled, but not before John Quincy suffered such lonely misery in Quincy that he became almost incapacitated. He told his mother he could not tell how much of his disorder was the sort "which required ministering to a mind diseas'd," but he acknowledged that he was in danger of surrendering to "despondency." Reunion with his wife was not the entire answer, as the young politician soon discovered. By 1806, as his career became more embattled, he felt that public life was no place for him, and he told Louisa that he was ready to resume law practice. She replied that though she herself was willing to make the change, she knew that such a prospect was likely to be no permanent escape for "a mind form'd as yours has been for a more brilliant sphere."

The difficulty for John Quincy lay in discharging his obligations in a world which family teaching and experience had exposed as fickle and unprincipled. John Adams only confused matters for the young Senator, wanting him to be prominent and at the same time telling him, "You have too honest a heart, too independent a mind, and too brilliant talents to be sincerely and confidentially treated by any man who is under the Dominance of party." The old man was certain that only his son had eluded the snare of partisanship. Thomas, who sensed that John Quincy was seriously depressed, also wrote, suggesting that his brother be less sensitive and more aggressive: "You lay the publick cause too seriously to heart and give too much domination to the crosses and vexations which are strewn along your path by the wayward nature of the times."

In 1808, John Quincy's lofty tactics and association with the Republican Party led his party in Massachusetts to repudiate him; the legislature elected another person to succeed him in the Senate. The insult was in doing so much earlier than at the expiration of his term. Senator Adams resigned immediately. It was hard for John and Abigail to argue against the rumor that desperation had driven their son into collaboration with the "enemy." Stories circulated that Jefferson's people would reward an apostate Adams with an escape—a diplomatic post far from the disgrace of Massachusetts. This encouraged his parents to be sharply critical of his dalliance with another faction. The letters of rebuke they sent were, said John Quincy, "a test for my firmness, for my prudence, and for my filial reverence." Indeed they were, for John and Abigail talked bluntly of integrity. The son could offer no good explanation for behavior strange indeed from one who professed to scorn factions. "I would wish to please my parents," he said, "but my duty I *must* do. It is a law far above that of my mere wishes." The young Senator retired from public life, insisting to himself that he had united with the opposition on matters of principle.

Under a shadow, both at home and in the public eye, John Quincy came back to Massachusetts in 1808, and bought a house on the corner of Frog Lane and Nassau Street in Boston. There he sat, embittered, ashamed, and restless, wondering if his sons had any better stuff in them than he had. John Quincy's family hoped that a professorship of rhetoric at Harvard, which he had accepted in 1806 under conditions which allowed him to continue as a senator, would be relief enough from the tedium of law practice to make him contented. There had even been talk that he would be chosen as president of Harvard. It did little good. For a time he seemed doomed to early oblivion, a more humbled figure than his father had been. This turn of events left John Quincy depressed and desperate, and his parents worried and ashamed.

By 1808, all three sons of John and Abigail had thwarted the hopes of their family, each for different reasons, but all evidently suffering compulsions acquired in their youth. Meanwhile, John was finding that his own tribulations were not finished, even if he was retired from public life. Family life in Quincy was costly, the Old House was usually full of children, grandchildren, nieces, and nephews, all needing so much physical and spiritual nourishment that Abigail complained she felt more confined than when her own children had been small. John was more than ever short of money, so that Abigail had to superintend the household closely, lest "the fruits of diligence will be scattered by the hand of dissipation." At least, the aging couple assured each other, while they might be surrounded by difficulties, they would go to their graves free of debt. In fact, John took every opportunity to satisfy his craving to own any land important to his or Abigail's forebears. His son Thomas spoke with awe and amusement of "the veneration which my father entertains for hereditary institutions, especially in his own family."

John felt one of his greatest triumphs was the purchase from relatives of the balance of the nearby Mount Wollaston farm, the vital part of which had been bequeathed to Abigail by her uncle, Norton Quincy. He was the last of a long succession of Quincys to own this estate, which had between 1625 and 1628 supported the early, notorious settlement Merrymount, presided over by Thomas Morton and his merry band, who were reported to have raised the only Maypole in New England. The land lay along the shore, east of the Old House. With this acquisition, John gleefully contrived such titles for his baronial role as Field Marshal of Mount Wollaston, Monarch of Stonyfield, Count of Gull Island, and Baron of Rocky Run.

In addition to his considerable real property in New England, John had money, 3500 pounds sterling, deposited with the London firm of Bird, Savage, and Bird. It had been moved there from Holland on John Quincy's advice. With debts mounting from the cost of running the family and his land purchases, John was planning to use his London money when, in 1803, Bird, Savage, and Bird failed. The collapse of the banking house staggered the elder Adamses and mortified John Quincy. He had been his father's agent since returning from Prussia and had written checks against the English account, paper that was now worthless in the hands of family creditors. Said John Quincy, "The error of judgment was mine, and therefore I shall not refuse to share in the suffering." At once he began plans to sell his own property to help cover the losses.

The remainder of 1803 was grim indeed. Abigail fretted so much over the fiscal troubles that she again became seriously ill. She talked of "catastrophe," of an affliction of such proportion "that I could scarcely realize it." No longer could she and John be certain of "what we may call our own." The former first lady of the Republic acknowledged that "to suffer want is a new lesson," especially after she had dreaded debt all her life. Facing a changed style of living, Abigail still tried to make the best of it. "If I cannot keep a carriage, I will ride in a chaise," she told Thomas. The vexing feature to both John and Abigail was that they could not pass an appropriate legacy to their descendants. They had looked forward to handing down to the next generation estates which had been their forebears' property since the earliest days of the colony.

It fell to John Quincy to rescue the family patrimony. He arranged to have another banking firm, King, Gore, and Williams, meet temporarily all the Adams obligations against the defunct Bird, Savage, and Bird. Then, using his own resources and credit, John Quincy purchased land from his father. The income from this sale allowed John to settle his debts and to complete his acquisition of parcels of the Mount Wollaston farm, those remaining in the hands of Norton Quincy's other heirs. John took about $12,800 from John Quincy who received for it title to much of the land around Penn's Hill. John's gratitude to his son was boundless, and John Quincy was assured of the legacy of the Old House and the area surrounding it. This prospect was so pleasing that neither father nor son seemed to realize that henceforth each of them would always teeter on the edge of financial disaster, land-poor and saddled by debt.

Meantime, two women who had entered the family by marriage watched these transactions while their own hopes lay broken about them. These were Charles' widow, Sally, and Thomas' wife, Nancy, both now obliged to live under John and Abigail's roof as dependants. Both became increasingly embittered. Sally had two small daughters to care for—although Abigail did most of the rearing of these grandchildren—while Nancy's husband seemed successful only in fathering her children. Both women grew vindictive and jealous of what they took to be the favored role of John Quincy and of his slowly improving station in life. What neither appreciated was the price extracted from their brother-in-law for his role in the family and for the success which began coming to him after 1809. He paid heavily for the eventual fulfillment of the hopes in him, but the cost was even higher for Louisa.

This remarkable woman was already showing that she was one of the most courageous members ever to grace John Adams' family. Of the women in the second generation, only Louisa's steadfastness finally brought her a measure of satisfaction and peace. While Nabby was equally stouthearted, her life afforded no such reward.

❧ 5 ❧

Marriages

Life briefly improved for Nabby after Colonel Smith took up the customs post his father-in-law, as president, finally secured for him. He claimed loudly that John Adams would never suffer for showing this confidence—though now that he had income, the Colonel summoned home to New York his sons who had been sent to the care of Aunt Eliza in New Hampshire in hopes the two lads might escape the pernicious influence of the Smith family. Once back with their parents, the boys' ruin was predicted by the Adamses, whose hopes for Smith's reform were dispelled when John Quincy tried to recover money the Colonel had squandered.

To requests for payment, Nabby's husband rarely replied and then with scarcely polite evasions. He refused even to try to repay the interest on his debt. Louisa was much stirred by Nabby's courage as a loving wife, recalling that Nabby had fallen from the "height of affluence and station" in happy London days to "the misery of poverty and obloquy." Her fortitude, sweetness, and devotion in the face of this were virtues Louisa frequently reminded herself to emulate. Colonel Smith was by Louisa's description a "gay, deluded boy." She said he was "born to be the charm and plague of doating women," while Nabby was the most exemplary of wives.

Nabby's marriage entered its gravest hour in 1805 when Colonel Smith's debts and other follies caught up with him. He was removed from the public's payroll and clapped in prison. This was due in part to Smith's indebtedness, the eternal testimony to his extravagant living and susceptibility to schemes for quick wealth. This weakness was painfully familiar to his in-laws, although it was not until after his death in 1816 that Smith's obligations were shown to total an astounding $200,000. What also led to his imprisonment, however, was Smith's stupidity in aligning himself with an old comrade, Francisco de Miranda, who undertook to liberate Venezuela from Spain, illegally using the United States as a base. To make matters worse, Nabby's eldest son was sent by his father to accompany this filibusterer, and the lad, Billy, was captured at sea by the Spanish and narrowly escaped the hangman. Despite her horror at this, Nabby stood fast, refusing to abandon her husband and seek asylum in Quincy, as her parents pleaded. Instead, she accompanied the Colonel to jail, where they shared a cottage on prison grounds, while her mate fulminated against the men who he claimed had unjustly wronged him. He demanded, in vain, that his lost office should be awarded to his brother, Justus, who was as undependable as the rest of the Smiths.

For the Adams family, the current plight of Nabby's husband was, as John Quincy put it, "the natural consequence of the principles and practice which have for many years been in unceasing operation." They also recognized that, should Smith be acquitted on flimsy evidence, which he was, he must flee New York City ahead of his creditors, leaving Nabby and his children to the care of relatives. Of these youngsters—two boys and a girl—and their "sorry prospects," John Quincy bluntly said: "Bred to nothing, possessed of nothing—having nothing to expect." It was this dismal future for her children which evidently was the greatest torment for Nabby and brought her as close to breaking as she ever came. Once she wrote so dejectedly that Abigail begged anew that she take refuge in Quincy. Nabby did so, until early in 1808 when Colonel Smith appeared to take her with him to his exile in remote upstate New York, near the town of Hamilton. Across this distance, Abigail sought to cheer Nabby, and urged her granddaughter, Caroline, to try to comfort her mother in her "many cares and anxieties to which you are yet a stranger and the innocent playfulness of youth is a great solace." The letters John and Abigail wrote to the Smith grandchildren display a special fondness for these youngsters. Deprived of the example and care of a wholesome father, they were doubly dear to Abigail, who was determined to set these youthful lives aright. The examples of John and John Quincy especially were put before the Colonel's waifs.

Then came a new affliction in Nabby's unhappy life. In early 1811 word reached Quincy that Nabby had developed breast cancer. Abigail wrote: "Let me know particularly concerning the state of it," and implored her daughter to come to Boston for medical consultation. Momentarily, Nabby resisted, saying she could not leave her husband, although, Abigail said, "the real truth is, I believe, she thinks the physicians would urge the knife, which she says the very thought of would be death to her." Eventually Nabby came to Quincy where, in the summer of 1811, the first examinations seemed hopeful—"the bunch is moveable"—and some doctors said this promised a cure. If hemlock pills were taken regularly, Nabby was encouraged to believe the growth would not enlarge. In fact, Abigail wrote Colonel Smith, some doctors said Nabby might not even have cancer.

The Colonel, though he was not in Quincy, was skeptical about this medical reasoning. If it was followed, he predicted the tumor would "dispose its malignancy thro' the whole frame," and then surgery would be useless. After urging prompt and more realistic treatment, Nabby's husband shifted the subject of his letter to a favorite topic, his belief that he had "foiled the base attempt of Mr. Jefferson to destroy me on the Miranda question." He then proposed to join his wife, a move rejected by the stricken Nabby, who did not wish her husband near her even when she finally agreed to a mastectomy. The operation was performed at the Old House in the autumn of 1811 after the family's close friend, the Philadelphia physician Benjamin Rush, had advised that removal of the afflicted breast was essential. Nabby resisted, but finally consented before an insistent family. Four surgeons were present for the ordeal, which Nabby is said to have endured with calm courage, despite the lack of effective anesthesia.

For a time the physicians offered hope. Colonel Smith was permitted to visit Quincy for a month, and Nabby settled into what her parents foresaw would be permanent residence with them. Colonel Smith, however, tolerated this for only a year; he then returned to Quincy in July 1812 and took his loyal but reluctant wife back to New York. There Nabby was soon reported bedridden in agony from what was called rheumatism. Meanwhile, her husband was absorbed in the sudden and short-lived gratification of a term in the federal Congress. Smith rushed off to Washington, leaving Nabby crippled and without funds to pay her bills. In June 1813 the elder Adamses had to be told the truth. Nabby was being overpowered not by rheumatism but by the rapidly metastasizing cancer. Upon hearing this, Abigail wrote in "inexpressible grief" to John Quincy: "Heaven only knows to what sufferings she may yet be reserved. My heart bleeds."

Nabby now showed a resoluteness that astonished even her relatives. She decided that she must die with the Adamses in Quincy. Summoning her son John, and joined by her daughter and a sister-in-law, Nabby—now a helpless invalid—somehow made the 300-mile trip to Massachusetts in fifteen days. On 26 July 1813 she reached home, fulfilling her only remaining wish, "that of getting to her father's house once more." Abigail broke down during the last stage of Nabby's illness. This was surely, she said, "the most trying affliction of all I have ever been called to endure." As Nabby experienced "spasms" through her abdomen and back, the occupants of the Old House turned to John Adams, who became the source of the strength everyone needed. Before Nabby died, Colonel Smith arrived and displayed such monumental grief that Abigail forgave him many of his sins.

On 15 August 1813, a Sunday morning, Nabby died in so touching a scene that Abigail was unable to write letters for a fortnight. Among the group gathered in her chamber, only Nabby and John were calm. Nabby called for a hymnal and asked her family to sing her favorite, "Longing for Heaven." Moments later she was dead. Abigail thought of the appropriateness in the lines of the hymn which promised their daughter that now, after a pathetic life, there was a world ahead where peace and love would be hers. When John Quincy heard the news he was so agitated that he wandered aimlessly about, thinking of "this excellent woman" who had been, he said, without blemish as daughter, sister, wife, and mother.

While his late sister may have been a paragon for the grieving John Quincy, he found his wife to be so complicated that he may never have fully appreciated her. At least Louisa claimed he never did, which made John Quincy something of a trial to her. In fact, Louisa was rarely comfortable with any Adams but her father-in-law. Her other favorite, Thomas, was already changing for the worse when Louisa reached America. Consequently, for most of her life she found solace among the Johnson family, where her powers for sympathy and her often astounding energy had room to flourish. In 1802 she began the custom of keeping at least one younger sister as a companion and nursemaid. This also provided modest financial relief for Louisa's mother, who was left destitute when Joshua Johnson died on 17 April 1802, never realizing his dream of wealth from the legendary thousands of acres around Augusta, Georgia.

As a young wife Louisa was especially grateful for opportunities to visit her relatives, for there she felt safe from the dubious stares of

Abigail, whom Louisa called a woman "equal to every occasion in life," a style the daughter-in-law coveted. Away from Quincy, Louisa was less inclined to see herself as "a mindless, hysterical fine lady, not fit to be the partner of a man who was," as she put it frequently, "evidently to play a great part in the theatre of life." The more bereft and troubled the Johnson family became, the more Louisa embraced them, drawing a kind of strength even as she suffered with them. All her life Louisa was quick to aid and defend her relatives, evidencing a mature strength which made the fact that the awesome Abigail wrote to her as if she were a helpless child all the more annoying.

But, then, Louisa was forever unsure about which side of her nature was stronger—the one that longed to be coddled and adored, as her father had done for her, or the one that yearned to help, admonish, and sustain others. Being now a daughter of the Adamses as well as a daughter of the Johnsons, and married to a man whose soul was baffled and tormented by his similarly complex personality, Louisa was pulled by conflicting moods and demands. However, she came to be an astute judge of herself as well as of her husband, whose awesome personality she had to struggle against to keep it from stifling her own. She knew that he was a gifted man shackled to a compulsive nature. Alternately pursuing fulfillment by external measures and then becoming deeply introspective, he gave himself no rest. Even when he was very ill, John Quincy would insist, to Louisa's dismay, upon exercising or meeting some obligation, pushing himself until he finally collapsed. Whenever he was treated by a physician, he expected immediate and dramatic results. If a doctor gave him a powerful laxative that affected him repeatedly during a day, John Quincy still would not be satisfied, deeming it "too gentle." It was not an unusual occurrence, for instance, when, despite a sore throat and a sprained foot, he insisted upon taking his two customary daily walks.

John Quincy's habit of keeping a diary, which he considered most valuable because it exercised his perseverance, allows the world to observe in detail how he battled with himself, a conflict which often left Louisa wounded as well. In this now justly famous diary, he would flay himself for the "pride and self-conceit and presumption" which, he said, "lie so deep in my natural character." Here, too, he bemoaned his frequent difficulty in applying himself, in keeping his attention upon his work. "I find it yet impossible to reverse the Doom of idleness and mental imbecility to which I have been condemned," he wrote. This "propensity to fly from the most urgent object of study to any thing else is to me the most pernicious of Enemies." He had "an instinctive horror of long and systematic pursuit of one purpose. This is one of the most essential defects in my character," he wrote in 1809. Especially noxious to him were the demands of domesticity. These "minutest and most miserable details of drudgery" he conceded left him little "command of my patience or of my temper."

Indifference to his health and slovenly dress persisted throughout John Quincy's life, even though (or perhaps because) his mother hounded him on these matters as long as she lived. When he was a United States senator Abigail insisted that he should eat "a hard bisquit and three figs daily between meals." And if he did not improve his appearance, she told him the world would ask "what kind of mother he had." His inability so often to persist in routine work or to concentrate may have indicated a resentment of a lifetime's indoctrination about duty, just as the powerful temper that he frequently allowed to show was a means of rebelling against restraint. John Quincy could not even turn to chess for relaxation, for there he complained he lost by mistakes he considered "gross and glaring" and which so angered him that "I sometime lose the calmness of my disposition to a degree bordering upon madness." Once, as minister to Great Britain, he created a scene in a Lombard Street bank when he fancied a clerk had insulted him. He confessed leaving the place "mortified and vexed with myself at having been irritated to intemperance."

Louisa had the better of her difficult husband when it came to problems in their family. Illness among those about John Quincy and especially an alarming event or accident would unhinge him, to use one of his favorite descriptions, while these occasions seemed to bring out Louisa's strength. This may have been another form of John Quincy's insecurity, as would seem to be the case in his remark after his son George had taken a bad fall: "I am afraid of my own weakness on this and the like occasions and I pray for fortitude from above." Grief affected him strongly. When, in June 1806, Louisa gave birth in Washington to a baby that soon died, the news overpowered John Quincy in Boston. As soon as possible he went to his room "and there yielded to the weakness which I had so long strove to conceal and restrain." It took several days "to subdue my feelings." And he told Louisa, "If the tears of affliction are unbecoming to a man, Heaven will at least accept those of gratitude from me for having preserved you to me."

All his life John Quincy struggled to contain his volatile nature, often regretting that "I have not enough of the Stoic in my Soul." But he also worked earnestly to teach Louisa to muffle her spirits, equally as explosive as his own. He told her that "duty and virtue" required that they "controul the impulse of our own feelings, to moderate those emotions which we cannot suppress." Otherwise, there was danger from the temptations and vexations which the Adamses saw lurking in the world. "To sink under evils, whatever they may be, is a proof of weakness," John Quincy admonished his wife. He was usually scornful of any woman's ability but his mother's, so that when Louisa displayed courage and strength, as she did surprisingly often, her husband was always astonished.

John Quincy Adams fell short of being super-human, a flaw for which he punished himself, and also his wife and children, repeatedly. Nevertheless, he could be very warm on occasion. He enjoyed swimming, fishing, wine, conversation about literature, writing poetry, and raising seedling trees. To Louisa he was sometimes charmingly erotic, at least when apart from her. Once he wrote her that "A very little cloathing you know, upon a Lady will answer all my purposes. . . . But then for that very little I am scrupulous in exacting it. I am still of the opinion that a Lady when she goes to bed at night should have something to do besides opening the sheets." On another occasion, he was so moved by the sight of the semi-clad ladies of Washington that he wrote a poem about it and sent it to Louisa, who was then in Boston. What she read, said a greatly amused Louisa, contained "the sauciest lines I ever perused," and she teasingly threatened to have them published in the Boston papers. The poem spoke facetiously of how nakedness now meant charm rather than shame, so that the feminine treasures should be displayed to all. The neck, the arms, the breasts shall be exposed, said the poet, "while a bare spider's web conceals, and scarce conceals the rest." He closed with a plea to all women:

> Fling the *last* fig-leaf to the wind,
> And snatch me to thy arms!

John Quincy could be very impulsive, especially when indulging his children by occasional and usually abrupt whims that contrasted markedly with his customary rigorous habits. This was one reason he so often ran short of money: "I never can set expense in one scale when the comfort or pleasure of my wife and children is in the other." At other times, however, when he saw what this policy cost him, he would fly into a rage, leaving Louisa and the children terrorized and uncertain even about his more liberal moments. He could be humorous on occasion, something observers seemed to find rare enough to be worthy of comment. His diary is spiced with quips, as, after a horse he was riding threw him twice in as many days, he wrote: "This is a second broad hint to me not to ride this horse any more—I think I shall not call for the third." He knew he usually failed in the small but pleasant courtesies which help to ease social interaction. He confessed to Louisa, "I have often wished that I had been that man of elegant and accomplished manners who can recommend himself to the regard of others by *little attentions.*" His failure here, he assured his wife, was not because he had "an unfeeling heart."

From the painful days when he first returned from Europe to attend Harvard until he died, John Quincy Adams' personality was powerfully affected by his knowledge of the qualities he lacked and the attainments that were beyond him. Yearning as he did to have mastery of self and others, yet actually able to command comparatively little of either, John Quincy increasingly took comfort in meditation and prayer as he matured. Especially did he solicit humility, "a lesson which I sorely want and which I pray God to give me the grace to learn." Yet greatly though he chastised and admonished himself, he mistrusted and feared the world even more, a lesson learned at his parents' knees. His cry in 1804 as an embattled senator to his father was a refrain of the Adams motto: "as much as I depend upon the dispensation of Providence, just so little is my confidence in the wisdom and virtue of men."

Actually, there was no haven for John Quincy, either in his own soul or in the comfort of others, save for those moments when he was so overcome that he allowed Louisa to console him. Once, when he was stricken by another recurring attack of melancholia, a physician urged him to relax, advice which John Quincy brushed aside without hesitation, as he had similar counsel from Louisa. The doctor, John Quincy scoffed, "recommends good living and generous wines to which the lust of the flesh is already too prone and in which I indulge myself more than enough. My own opinion and experience both prescribe a curb to the appetites and not a spur. I distrust therefore the advice which recommends pampering."

Aptly, then, Louisa's trials as his wife were caught in a cry she once uttered to her husband: "I can neither live with or without you." One of her greatest distresses arose from the different views she and John Quincy had about raising children. Feeling, as she often did, closed out of her husband's life and thoughts and even out of the Adams family circle, Louisa put her considerable vigor and passion into child-rearing, with the result that most of her conflicts with John Quincy came over questions about the children's welfare. These differences, however, were only a part of Louisa's problems as a parent. Her difficulties and her fancied failures in rearing her youngsters were always in mind during her long life.

Louisa's troubles with child-bearing were not uncommon in her time. After four miscarriages in Berlin, she had carried her first son, George Washington Adams, to full term in 1801, only to be herself mishandled woefully by a midwife at the delivery. Thereafter, miscarriages continued to plague Louisa for many years, although they were interspersed by the births in Boston of two more sons, John in 1803 and Charles Francis in 1807; a daughter, Louisa Catherine, was born in 1811 in St. Petersburg when John Quincy was minister to Russia. Even at such successful moments, matters never seemed to go well. Louisa found herself virtually alone when John was born, for her husband had let a higher duty compel him to be in Quincy visiting his parents. At birth Charles Francis seemed lifeless for an unbearably long time, while the baby Louisa lived for only a year, though easily long enough to ensnare her mother's heart forever.

Concerning each of her sons, three very different individuals, Louisa was keenly perceptive. Despite her passionate regard for them, she was almost disinterested in her capacity to study them critically as she strove to care for their souls and bodies. She had little patience with physicians, especially where treatment of children was involved. Her homely medications were not subtle. When George was small she found it advisable regularly to give him two teaspoons of castor oil mixed with a black powder she admired. When it was feared that the lad was afflicted with worms, he received on alternate days five drops of spirit of turpentine on a lump of sugar. For little John, who suffered from hives, she mixed brown sugar with sliced onions and water, pouring off the juice for teaspoonful doses supplemented by an emetic, and then rubbed goose oil on his feet and stomach should the case prove stubborn. Yet the gentle-souled Louisa hesitated more than was then fashionable to employ the stern remedies of purgatives, emetics, bleeding, and leeches.

Louisa undoubtedly shared the Adams family's hopes for the new generation. She was as fiercely eager for her sons' success as any family member, although her means toward that end were softer. Louisa had taken to heart "the many prophecies of his future greatness" she heard about George when the little fellow was first presented to his grandparents. However, as the years passed, events made her remember the French diplomat who had examined George's horoscope and announced that he would be in dreadful danger at age twenty-eight. That was George's age, as it turned out, when he died, a probable suicide, after fulfilling his mother's tearful prediction that he was born "to taste the ills of life." From childhood on, there were warnings that Louisa was correct. George Washington Adams was a hyperactive child, often ungovernable, who stoutly resisted parental efforts to curb him. He liked visits to the Old House, where the grandparents had at last forgiven John Quincy for his impulsive decision against naming his first-born after John Adams. Abigail had watched John suffer from this slight, so sensitive was he in 1801 to invidious comparisons to George Washington. It would have been much easier, she said, if the baby had been named after Thomas Jefferson. With this in the past, little George at the age of six was soon boasting of how kindly he was treated by his grandparents, where French lessons were not imposed on him. Abigail reported the lad's taunt that "his mother tried to make a good boy of him, but she could not." She found her grandson good at making but not keeping promises, a sign which alarmed her.

Even after his parents had left him behind in 1809 when President James Madison sent John Quincy to Russia as minister, George continued to throw off restraint, and particularly any responsibility for self-discipline. His father had failed with harsh punishments, to break the son's wilfulness. George had responded by roaming away from home until hunger drove him back in near-collapse. Thinking that Louisa's motherly solicitousness offset his paternal severity, John Quincy kept her from seeing George when the boy was being chastised. One reason the father was willing to go to Russia without his sons was his fear that Louisa was too weak a parent, but he also fretted over his own incapacity to rule the children successfully.

Thus George grew to young manhood out of the presence of his parents, who never left him long without letters of exhortation, spurred by news from relatives in Quincy that George still had a brain "overcharged" and was "always in trouble and in punishment." The boy's warmest friend, his grandfather John, tried to be hopeful, saying, "George is a treasure of diamonds. He has a genius equal to any thing, but like all other geniuses requires the most delicate management to prevent it from running into eccentricities." But Abigail's sister Eliza, with whom George lived during much of his formative years, worried more about persisting disobedience in the boy and a tendency "to *fiddle away* time—no object in view—this is a sad thing." When John Quincy wrote incessantly concerning his son's morals, Eliza told him to worry more that George's waste of time would "shade his fine talents."

The second son, John, was born in 1803 and lived until 1834. Here, as in George's case, the family papers are meager, perhaps because John, too, was a sorrow to both his parents. Known for his fiery temper, which contributed to his expulsion from Harvard just before he was to graduate in 1823, the second John Adams always remained a burden to his parents. He refused to return to Quincy in 1829 after his father failed to be reelected president, but instead stayed in Washington where he managed some family property in a hapless manner and succumbed to poor health and intemperance. Aside from his dismissal from college, John's most memorable contribution to family lore was when, as a two-year-old, he fell into a rain barrel at the Penn's Hill farmhouse where his parents were then living. He nearly drowned, while John Quincy was reading aloud in the parlor and Louisa patiently listening.

Louisa never forgot the boy's blackened face as one of her sisters stripped him to rub hot rum over his body. She also remembered that during this crisis the most immediate attention and affection she received was from John Adams, who hastened from the Old House to reassure her. "God bless him for all his goodness to me," Louisa breathed, and little John seemed thereafter to become wholly dependent upon his grandparents, tormented at any need to part from them. The lad was "too much attached to the old mansion," his grandfather observed in 1807, "but that is flattering to old hearts." John's grandmother, however, was more concerned that Louisa and John Quincy hear of how their son's imperious ways were developing. "He has a spirit that I would certainly avoid to break," Abigail reported, "but which must be very carefully managed." In their parents' absence, the boys were divided between the Old House, where John, asthmatic and very fat, sat perpetually on Abigail's lap, while George, who was equally difficult, was kept with Aunt and Uncle Peabody in New Hampshire.

Young John maintained a determination to do just as he chose, often being secretive with his parents. It distressed him immensely when he was obliged to leave his indulgent grandmother in 1815 and travel to England to join his family in John Quincy's latest diplomatic assignment. Abigail sent along careful recommendations out of her efforts to inculcate in John "virtue and usefulness": "John is the most ardent, volatile, active being you can conceive, full of the kindliness of gay spirits, with a good capacity wanting a watchful guide and governor." When the impetuous and disobedient John angered Louisa, Abigail was quick now to counsel firm action. "He must be accountable and submit to the regulation of the family," said the old grandmother, adding that John "loves to be his own master." This was, for any Adams, a grave charge.

Louisa's tie with her third son, Charles Francis, was sturdy. Unlike his brothers, he was never obliged to leave his mother before he went to college. Eventually, his success and his own thriving family brought much solace to Louisa, who fixed her intense feelings upon him after the failure of his brothers. Charles Francis was born on 18 August 1807, in a breech presentation, which convinced Louisa that "the little gentleman" had a good future. "He was born to be lucky," she said. Named for his late uncle and for his father's mentor, Judge Francis Dana, Charles Francis had the full attention of both parents in his early years, since he was not left behind when they went to Russia.

About Charles Francis there is much surviving material. Even his earliest letters, in great numbers, remain—many of them showing the voice of John Quincy in the background dictating the message. Before he was seven, Charles was keeping a letter book and listening to his father's emphasis on orderliness in all affairs of the mind. The lad's response was not altogether pleasing; his father noted, "Like my two older boys I find him a Child not easy to manage." Soon thereafter, Charles wrote to his brother John about the ordeal of copying two mottoes at John Quincy's direction: " 'Recreation ought to be allowed.' " and " 'Never neglect your employment.' " "Papa asked me which I liked best," he wrote, "and I said, the First, Was not that funny?" There is no report on what John Quincy replied, but Charles prudently added that "now I say, I like them both just the same."

Charles' letters to his Grandfather Adams were delights to the old man, who begged for more. At the age of eight, the lad reported from England that "It is the fashion at Ealing for Ladies to ride on donkeys which is the genteel name for asses." A year later he told John Adams: "Mama says she thinks my letters are like the mountain that brought forth a mouse, therefore, Dear Grandpapa, take care when you open this letter that the mouse don't spring out upon you." Louisa had feared, and her husband had hoped, that Charles might be a prodigy because of his grasp of language. He became indifferent to study, however, when he was reunited with his brothers in England at age seven. He, too, had learned the severity of his father's ways, for when John Quincy left Louisa and Charles in St. Petersburg while he was in Belgium ne-

gotiating the peace treaty that ended the War of 1812, his instructions
were that Charles was to be given the father's love "only if he is good."
As his mother struggled with her precocious lad's "strange mixture of
stubbornness, arrogance, and sweetness," she found that an excellent
punishment was to forbid Charles to write to his father. Out of such
an extraordinary background, this lad of remarkable qualities twenty-
five years later took over as head of the Adams family.

John and Abigail's financial debacle in 1803 had so reduced John
Quincy's means that he could not afford a suitable Boston home for
his wife. Consequently, before departing for Russia in 1809 Louisa had
to keep house mostly in Quincy. During the summers of 1805 and 1806,
she especially enjoyed romping with her sons in the farmhouse at Penn's
Hill where John Quincy had taken his family after buying it from his
father. This dwelling was about two miles from the Old House and
Abigail's watchful gaze, but not so far that dutiful visits could not be
paid during daily walks.

   This little home was, of course, incalculably different from the handsome environment Louisa had shared in London with her own parents,
but at least she had her boys at hand in the healthful countryside and
ocean breeze. Years later she wrote of the hilarious escapades of life at
Penn's Hill: trying to milk a cow for the first time; the day unexpected
callers from Baltimore found her covered with flour, baking a cake;
her efforts to learn to drive a carriage—how calm she had remained
when the horse bolted at the sight of a red wheelbarrow and she and
the boys were overturned. To her husband's expressed hopes that she
would be reconciled "to so humble a residence" as the Penn's Hill house
she responded by proudly assuring him that, if Abigail could have once
managed in it with four children, "I can certainly live there with two."
After all, she said, smarting still from the scorn she had received in
America as a fancy English lady, "I am not *More timid* than the rest of
my sex."

   Although John Quincy lived in Quincy only for intervals, he wanted
to use that time "to do all in my power for the preservation of my
family," which included taking five-year-old George for a long stroll
across the land, to familiarize him at this early date in his life with what
John Quincy called the "scene upon which my own earliest recollection
dwells." In his diary he recorded: "I feel an attraction to these places
more powerful than to any other spot upon Earth." This rural vista

reminded him of the ideals early implanted by his parents, "sentiments and opinions which I most cherish and which I should wish my children to possess." However, while he knew how much strain there was upon these ideals when they were taken from the pastures of Quincy into the arena of public life, John Quincy was still increasingly drawn into that larger world. His return to public life meant renewed anguish for Louisa.

To his wife's and his parents' distress, he was not content to bury his failure as a senator in teaching rhetoric at Harvard and in the practice of law. Instead, without consulting Louisa, he seized upon legal business as the pretext to return to Washington and to seek a diplomatic post from the Jeffersonians. This alarmed those at home who believed that when an Adams entered the dangerous grounds of public life, it must be as an independent participant, called to demolish wrong and erect good. Traveling south in the snow of January 1809, John Quincy prayed "for perfect self possession, cool persevering, inflexible energy, and the unalterable consciousness of integrity." He carried with him his mother's plea to resist the urge to find political work. "I do not wish to see you under existing circumstances any other than the private citizen you now are," she said. John Quincy later denied to his mother that he had known anything about his nomination as minister to Russia, and when the Senate rejected his appointment, he claimed to be relieved. It would have meant much trouble, he said, trying to hide the fact that he still pined to escape Boston, where he was coldly treated as a turncoat for his associations with Thomas Jefferson's party. Even exile to the remoteness of St. Petersburg would be welcome.

Consequently, John Quincy was thrilled in the early summer of 1809 when talk reached Boston that President James Madison was renewing the effort to name him minister to Russia. The only problem was that according to a persistent rumor John Adams was trying to aid his son in getting a public job. John was furious, and ordered that the St. Petersburg post be rejected if it were proffered. However, when the nomination passed the Senate, John Quincy accepted the appointment, disobeying his parents and ignoring Louisa's opinion. She and all his relatives were outraged, but there was little choice for the family but to adopt the new Minister's view that he had a summons from the nation no honorable man could reject. On 6 August 1809, leaving two sons behind and taking along a still-protesting Louisa and the infant Charles Francis, John Quincy sailed away from his American disasters. In Russia he hoped once more to be dutiful and independent.

Louisa felt that her husband had been badly misguided in his determination to leave America. She found it difficult even to talk about the
disruption of their family, a step he had chosen not to share with her
until he had made all the decisions. When the Senate had first rejected
his appointment to Russia, Louisa had been jubilant, claiming she knew
"how unfit I always have been for the Wife of a *great man* and a Politician." Then, swiftly, she was hustled away from her two oldest boys for
an absence lasting six years, during which Louisa's suffering and worry
amid a harsh life help account for the quality of anger, frustration, and
guilt which thereafter always threatened to cloud her outlook.

The trip began wretchedly, for as the moment to leave for St. Petersburg approached, Louisa was not permitted to be alone with John
Adams. Her husband feared her pitiful pleas that little George and
John go with her might so move the old gentleman that he would order
John Quincy to comply. It was grimly appropriate that she and tiny
Charles Francis came into St. Petersburg just before the ice closed
around the city in late October 1809. At once Louisa felt herself imprisoned in an alien land. The Adamses were too poor to move properly in the social whirl of the capital, so that John Quincy often attended functions in the Czar's court while Louisa, who was usually
indisposed anyway, stayed alone.

Slowly John Quincy rebuilt his self-esteem as he shrewdly used his
talents in St. Petersburg to advance his government's cause—freedom
on the seas, a stronger claim to the far northwest of the North American continent, and, most important, a reasonable end to the War of
1812. He soon was one of the most skillful students and practitioners
of international politics in the Napoleonic era. Most of this was lost on
the desperately unhappy Louisa who thought of her sons in Massachusetts and her family in Washington. The strain between her husband
and herself grew more severe as each perceived the other as indifferent
and cold, although Louisa became pregnant in the early winter of 1810.

Realizing the plight of her son and his wife, Abigail attempted to
help them by trying to bring them home once more to private life. She
discussed their situation with Louisa's mother, who had earlier visited
at length in Quincy, thus sealing a bond between the two. John Quincy
was too independent ever to succeed in politics, Abigail said, and she
hoped for that reason that he would never be president. He would be
unable to bear the office in the face of a fickle public and a licentious
press, she predicted with remarkable accuracy. Meanwhile, Louisa wrote
of her fondness for the quiet of Quincy, saying she had left with her

husband only because of "the state of cruel anxiety and uneasiness in which Mr. A. passed his life in America."

Unaware of John Quincy's growing diplomatic success, Abigail went so far as to beg President Madison to summon him home. Thanks to her eloquent pleas, an admirable way seemed to open for the President to recall his Minister. John Quincy was appointed in 1811 to the United States Supreme Court. At this point John Adams entered the drama. Knowing of his son's distaste for the law and the bench, John tried to counteract this mood by telling him that the Supreme Court appointment meant "the preservation of your family from ruin." Refusal to accept it would create "a National Disgust and Resentment." Both parents called the opportunity a summons from Providence, returning their son to the nation with pride—"he is not called home to bring his talents in a napkin," Abigail reassured Louisa.

Abigail's excitement over this appointment was extravagant, betraying once again her passionate ambition for her family. President Madison's generosity gave her "a confidence and firm belief" that John Quincy was meant by Heaven to be "a great blessing to this nation." She said she did not wish John Quincy ever to be president, so that the judicial appointment was "more gratifying than if he had been called to the highest office in the land." The family's fervent thanks were privately conveyed to James Madison. Indeed, one wonders what might have been the course of Adams family history if John Quincy had bowed to his parents and chosen to sit on the Supreme Court. His father insisted here his restless son would at last find happiness in a post where he could be both statesman and scholar.

But John Quincy refused the seat, seizing upon Louisa's pregnancy as justification, since a dangerous journey would be required of her if he returned to join the Court. John Quincy was relieved at having so indisputable an excuse for rejecting a post which he did not relish. His real reasons were mixed—the pay was low; several men had already turned down the position; his ambitions were set on higher things. He surely despised any career daily associated with legal matters, which he found taxing and dull. John and Abigail did not refer to his repudiation of a family summons when the baby, whose expected birth John Quincy had used so fortuitously, died on 15 September 1812 before she was two years old.

Nor did John Quincy mention how different his life might have been had he returned to America, but his grief over the youngster nearly unhinged him. His letters home in the autumn of 1812 cry with anguish: "The desire of my eyes, the darling of my heart is gone." There was, he said, "nothing upon earth that can administer relief to my affliction. I cling more than ever to the bosom of my family." He wondered if the little girl's painful death was meant to atone for his offenses, and he called upon what must have seemed familiar strains in admonishing his sons in America that loss of a sister meant "a new obligation upon you to contribute every thing in your power to the consolation of your parents . . . which you can only do by your steady and continued improvement in piety and virtue."

The death of her tiny namesake so crushed Louisa that she even stopped writing for a time, trying to absorb this and another grievous loss, for news arrived that Mrs. Joshua Johnson had died in Washington. Losing her mother made Louisa the more brokenhearted over the distance between her and those she loved. At times she wished she would die and be left peacefully in St. Petersburg, buried at the side of her daughter. These trials were made worse because she and John Quincy found it so difficult to talk about their situation. Certainly it was no comfort when, during one of Louisa's moments of severe depression, John Quincy, with monumental tactlessness, directed her to read a volume on the diseases of the mind.

Louisa's consolation came from her deepening Christian faith; she prayed continually for help to "bear the dreadful afflictions with which thou has thought fit to try thy servant." Even so, she yearned for death and had to grapple with "the evil propensities of my nature" which brought her to contemplate suicide. If only she could have tenderness and "a little indulgence" from her husband, Louisa wrote in her journal, but in him she seemed now to find only "harshness and contempt." At this point, she even longed for Abigail—"a comforter," one who "would pity sufferings which she would have understood." Her dreams were filled with her father's presence, while her thoughts strayed to women of her age, "when the passions are supposed to be dead," who were stirred into sexual liaisons with younger men. At least, she noted, her life could be made even worse were she to give way to such an impulse. Russia she called a "horrid place."

John Quincy, too, lonely, depressed, and with much time on his hands in St. Petersburg, retreated into himself to brood over not only his fate but what that of his sons might be. Nothing so tore at his complex emotions as fears and affection for his surviving children, who were the most sensitive, vulnerable projections of his greatest anxieties about himself. Now the distance from them vexed him as much as it did

Louisa. He claimed that his love for his children was "inordinate," that
when they were in peril his "distress of mind" was so great as to be a
"gross and culpable want of Fortitude, for which I ask the forgiveness
of Heaven, but which is its own punishment." Frank and quick to concede that none of his sons seemed designed *"to tug* with the world as I
have done," he watched with perturbation "what their present qualities
and propensities will bring them to." Needing to see each son succeed,
to bring honor to the name his own father had established, John Quincy
had scant success when he tried to pledge himself to be contented if
only his boys went through life with uncontaminated morals.

In this quality Abigail and John had succeeded with their eldest son
to a degree they might not have foreseen nor entirely desired. John
Quincy's burden as a parent was most in evidence in the frenzy with
which he reacted to any intimations of sin on his sons' part. He even
outdid his own parents' fervor in seeking to keep his children from
succumbing to human frailty. There was no more graphic nor moving
example of this concern than in his letters from St. Petersburg, 1811–
13, to his oldest son George, then not yet a teenager, letters which
implored the lad to recognize the evil in himself and the world. The
key to life, John Quincy said, was realizing that "Heaven has given to
every human being the *power* of controuling his Passions, and if he
neglects or loses it, the fault is his own and he must be answerable for
it." These letters, lectures on the Bible, disclosed how determined
John Quincy was that he and his sons use such power, and how fearful
he was that their passions—his own included—would break forth.

For John Quincy, the Bible was the chief bulwark of man's weak
nature before a world of temptation. Not only did he read the scriptures daily, but he tried to have his children adopt the same practice so
that they might be "fruitful of good works." He particularly directed
his zeal toward Charles Francis, especially when the mission to Russia
cut him off from the sons remaining in Quincy. With few diplomatic
duties to distract him, he became fascinated as he tutored Charles
Francis: "I find as with his elder brothers a difficulty in fixing his attention," he said before his little pupil was even three years old. "The
sugar plums yet serve a guard against the sentiment of toil, but to dis-
pense them with efficacy is a delicate task." John Quincy sought to contain his irritation when the child doggedly refused to heed his father,
willing, in short, said the exasperated teacher, to do anything "but
naming the letter to which I point." He prayed for more of what he
called the essential qualities in teaching, "Patience and Perseverance."

By the time Charles Francis was five, John Quincy was spending three hours a day instructing him in English, French, and German. Soon, it was a six-hour work schedule. Arithmetic was added and the teacher noted that his pupil "has a thirst for learning beyond his years" combined with "a singular aversion to being taught. He chose to learn everything in his own way." When Charles Francis had driven his father nearly to despair by insisting on "reading or pronouncing words wrong when he knows perfectly well the right," he would suddenly confound his parent by wanting to continue reading at the end of a long day's work. He was then five years old.

John Quincy refused to heed urgent advice from Louisa and from Quincy relatives that he might be pushing Charles Francis too soon and too hard. Instead, he considered it a personal achievement whenever his son behaved like the prodigy the father craved to see. All this was interrupted when John Quincy left Russia to lead negotiations which ended the War of 1812. When he and the child were reunited a year later in England, the spell of scholarship had evidently been broken, for by then Charles Francis displayed at school the qualities of an ordinary student. John Quincy conceded to his mother she might be correct "that we must be content to take children for children."

In spite of the time given to his son's education, John Quincy had used every opportunity in St. Petersburg to encourage negotiations for peace between the United States and Great Britain. President Madison showed his gratitude by promising that John Quincy would be made minister to England once the war was over. Now at the fore as an American diplomat, John Quincy Adams was the plausible choice to head the American delegation in the peace conference at Ghent in Belgium. In late April 1814, John Quincy took leave of Louisa and Charles Francis and began one of the great moments in Adams family's public story, when he led Albert Gallatin, James A. Bayard, Henry Clay, and Jonathan Russell to meet the British spokesmen headed by Admiral Lord Gambier. On Christmas eve 1814, John Quincy's courageous tactics and the able assistance of his associates finally created a brilliant treaty, the Peace of Ghent. When this news reached Quincy, it brought old John boundless pleasure.

Meanwhile, the distance which his duties in Ghent put between John Quincy and Louisa had its own pacifying effect. Louisa had the opportunity to be a loving parent to Charles Francis, sparing him the rigorous demands made by his father. Also, John Quincy made an acknowledgement that his wife was an intelligent person, for, to her astonishment, he disclosed to her his property holdings and financial assets and liabilities. Then, too, when he was away from her and could put his thoughts into writing, John Quincy could express his suppressed affection and tenderness, which came to her often quite charmingly. One letter closed, "Adieu! Love me and pray for me as I do without ceasing for you." Another spoke of "ardour beating at my heart." In turn, Louisa found herself thinking of those times when John Quincy had been attentive, and told him that she wept at remembering "how often you used to come to my bedside and kiss my hand." To this, her husband replied that he yearned to kiss her hand, adding, "I hope it will not be at your bed-*side.*" When in her letters Louisa lapsed into depression, John Quincy did not scold but tried to cheer her.

The encouraging news Louisa awaited, of course, was that she must hasten to Paris to join her triumphant husband before they took up residence in England and where she hoped to be reunited with George and John, summoned from Massachusetts. The Empress of Russia said that never had she seen a woman so altered for the better as Louisa when the joy of this news finally sparkled in her eyes. In February 1815, Louisa left behind her experiences in Russia which had for a time taxed her strength—guilt at abandoning her sons, grief at the death of her daughter, and attempts to understand John Quincy's emotional problems and increasingly cold manner. With Charles Francis she made a truly heroic adventure for those times—a trek by stagecoach, alone except for undependable servants, across a winterbound Europe still troubled by Napoleon. The journey allowed Louisa to display her mettle.

The next two years were happy for Louisa. She accompanied her husband to London after a gratifying reunion in Paris. She shared his pleasure at receiving honor and duties similar to those John Adams before him had held, when John too had been fresh from the peace table. These parallels were evident to the family in Quincy who at last had reason to be content with John Quincy's public career, even though he was still working for the Jeffersonians. There was also tribute for Louisa's courage on her carriage trip, praise which she waved aside. "I have really acquired the reputation of a heroine at a very cheap rate," she wrote to Abigail, while setting to work to arrange a happy home in Little Ealing outside London for a united family now that George and John were with them.

The surroundings were wonderful, a charming residence called Boston House, with its coach building, stable, and dairy, as well as gardens for fruits and vegetables. There was even a church pew which came with the rent, and Louisa attended daily Anglican services, her favorite form of worship. In these English years, family life seemed at last to move in the way she wished. The boys were frequently home from school, allowing the family to sit by the fire for long talks. John Quincy often joined in, to Louisa's delight. Not only did their sons have many friends, but Louisa, too, found pleasing companions, especially in the daughters of the boys' academy master. The general happiness arising from this was evident in a teasing poem John Quincy wrote to Louisa and a friend, entitled "On their Laughing during Divine Service." More important, certainly, were the poems John Quincy and Louisa addressed to each other in celebration of twenty years of marriage, lines by both which were full of love and admiration. John Quincy announced he would marry Louisa again if that were possible. When her husband was nearly blinded by an eye disorder, Louisa cared tenderly for him and read aloud and wrote endlessly at his dictation while he chafed at his helplessness.

John Quincy also became a more relaxed father. "Spent the evening at sport with the children" was not an uncommon diary entry for him in England. However, regardless of what new confidence he might have in himself in view of his recent achievements, John Quincy could not be entirely comfortable as a parent even in the English interlude. "Among the desires of my heart the most deeply anxious is that for the conduct and welfare of my children. In them my hopes and fears are most deeply involved," said he, adding, "none of my children will probably ever answer my hopes." He prayed that "none of them ever realize my fears."

The extraordinary extremes of mood in John Quincy were especially evident in his poetry, for life in England was unhurried enough to afford him time to write. In fact, he became quite a zealous poet, finding nothing so intriguing as composing lines for Louisa and her acquaintances. Walking, sitting, even when arising from bed, he found himself composing verse. Some results were so amatory that John Quincy was shy. "I scruple to show the lines now they are written," he told his wife. Perhaps he recalled that five years earlier in Russia he was busily condemning men who let their character be fouled by a "Passion for Women." Was it a "constitutional" tendency in some males, he wondered, adding that he could not offer an opinion on the latter point. However, he had been certain about one thing: "I can never consider the disgrace of a goat as the honors of a man."

While Louisa was certainly circumspect, she was also an ardent creature, and her talk of sexual relations was more consistent and good-natured than her husband's. She acknowledged that men and women were slaves of passion, creatures of these impulses, but felt that the attraction between the sexes was basically innocent, as sent by God. When it was unduly indulged, then the "grosser nature" of humanity was in command and evil would surely result. Louisa's sons were amply warned about such matters by her, there being little evidence that John Quincy discussed sexuality with them. Louisa did not hesitate to condemn the infidelities of Daniel Webster when Charles Francis later studied law with him. She had learned one lesson, she said—"that the wisest men are the greatest fools," and she hoped the boys would keep silent lest Mrs. Webster's feelings be devastated by gossip about her husband.

Once, when Louisa somehow encountered a book of pornography, she threw it away in disgust and seized pen and paper for new admonitions—her sons must not look at these foul books. "If the purity of a young mind is debased by loathsome and disgusting pictures of nature or of vices, our actions and our thoughts will become gross and indelicate, destroy all relish for moderate and virtuous enjoyment, and render us unfit for any society but the lowest and most degraded." As her son John was about to become eighteen, she told him not to be too confident with women: "You are not yet perfectly aware of how dangerous women become when they choose and how very difficult it is for a young man to withstand their allurements." Even a homely woman could know how to arouse a male, Louisa warned, adding, "I am sorry to say that there are many of my sex who address themselves alone to those passions which are the most easily excited and at your time of life the most difficult to control."

John Quincy was held up to his sons as one who had triumphed over his passions, especially the terrible habits of tobacco and intemperance, from which Louisa assured the boys their father "has weaned himself." In short, Louisa often used many of the same exhortations which Abigail had made into an art. Where Abigail urged her grandsons to obey their parents—"In this way only and by these means only, will you be safe in the awful world that surrounds you"—Louisa added a touch of drama: "Imagine me near you at the moment of trial and fancy you see the blush of shame and indignation burning on my cheek, should you sink into vice, or even run into error."

One reason, of course, why Louisa's stay in England, 1815–17, may have been the most satisfying time of her life was that she had, for the first time, complete possession of her sons. There was no competition here from grandparents and Quincy. Louisa savored each day. Both George and John had been loath to leave Massachusetts, although their grandparents sent them on the voyage with ample warnings and admonitions—especially George, who was warned that his father was sure to put sterner stuff in him than grandparents, aunts, and uncles had been able to do.

Predictably, when George was reunited with his parents, they undertook to compensate for time lost with this eldest son. On a typical day in England, before it was clear that George could be trusted to be enrolled in school, his father had him out of bed by 5:00 AM to read aloud five chapters from the Bible in French, and then as many from a Latin Bible. Thereafter, came classical studies, with special instructors arriving for Greek, Italian, and penmanship. Late in the day, the father took the boy on two long walks, before and after dinner. Meanwhile, George had grown attached to imaginative literature and to poetry. In the world of plays, novels, and poems he seemed to find a haven.

Only in this English interlude, where he seemed remarkably at peace with himself, did John Quincy concede that his children would be acceptable even if they proved not to be extraordinary persons. "I comfort myself with the reflection," he told Abigail, that my sons "are like other children; and prepare my mind for seeing them, if their lives are spared, get along in the world like other men." This contentment lasted just as long as the stay in England. When John Quincy brought his sons back to the United States, there to see them enter Harvard and thence into careers, the old anxiety and expectation reappeared and the exhortations became even sterner, making Louisa as uncomfortable as her boys.

For the moment in England, however, these parents lived in cautious hope, with Louisa particularly grateful for having survived the early tribulations life had brought her as an Adams wife and mother. She soon found herself in better health than at any point in her married life, reporting to Abigail, "I am only afraid of growing too fat." These days were fixed especially in the family's memory by the domestic concerts which Louisa conducted, singing and playing harp or piano, with music drawn mostly from the works of G. F. Handel, John Quincy's favorite composer. He was particularly moved when his wife sang, "Oh! Had I Jubel's Lyre" and "Comfort Ye, My People."

These easy days of poetry and song ended with the Adamses' return to the United States. Louisa recalled her moods of anxiety in the dreadful early days in Boston and Washington. She foresaw losing her husband once again to a hard public life and, beyond that, to the compelling presence of his mother. The change drove her at once into claims of illness and into tearful states of weakness which the family called "fainting fits." These neurotic indispositions became Louisa's refuge increasingly over the next decade as she struggled to play the part assigned to her in public and private. Her husband became testy and distracted under the mounting responsibilities the Jeffersonian party asked him to assume in national politics. He could not withdraw from this life, nor could he enjoy it. He was his father's son in more ways than he could perhaps appreciate. Only Louisa understood what happened; but after 1817 began a twelve-year period during which she felt she held little importance in her husband's life and had few ways of helping him.

6

Quincy

On 18 August 1817, John Quincy and his family were reunited with the Adamses at Quincy. For the moment, a blissful state reigned in the Old House, from having wanderers return in such glory. Abigail was frail now but so cheerful that she even praised Louisa, finding her "looking better than I ever saw her, and younger I think." But, all too quickly, the new Secretary of State took his wife to Washington, leaving Abigail and John to tend the grandsons who were off to school.

Despite advancing infirmity, Abigail responded to any demand upon her as matriarch, even seeing to serving more than twenty guests at Sunday dinner. She was still able to handle tragedy in the family. One case involved Thomas Greenleaf, an Adams relative and family intimate who struggled with depression. He lost his battle, put a pistol barrel in his mouth and, as Abigail bluntly stated, "blowed his head to attoms." His suicide note asserted simply that a comfortless life was unbearable. With their Greenleaf kinsmen in shock, Abigail took command, persuaded the family to do without a funeral. "To look upon him was out of the question" and a ceremony would only grieve the family more, she said, directing that the body be quickly buried in Boston. This was October 1817, and Abigail herself had only another year to live.

The grim story of Thomas Greenleaf was more characteristic of the events of Abigail Adams' last days than the triumphant return of her once fugitive son. In fact, John Quincy's embarrassing departure to Russia in 1809 had seemed to start a chain of trouble, from which there was little surcease thereafter. After Nabby's tragic death, her eldest son,

William, in particular, caused continuing concern for his grandparents. His narrow escape from execution in South America seemed to have taught him nothing. When Nabby begged John Quincy in 1809 to take William to Russia as a secretary, her brother had reluctantly agreed, bowing to the family's argument that such a venture might be the lad's making. To the contrary, William proved one of his uncle's greatest trials. He not only ran up sizeable gambling debts, but he was also obliged to marry Louisa's sister-companion in St. Petersburg, Kitty Johnson, in embarrassing haste, a calamity Louisa had foreseen but had been unable to avert. The young couple was dispatched to the United States with their child, leaving William's gambling debts to be paid by John Quincy. Nabby died without being told of this new shame.

Like his father, the younger William Smith never gave up seeking an easy route in life, even imitating the Colonel's clamor for public office. By 1817 not only was he entangled in his father's debts, which he grandly undertook to assume despite their staggering size, but also with overdue commitments of his own. Soon sheriff's representatives were at William's door in Washington, while he tried to sponge from his uncle John Quincy, now secretary of state, who resisted sternly. John Quincy reminded his nephew that he knew how William had bilked the government of more expense reimbursement than was rightly due him upon returning from Russia. Poor Abigail could only weep over her grandson, now far beyond listening to her admonition. "Every heart knows its own bitterness," she said, giving thanks that Nabby was not alive to witness this latest family indignity.

In his last days, Colonel Smith was less trying to his in-laws than his son had become. Offended by failure to win a second term in Congress, shackled to fearful debts, and in deteriorating health, Nabby's husband grew old not far from Hamilton, New York, in Smith Valley, on land claimed by his brother. He lived to see his daughter Caroline, who was now doubly dear to Abigail and John as replacement for Nabby, married to a wealthy young New Yorker, John Peter de Windt of Fishkill. The ceremony enlivened the Old House in September 1814 with a pomp which would have pleased the Colonel, who always fancied regal living. Caroline's groom had arrived for his bride in an open carriage pulled by four horses. An astonished Quincy clergyman was paid fifty dollars for performing the ceremony, and Caroline Smith de Windt rode away to become the mother of many children and behave so haughtily as to make her another annoyance to her Adams cousins.

Colonel William Smith died of a liver complaint at age sixty-one on

10 June 1816, just after his two brothers had died. These gentlemen, Justus and James Smith, had exasperated the Adams family in their own rights, having helped mislead Charles, refusing to pay debts to John Quincy, and seeming to take advantage of their brother William's weaknesses. Since they were rumored to be more successful than Colonel Smith, Abigail rushed advice to John Quincy—ought he not to try one last time to get his money back from Nabby's husband? But even had the Colonel lived there would have been no relief, for his brothers were nearly as mired in debt as the Colonel.

In death Nabby and her husband were parted, just as they had been so often in life. She had been placed in the Adams vault among her relatives in the old Quincy burying ground, while Smith was interred next to his mother and brothers in the town of Sherbourne, New York. He did not even own the cemetery plot where he was buried. The Adamses' adoration of Nabby had helped to soften their grudge against her husband when he, too, became helpless and died. Much emphasis was placed upon his bravery as a soldier—his last years had seen him limp proudly from Revolutionary wounds. Abigail was even willing to speak about her son-in-law's "noble and generous spirit," but he lacked judgment, she admitted, accepting in part Smith's own defense that "he too often became the prey of the artfull and the designing."

Abigail was somewhat less generous later when she learned that Smith's debts exceeded $200,000, an amount "vastly beyond any sums which I had any idea of. I never knew anything of his affairs," Abigail said to Louisa, adding that Nabby had been equally in the dark, knowing only enough "to make her life a scene of anxiety, patience, submission and resignation." Nabby's husband's career and now that of his son and namesake reduced Abigail finally to a simple indictment of the Smith connection: "I am ashamed." A more charitable John Adams observed that "the vicissitudes of human life have not been more exemplified in the biography of Napoleon than in that of Col. Smith." Let the Adams family seek to be a little blind to his faults, John recommended, for no one could ever really know all the evil or good due to Smith. However, the former President could not resist remembering that his son-in-law "undesignedly did more injury to me in my administration than any other man."

By this time, however, there was cause for dismay much closer to Quincy. At the time Colonel Smith died, Thomas Adams was in circumstances so delicate that Abigail wrote about her youngest son's predicament, "I have to feel more than I can express." Since 1810, Thomas and his family had been living in a measure of independence, using the other ancient farmhouse at Penn's Hill, where John Adams had been born. They had left the Old House after some strain, evidently, for while Nancy wrote to Abigail in fervent thanks for all the aid granted by the older couple, she also begged forgiveness for so often having isolated herself from the Adams circle. She hoped that John and Abigail would "draw the veil of charity" over the failings they had discovered in her.

Once on their own, Thomas and Nancy continued producing children. There were seven in all, the last born in 1817 when Thomas was forty-five and Nancy forty-three. However, life at Penn's Hill was no easier, apparently, for Thomas' woes increased. He continued to suffer from wretched health and poor luck, including once being badly injured when thrown from a horse. Rheumatism was mentioned—but it was only part of the problem, for it is impossible to read the cautiously worded exchanges among the Adamses without realizing that Thomas was, with increased frequency, disabled by bouts of drinking. This situation was made potentially all the more explosive since, when John Quincy departed for Russia, Thomas had become business manager for the entire family. He was still dependent on his parents, having only a meager legal practice and a modest judicial role in common pleas procedure for Norfolk County.

Thomas began to act impulsively, as when, without consulting John Quincy and Louisa in Russia, he announced in 1810 that he would take their sons, George and John, into his home and rear them away from the Old House. Learning of Thomas' intent, Abigail moved quickly to avert it by warning John Quincy in the guarded Adams language that his sons "would be too much indulged" in the care of Thomas. When he learned of his brother's behavior, John Quincy rebuked him sharply, which added considerably to Thomas' misery. His excuses and complaints grew more passionate, so that John Quincy turned from his own woes to urge his brother to have more fortitude—"childhood has only the privilege of screaming at the scratch of a pin," he told him, and added, "Let me entreat you my dear brother for your own sake and that of your family, to reflect in future before you undertake." He must defy depression. "Your lot is not so dismal as it paints itself to your imagination."

Thomas was not convinced and he continued to talk of being borne down by "a sickness and a sorrow at the heart," a weight increased by the death of a daughter. Thereafter, many of Thomas' letters are miss-

ing, making it difficult to chart his deepening despair. He was much
away from home after 1812, ostensibly on legal and judicial business.
He returned from each trip seriously indisposed. In the spring of 1814
one such attack disabled him for a month. Said Eliza Peabody: "It is a
melancholy thing for the head of a family to be so frequently taken
from his business." Was Thomas grateful, she wondered, for all the aid
John and Abigail continued to give him and his brood? By 1815, John
Quincy was clearly alarmed, not only for the family's sake at this prospect of another tragedy in their midst, but also because he had given
Thomas authority over his finances and property. Here was a problem,
John Quincy told his father, "which of all others gives me the most
concern and uneasiness."

The difficulty between the two brothers became so grave by 1818
that John Quincy even questioned Thomas' honesty. Thomas now appeared to be gambling as well as drinking heavily. In June 1818, he
asked to be relieved of much of his responsibility as agent for John
Quincy, an assignment for which he had been quite handsomely paid.
The elder brother recorded some of the sad story in his diary, speaking
of his deep concern for his brother and his wife and children and observing how "painful" it had become to write to Thomas. Because
Thomas was so easily antagonized, letters to him required "much meditation and much delicacy." In December 1818 John Quincy stated that
it took nearly a day to compose a short letter to Thomas.

In the remnant of this letter which survives, John Quincy told his
brother that he feared to say anything lest he "would alienate still more
your affection from me, which I would deeply lament." The delicate
subject was gambling, and John Quincy said that if his rebukes "have
unnecessarily wounded your feelings, I am sorry for it." Speaking with
unaccustomed gentleness, John Quincy begged his brother to seek "the
spirit of truth and self-knowledge." This fragment, which starkly suggests the deepening plight of Thomas, is one of the last existing between them. Exactly a century later, one of John Quincy's grandsons,
Brooks Adams, recalled this sad episode in his family's story: "poor
Tom," said Brooks, "he was not very brilliant. There never was harm
in poor Tom except that he would drink. He nearly drove John Adams
and J.Q.A. into insanity with it."

Thomas' troubles were by no means the extent of John and Abigail's
grief. A terrible blow fell in 1811 when their closest friends in Quincy,
Abigail's sister Mary and her husband Richard Cranch, died within
hours of each other. The awestruck family could breathe: "Together

freed," as Mary was carried off by consumption, while Richard suffered a stroke. He died on 15 October and Mary followed on the next day. Abigail and her remaining sister Eliza took comfort in the belief that the Cranches had flown as one to immortal bliss. Meanwhile, John sought in his quiet way to face the loss of Richard Cranch—"Never shall I see his like again. An invariable friend for sixty years." Then came still more bad news. The death of Louisa's mother occurred on 29 October 1811 during an epidemic of fever which also was fatal to one of Louisa's brothers-in-law. This seemingly endless family tragedy brought John additional concern about Abigail's frail health. "Your mother," he wrote John Quincy in Russia, "takes so deep an interest in all such distresses of her Friends [family] that I am not without apprehension for her."

Meanwhile two grandchildren had also died. John Quincy and Thomas each lost a daughter, one, of course, the grandparents had never seen, for she had been born in St. Petersburg. At this Abigail turned to sister Eliza in bewilderment. Why were so many afflictions besetting the family? Eliza herself had little to say except to remind Abigail of other blessings and that no one could escape anguish in this world. John emerged stronger, although he could not help wondering why the blossom-like granddaughters were "blasted" and he was spared. He brushed the question aside as one "we are not permitted to ask," going on to try to rouse a dangerously depressed John Quincy. Despite his own sadness, the old gentleman wrote: "Grief, sorrow, and mourning for irremedial events are useless to the living as to the dead except as they procure reflection, consideration, and correct errors or reform views. Remember the best of philosophy. The beloved are chastened."

Despite omnipresent illness and death, there were still young Adamses to be superintended and admonished, responsibilities which distracted John and Abigail from their grief and worry. John reported with evident glee to John Quincy, then in England, how he was "engaged in an edifying correspondence with your sons." And an aged Abigail continued to moralize, telling her grandsons: "I shall be glad to see you growing better and better every day. This is your duty. This is what you live for." The experience of the children had made Abigail, if anything, even more skeptical of the world. At a time when she felt especially hopeless, she told Louisa and John Quincy: "Caution, advice, even precept and example, are but weak barriers against the ardour of youth, the warmth of passion, and the allurements of pleasure."

Abigail did her best, however, showering upon the grandsons the trusty maxims which her own children so often heard. "To be happy you must be good, for happiness is built only upon virtue," she admonished an eight-year-old Charles Francis. All the grandchildren seemed drawn to Abigail upon their return from England in 1817. Even such world-travelers as Charles Francis and his brother John begged not to be sent away from her to school in Boston. "I feel not a little mortified," Abigail told their mother. "I must wean them by degrees." Abigail's tribulations had taught her to detect trouble ahead. She had a shrewd premonition, for instance, about John Quincy's eldest son, George, who she sensed would readily become "the prey of any artfull designer." George's was a mind, she warned his father, which "requires no common share of vigilence and attention to direct it right."

More pungent and charming advice to George and his brother John came from their grandfather. John Adams was still concerned about the evil world into which grandchildren must go, as he had been for his own children. Now, however, he seemed better able to capture the imagination of youngsters. "I had rather you should be worthy makers of brooms and baskets than unworthy Presidents of the United States, procured by Intrigue, Faction, Slander, and Corruption," he said. On one occasion he even went so far as to tell young George: "Follow, my dear boy, the good impulse of your mind and the amiable dictates of your heart," advice, as it turned out, the hapless George may have taken too literally. At any rate, John acknowledged that he did not "push" the lads the way he feared John Quincy was doing.

Abigail's passion for superintending the lives of others extended to Louisa's sisters, who regularly turned to the lady of the Old House for counsel after their mother died. There was little about the private affairs of others which Abigail did not know, nor did she hesitate to give advice. Usually her clients were willing and docile enough, if not eventually obedient. However, with Sally, her son Charles' widow, and their children, Abigail really had more than she could manage—not that this entirely surprised her, since her daughter-in-law was, after all, part of the notorious Smith clan. "What queer fish they are," was Abigail's comment after watching Sally Smith Adams suddenly take on airs when a wealthy resident of Utica, New York, Alexander B. Johnson, married her daughter Abby. Now that she had unexpectedly acquired a rich son-in-law, Sally announced she had loathed Quincy and the Old House during the many years when the Adamses had given her and her daughters shelter. Abigail tried to reason with Sally but gave up when the younger woman became, as Abigail put it, "strangely bewildered."

Eventually, Sally allowed herself to be alienated from all her family. She died of stomach cancer in 1828.

After Sally Adams had left Quincy to live with her married daughter in New York, Abigail had a last chance to improve upon the sad memory of Charles, for Sally left behind her daughter Susan to be supervised by the grandparents in the Old House. This young lady was so difficult to handle that Abigail felt she would never succeed in curbing the girl. A crisis came when Susan decided to marry a sailor, Lieutenant Charles T. Clark, in 1817. Abigail firmly opposed the step, but the wily Susan coaxed permission from her far-away mother. Susan soon found that her grandmother knew best. Lieutenant and Mrs. Clark went to Philadelphia to live, but only briefly. He was promptly sent to sea and a pregnant Susan had no alternative but to reappear at the Old House where she became so depressed that Abigail had full charge of a baby great-granddaughter.

Her husband was soon incapacitated by ill health, so that Susan left Quincy once more to be with him in Washington. Again, however, her hopes of freedom from the Adamses evaporated. Charles T. Clark died in 1819, and the new widow's family were alarmed by rumors that she was pregnant again. Instead, she was merely destitute. She had no choice but to implore John Adams to grant her and her daughter shelter in the Old House. When Susan returned, Abigail was no longer alive to help handle her, but John still welcomed her—with the stern reminder that under his roof she must live in peace and hold her tongue.

Susan remained in Quincy until after John himself died, actually helping him considerably in his final illness. Afterwards, she left Quincy forever, going to Utica to be with her ailing mother, Sally Adams. There she met William R. H. Treadway, whom she married in 1833. Her second husband lived barely longer after marriage than her first. Treadway died in 1836. When her daughter married Judson Crane in 1839, Susan at last found haven until her own death in 1846. She and her daughter kept in touch with the Adamses only when it was time to collect the quarterly payments John Quincy and Charles Francis dutifully sent them as provided in John Adams' will.

Still the Adamses' involvement with the Smiths did not end. While worrying over Susan's love life, Abigail had also been trying to help Nabby's second son, John Smith, become secretary to John Quincy when the latter moved to England from Russia. Having just rid himself of John's brother William in St. Petersburg, John Quincy tried to dissuade his energetic mother: "I must be responsible for the conduct public and private of no more nephews." But it was too late. John won the appointment and surprised the family by his decent deportment.

Abigail herself began finally to falter, worn in body and with a spirit made ragged by seeing so many descendants fall into grief. When her remaining sister, Eliza Peabody, died suddenly in April 1815, Abigail was crushed. Sorrow, she said, had "broken me down . . . and now I am left alone." She was the last of her parson father's children to survive, and she called herself the least worthy. More and more in her last months of life, Abigail turned to John for comfort, never ceasing to marvel at the wonderful changes which retirement and advancing years were bringing to her once combative and impatient husband. Even his physical alteration struck Abigail favorably. "Age has softened his features and shed a mild lustre over them," she reported to John Quincy, while Eliza Peabody had been similarly struck, telling Abigail, "When I look upon him, it always reminds me of that line of the Poet, 'Eternal sun shine settles on his head.' "

John himself never ceased giving thanks for his years of retirement, even when he looked about Quincy in vain for familiar faces gone to the grave, such as old Cotton Tufts, faithful cousin and family retainer, who had died soon after Eliza Peabody. John maintained a consoling philosophy, refined from an outlook he had tried to embrace all his life. "I believe you have often heard me say," he reminded John Quincy in 1814, "that human nature cannot bear prosperity. It invariably intoxicates individuals and nations. Adversity is the great reformer. Affliction is the purifying furnace." The old warrior undertook to persuade John Quincy to recognize the dilemma before the Adams family which John himself now saw clearly. The son must master his ambition before it was too late, and accept the merit of virtuous retreat. Certainly an Adams was by nature eager to serve the public, John observed pointedly, his own record in mind. But the fact was now equally clear that an Adams must condemn the public process as malicious, fickle, and irrational. A man of principle could never enter openly into the affairs of a republic. "Poor human nature," said John to his son, "disputes in religion and dissensions in politics without end."

Repeatedly, John passed on to John Quincy the wisdom he had learned from his own excursions into public affairs. Better indeed for the son to retire to home life, even if there "you should be obliged to live on turnips, potatoes, and cabbage, as I am. My sphere is reduced to my garden, and so must yours be." As for gaining the high repute even of a George Washington, what was it really worth, the elder statesman asked the younger? "No more than the grunting of a bagpipe." Nevertheless, John was shrewd. He now understood his son better than John Quincy did himself, even in the recesses of his cherished diary. The father put it bluntly: John Quincy had such an ambitious spirit that he would never accept a secluded life. "No, your nature will not admit it," nor, John added, would Louisa's. "And what is harder to resist than your own or your wife's nature?" The only way he saw that an Adams might properly be drawn into public life was if the world came to such a pass that it needed one of John's family to lead it. Such advice the son knew was sound, but he could not follow it. Nor could his aged father cling to it when John Quincy returned to America in 1817 as a universal success in diplomacy and became secretary of state.

John could not help being pleased—a son with triumphant diplomatic attainments in St. Petersburg and Ghent, who stood now in a fair way to attain for the family the popular esteem John had always craved. It was doubly gratifying to him that his eldest son made his way by using John's own talents and principles. What could John do besides become an enthusiastic supporter of John Quincy's active career? He sought to help him remain realistic about the world's fickle and foolish ways. His advice to his son took strength from his discussions of the sorrows of public life with another retired statesman—Thomas Jefferson. Jefferson and John Adams had patched up their differences after New Year's Day 1812, when John undertook to renew by correspondence the friendship which had flourished long before in Philadelphia and Paris. This now famous exchange of letters brought some of the happier moments in John and Abigail's last days.

In other ways, too, John could not entirely stave off his old longing for the good opinion of the public he so often scorned. Gently he might chide his cousin, Nicholas Ward Boylston, for being concerned about monuments to their clan. "I regret the Boylston Street, and the Boylston Market, and even the town of Boylston." To John Quincy he might talk disgustedly of the pride in "pedigree" which beset so many: "Quincy church yard will furnish you with proof that your blood has not flown through scoundrels, and that is all I desire." But the old gentleman could not resist addressing the written record of the Adamses, which someday might be divulged to the world. He began the preservation of his family's papers by attacking the mountain of accumulated manuscripts, getting these precious relics in order for his descendants. "Trunks, boxes, desks, drawers locked up for thirty years have been broken open because the keys are lost. Every scrap shall be found and preserved for your affliction, for your good," he told John Quincy. It would be a tormenting inheritance, he predicted: "the huge pile of family letters will make you alternately laugh and cry, fret and fume, stamp and scold, as they do me."

In old age, John seemed more able to yoke patience to his sharp opinion of society. He was certainly amused and perhaps even pleased that doors in Boston and Quincy were now open to Abigail and him after a lifetime of being "buffetted" and "neglected." Even the Otis family, once bitter political foes, gave a great Boston dinner to honor John. "I am invited into all societies and much caressed," he said. Increasingly, he met the sad events of his twilight years with a profound resignation. As he said to John Quincy, "I foresee what all ages have foreseen, that poor earthly mortals can foresee nothing, and that after all our studies and anxieties, we must trust Providence." He preferred to talk of love and fellowship, although satisfactory words failed even his brilliant tongue. "Oh! how can I express my feelings for you and your family," he wrote to John Quincy, in England, "The whole must be included in the words John Adams."

All his life, John Adams had been a deeply religious person. Until the fourth generation, most Adamses were drawn in one fashion or another to a spiritual life. Clearly both John and Abigail were sustained in their later years by faith, although they certainly differed significantly in the manner of their faith. With her sisters, Abigail retained from childhood a personal Christian outlook, while John's God was more general and remote. Once, the elder Adams sought to draw John Quincy into a debate upon theology after reading the latter's letters to his sons endorsing the Bible and its divine advice. "The Biblic Rule of Faith!" John said to his son. "What Bible? King James's? The Hebrews?" There were 30,000 variations. Which one, he prodded? In this moment, John was violating a solemn pledge never to dispute religious matters, but he could not resist trying to provoke John Quincy into reconsideration: "An incarnate God!!! An eternal, self-existent, omnipotent, omniscient author of this stupendous universe, suffering on a cross!!! My soul starts with horror at the idea, and it has stupified the Christian world. It has been the source of almost all the corruptions of Christianity."

To have his grandchildren "all about me" was the old man's final idea of happiness, for he cherished every moment even, he gleefully noted, as they "devour all my strawberries, Rasberries, cherries, currants, plumbs, peaches, pears and apples, and what is worse, they would get into my bedchamber and disarrange all the papers on my writing table." And when his namesake, John Adams, John Quincy's son, persisted in wishing for a naval career, the grandfather told the lad that, so long as he covenanted with God to be a just and humane man, "I consent that you should leap at the moon and seize her by the horns when necessary, as your grandfather and father have done before you." In old age John Adams overcame his despair of 1798 when he had regretted having children and prayed for no more grandchildren. In 1815, Nabby's daughter, Caroline de Windt, presented the old couple with another great-grandchild, which prompted John to call down Heaven's blessing upon all generations and to acknowledge that he sorrowed over "the loss of every sprig" which had fallen from his family's tree: "It is very ungenteel in these days of politeness and civilization to have so numerous a posterity; but I cannot help it and would not if I could." It clearly satisfied John that he was "descending so smoothly" the hill of life he conceded he once "ascended so roughly."

For Abigail the descent was not so gentle. She was often lonely, which pained her gregarious nature almost as much as the arthritis and rheumatism which badly bent her frame. After Thomas finally moved his large family to Penn's Hill, Abigail and John were left in the Old House with Louisa Smith, one of the waifs of Abigail's scandalous brother William. Louisa Smith had become practically a daughter to them and was now a faithful nurse to both John and Abigail. When she felt well enough, Abigail was always eager for the Sunday evenings when she threw open the Old House, holding court for all villagers who cared to call. Mostly, though, after Nabby's death, John and Abigail were alone. However, they were never unoccupied. John's determination to read or to be read to from his capacious library kept the elderly pair eyeweary, since Abigail often read aloud by the fire, once John's vision had faded for the day. He usually arose at 5:00 AM and opened his latest volume, leaving Abigail abed despairing, certain that "the President," as she called John, would catch cold in the bitter chill of the drafty mansion.

There, in the quiet of the morning Abigail could take her turn thinking about her family, and the promising and thwarted lives it encompassed. "What delight we garulous old people take in living over again in our offspring," she wrote. Even after so many family hopes had been spoiled, she persisted, leaving John to observe, "she makes me tremble by her uncontrollable attachment to the superintendence of her household." A mellowed Louisa Adams wrote to her mother-in-law: "It has often been to me a source of wonder how you write to so many in one family, and yet never appear at a loss for subject." It even astonished Abigail at times. "At the age of seventy," she conceded, "I feel more interest in all thats done beneath the circuit of the sun than some others do at, what shall I say, 35 or 40?"

On 28 October 1818, a year after her eldest son had returned to become secretary of state, Abigail died. In 1814, she and John had celebrated their "day of jubile," their fiftieth anniversary. Even then Abigail sensed that John, though older, would be more durable. She prepared her will; she wished, she said, to be even-handed with her sons. To nieces and granddaughters she gave money and distributed her jewelry and clothes. For the wife of a Quincy farmer, Abigail had an enormous wardrobe. She expressed contentment with her life: "I ought to rise satisfied from the feast." In October she developed typhoid fever, and though she fought bravely to recover, she died early one afternoon. Neither of her surviving children was present. In Washington, John Quincy was so caught up in negotiating with Spain's minister, Luis de Onis, on questions of America's geographic boundaries that he allowed public duty to prevent his answering the summons from Quincy. Abigail's other son, Thomas, had become too debilitated in body and spirit to attend a parting scene. Her niece Lucy Cranch Greenleaf stood beside Abigail at her death. Abigail's last words were of almost professional admiration. Lucy, Abigail murmured, had been "a mother to me."

The family marveled at old John's serenity. He said simply as he stood by Abigail's bed, "I wish I could lay down beside her and die too." At the moment Abigail surrendered, John's already palsied frame shook so that for a moment he could not stand, but then he turned to Louisa Smith, the niece whom Abigail had adopted, and took the distraught woman into the next room to comfort her. When friends and other relatives made sympathy calls, John kept them enthralled with stories of Abigail's remarkable life and talents. Perhaps the secret to this latest display of steadfastness and tranquility can be found in his observation to John Quincy: "The bitterness of death is past," he said. "The grim spectre so terrible to human nature has no sting left for me." He and Abigail would now be separated for less time than many of the intervals endured when they were younger, he predicted, adding that it had been more difficult to part from her when he left for France in 1778. John leaned on his great staff to walk the distance to the church behind the body of his wife, a trip they had taken together so often. Among the pallbearers were the governor and lieutenant governor of Massachusetts and the president of Harvard College.

John Quincy atoned for his absence by extravagant tributes to his mother. Hers was "a life of one most truly and emphatically little lower than the angels;" if she had any "mortal infirmity," it was imperceptible, and "in passing from earth to heaven she can scarcely be conscious of a change." John Quincy asked of Thomas: "In all this my brother is there one word of exaggeration? No! No!" Thomas had not been allowed to see his mother when she was dying, and evidently much of the time he was in no state to appear before anyone. His wretched condition was evident in his cries to "our Saint-like mother" and his prayer that her death "may prove to me individually a lasting lesson of humiliation and amendment of life."

As time passed, John Quincy mounted tribute upon tribute to Abigail. "There is not a virtue that can abide in the female breast but it was the ornament of her," he said, adding, "Never have I known another human being the perpetual object of whose life was so unremittingly to do good." She had "little less than heavenly purity"; hers was a spirit "nearly approaching perfection." Then, in a revealing glimpse, this eldest son said that he had always delighted in returning to his mother's house, for there "I felt as if the joys and charms of childhood returned to make me happy." However, Louisa wrote to a family intimate to explain that, while John Quincy had learned in ample time of his mother's grave illness, he deliberately remained away from her bedside. His negotiations with Spain were so demanding that, had he left, he felt he should have had to resign as secretary of state. "This is for you only and perfectly confidential," Louisa warned. In fact, it was Louisa who paid the most perceptive tribute to her mother-in-law. Abigail, she said, had been "the guiding planet around which all revolved, performing their separate duties only by the impulse of her magnetic power."

Struggling to close his life without Abigail, John sought to be cheerful, although he acknowledged that "the world falls to pieces round about me," as he tried to maintain the Old House where twenty dependants were gathered now, including Thomas, his wife and six children, all of whom came for refuge in 1819. In finding it necessary once more to surrender whatever fiction of independence he had nourished, Thomas was so depressed by his return to his father's house that he could not speak about it. His title as "Judge," inherited from days when he was a minor county functionary, was now more derisive than ever. Nancy, Thomas' wife, was reduced to asking her father-in-law to "see you alone without exciting suspicion" so that they could talk about her husband's dissipation which she knew endangered the "harmony of the family circle." She also tried to distract her father-in-law from her husband's sins by telling him about the drinking escapades of Billy Shaw, John's nephew, who also was sheltered from time to time in the Old House.

Trouble broke out over Thomas soon after Abigail died, and John Quincy could not avoid being involved. He had been obliged to scold Thomas more and more as the years passed, so that a confrontation was inevitable when John Quincy visited Quincy in the summer of 1819. Thomas' drunkenness was a sore affliction for the stoic calm that John Adams had cultivated. When John Quincy was not soothing his father, he was confronted with a pathetic Nancy—"Much of her conversation with me I would wish for her sake to forget," the brother-in-law noted in embarrassment as he listened to the woman's suspicions about her husband. Then there were family advisers such as Josiah Quincy who urged that Thomas be removed from the Old House, and the village, if possible. Someone else suggested that Thomas be handed a government job, to John Quincy's dismay.

Faced by all these accusations, Thomas ran away one day after a rebuke from old John. For six days there was no sign of him, although John Quincy searched around Quincy and in Boston. Then poor Thomas reappeared, but in such condition that he could talk to no one but was put to bed. That night, John Quincy was sleepless as he pondered what to do. At dawn, he knocked on Thomas' door to begin the necessary talks, but Thomas was "not yet fully recovered." Eventually John Quincy consented that Thomas should be his ward, granting him a much increased "allowance," and agreed to forgive his large debts "upon a single condition to be performed by himself for his own benefit, and as long as he performs it." The price, of course, was no more drinking. Thomas promised full reform at the feet of his older brother, who probably never understood how his own rigid self-command was beyond the reach of some.

This crisis closed, in 1819, with John Quincy accepting the expenses of the entire Old House ménage and old John reconciled to giving Thomas another chance, the sinner loudly pledging to mend his ways. Having restored calm, before he returned to Washington the Secretary of State took a solitary walk over the hillside to the ancient village burying ground where his ancestors and now his mother rested. It was a spot he visited often, but this time he stopped at length before Abigail's tomb to consider how costly her absence was to the family, and what burdens had been placed upon him: "I implored the divine blessing that the cup of affliction might be administered in mercy and that all my duties resulting from the emergency might be faithfully, diligently, and discreetly performed."

John Quincy's wife disagreed with her husband's solution to the problem of Thomas. Always perceptive about the troubles of others, Louisa saw that Thomas needed less rather than more dependence, and that removal to some distance from the Old House was essential. Otherwise, she was certain that Thomas' "disease"—the family were loath to commit to paper words somewhat more blunt—would remain "incurable." Despite her views, Thomas and his father were made to remain together, leaving Louisa mourning what a "bitter affliction" it was to "the poor old gentleman" to see daily the testimony of parental failure. John Quincy remained unmoved, and somehow the uneasy association in the Old House continued. Thomas and his family lived in the mansion until John Quincy claimed the place for himself in 1829.

Few letters of Thomas Boylston Adams survive, especially from this period. Those which were allowed to remain in the family papers show a man habitually depressed, but there is an occasional flash of the sweet charm for which he had been known in his younger, happier days. Thomas' sharp-eyed nephew, Charles Francis, could not keep from liking his uncle, despite the trouble he caused and the sorrow he brought to the family patriarch. "It has often made me grieve to think that this man should make himself a ruin to others and to himself," Charles Francis observed in 1824, adding that Thomas actually had all the qualities needed "to make an excellent member of Society," save for his "being under the influences of this fire which he perpetually takes." Thomas had gotten so that he often required alcohol to summon the courage to talk with people, and when his condition then kept him from doing so satisfactorily, the poor man would rush off complaining about how uncongenial the family was.

In the face of this tragedy, old John usually contained himself and rarely fell into self-pity or recrimination. "I believe there never was a man in a more romantic, dramatic, tragic, or comic situation than is, at this moment, your affectionate father," he wrote John Quincy. He wrote Louisa that he was coming to believe in that power of life or death over children which Greek and Roman parents had enjoyed. However, in 1824, as Thomas began another drunken spree by lunging uncontrollably around the Old House until he jumped onto the stagecoach for Boston, Charles Francis found that the old gentleman wanted to talk about "the fate which has thrown so much gloom over our house."

These sorrows, he said to his grandson, and particularly Thomas' plight, were sent to check the family's pride. Had all his sons flourished, John predicted, the world would have been obliged to crush such a family, "but now while the World respects us, it at the same [time] pities our misfortune and this pity destroys the envy which would otherwise arise."

Mostly, however, John Adams radiated cheer to the family of his eldest son as he extolled the virtues of peaceful life in Quincy. He joshed at his son's pleas that he was too busy to come to Massachusetts: "I would not care a farthing about Spain. I would leave the census to my clerks, and throw the weights and measures up into the air." However, when the Secretary of State managed to visit the Old House John moved his son to tears by taking his hand and beseeching him to continue faithful to his duty, and John himself was deeply stirred when he lived to see his son become president. The aged parent's seeming indestructibility helped John Quincy justify making fewer trips from Washington. After all, in 1820 John Adams had gone to Boston to participate in a constitutional convention for Massachusetts, reporting delight in the endless "luxurious dinners in the best of company in the world" and in his numerous visits "to widows as if I was looking out for another wife." In 1821, he had traveled a long distance to visit a Boylston relative, and had reacted quite calmly when the news followed him that a fire had ravaged the roof of the Old House. He continued the trip. The next year he had delighted in a large feast of venison and wild turkey at nearby Jamaica Plain and was taking carriage rides about his Quincy lands until a few days before his death.

Fortunately for posterity, John Quincy decided that his father's marvelous old age should be recorded, and he commissioned a portrait by Gilbert Stuart, one which he hoped might prove a picture "of affections and of curiosity for future time." This portrait (the frontpiece to this book) now seems the most charming and moving of the many Adams family portraits. Not long afterwards John died, lying in the big upstairs room which Abigail had added to the mansion a quarter of a century before. It was the Fourth of July 1826—exactly the same day on which his friend Thomas Jefferson also expired. "He died as calmly as an infant sleeps," one grandson reported to John Quincy. His final request had been to hear William Ellery Channing's sermon on walking by faith, and after this was read to him, John Adams pronounced himself satisfied, saying that he would see his beloved family again in another world.

7

Washington

When James Monroe became president in 1817 there was little doubt in his mind who should be secretary of state in the new administration. Having himself been in charge of foreign policy in Madison's presidency, just concluded, Monroe recognized better than most the brilliant role John Quincy Adams had taken in the international scene since 1809. Adams' experience and his innocence of recent involvement in American politics made him preferable to Henry Clay of Kentucky, who wanted the appointment. General acclaim greeted John Quincy's selection. While State Department duty often kept him from visiting his father, it led to one of his greatest triumphs, for his record as secretary of state is possibly the most brilliant in American diplomatic history. He managed this with the aid of only eight clerks. The tiny size of government bureaucracy then suited John Quincy's diligence, knowledge, and experience.

The issues facing him were complicated. There were important questions left over from the peace conference at Ghent. Should Great Britain and the United States continue maintaining an armed navy on the Great Lakes boundary between the two nations? How might the two peaceably share use of the North Atlantic coastal fisheries, an issue in which John Adams had taken a keen interest long before? Then there were troublesome problems involving the location of a boundary line between the United States and Canada west of the Great Lakes, as well as indemnification for slaves reputedly abducted by the British during the 1812 war. Succeeding in settling most of these within a year, John Quincy then turned to America's festering dispute with Spain over

Florida's boundary. In the negotiations which kept him from visiting his mother during her last illness, John Quincy negotiated the Transcontinental Treaty of 1819, which took the nation's frontier to the Pacific—an achievement scholars have called the greatest diplomatic victory won by any one person in United States history.

Soon John Quincy was applying his diplomatic genius to the formulation of the Monroe Doctrine, which was completed in 1823. This policy sought to limit European influence in the Western Hemisphere. The Secretary of State also worked toward assuring free maritime commerce for all nations, suppressing the international traffic in black slavery, encouraging wider reciprocal trade agreements for the United States, and clarifying the nation's Northwest boundary. Amid these prodigious efforts, John Quincy also managed to produce one of the most significant, if little recognized, scientific treatises in American scholarship, a work that grew out of his compliance with Congress' request for an analysis of worldwide systems of weights and measures. His superb *Report on Weights and Measures,* completed in 1821, had taken most of his waking moments not given to diplomacy. He was usually stirring before 4:00 AM and abed often only after midnight. Since vigorous early morning exercise, Bible study, and diary writing also were sacred duties, John Quincy was left little time to attend to his family in Washington, to say nothing of his aged parents in Quincy.

Yet these family responsibilities came near to driving the overworked Secretary of State to distraction—he feared at one point that he was losing his sanity. All of the domestic joys and sorrows, particularly those involving his sons, invariably stirred him, no matter how captive he was to official business. There was little about himself of which John Quincy was unaware, for he was usually his own most severe critic. "My natural disposition is of an over-anxious cast," he once told Louisa. "[M]y struggles to accommodate myself to circumstances which I cannot controul have given my constitution in less than fifty years the wear and tear of seventy." Even worse, it was a concern which he sought to bottle up "within my own bosom."

All this was, of course, no surprise to the perceptive Louisa, who was justified in saying, "I know my husband's character." It was she who recognized from the start that the old family dilemma would haunt John Quincy's career; and that, despite his proper Adams feeling of horror attendant upon serving the world, "he would not live long out of an active sphere of publick life . . . it is absolutely essential to his existence." It annoyed Louisa that portraits of her husband presented him appearing "disagreeable." His mouth seemed contemptuous, she thought, while it was actually only excessively earnest and serious. Instead, Louisa tried to remember those rare moments when her husband was high spirited, and particularly when he gave more attention to his children than his traditional, grave inquiry after their health each morning at breakfast. Such moments, she said, showed "how pleasing he can make himself when he has a powerful stimulus to excite him."

More often, however, John Quincy "looked all the ill temper that I suppressed in words," as he put it. ". . . I am a man of reserved, cold, austere and forbidding manners," a quality of which he said, "I have not the pliability to reform it." Not only did Louisa suffer from this manifestation of her husband's terrible load of anxiety, self-doubt, and compulsiveness; his children were just as deeply affected by it. At the age of seventeen, the acutely perceptive Charles Francis observed that his father was nearly impossible to draw near to: "He is the only man, I ever saw, whose feelings I could not penetrate . . . I can study his countenance for ever and very seldom find any sure guide by which to move." To this son, John Quincy Adams seemed perpetually to wear an "Iron Mask."

Eventually, this facade was shattered by worry and grief over their eldest son, George. So methodically were George's papers destroyed after his death in 1829 that only glimpses remain of this child's flight into erratic behavior and a fanciful world, a tactic which apparently was his way of trying to stay beyond his father's reach. When he was sixteen and back from England to begin studies at Harvard, George described a dream in which, while receiving smiling encouragement from a young lady, "I saw the form of my Father visible to me above, his eye fixed upon me." His amorousness "relapsed in the most cold indifference," while the parental voice "rang at my ear—Remember George who you are and what you are doing." After awaking George recalled that he "sank into a gloomy torpor." This was increasingly his response to life.

Even before he entered college in 1817, George's career was planned for him. It was, of course, decreed that he study law and manage the family's estates, as befitted the eldest son of the new generation. After a lackluster college career, George graduated from Harvard in 1821, his part in the ceremonies being an oration upon "the influences of natural scenery on poetry." This contribution his father failed specifically to comment upon, preferring generally to condemn the exercises as the poorest he had ever witnessed. Little was left to George's choosing, not even his own celebration at graduation, a time at which he had wanted a small party only for his parents, brothers, and Uncle Thomas' family. John Quincy could not leave it at that, but summoned the governor, the college president, and other dignitaries who, it was felt, should be acknowledging an Adams event. This beginning of a career was not auspicious, for Louisa was certain that her son, torn between doubt and enforced ambition, had begun to drink, which bothered her as much as his addiction to tobacco. George's tendency to flee reality also alarmed his observant mother, who saw him blinded by "the baseless fabrick of the visions which continually bent his imagination where all is Poetry, fiction, and love."

Some of George's most earnest attempts at writing poetry went into conventional lines devoted to his mother, who nevertheless kept worrying for the boy. "Bitter experience is the only lesson which can make any impression on some minds," she said. The poems George sent her left her less moved than fearful as she read his tributes to her, one of which went:

> In that whole past before my sight
>   One brilliant beam I see—
> A gentle, mellow, lovely light
>   Thy kindness proved for me. . . .
>
> What ere my fortune be
> Heaven grant that I for ever may
>   Devote myself to thee.

Before George finished Harvard, his two younger brothers joined him there, but their efforts did nothing to cheer John Quincy. First there was John, who had already disquieted his parents by his temper and what his mother called "his pretensions." He became an incessant smoker, often neglected college exercises and duties, was scolded by authorities for liking to make "a festive entertainment" and for creating much noise. Louisa begged her third child, Charles Francis, who was four years younger than John, to keep the elder brother "out of mischief." He was to do this, "not by contradiction but by persuasion and sound reasoning," she said. She saw her second son with painful clarity, "a hotheaded, noble, rash *boy.*"

Charles Francis, however, soon created his own share of concern for his parents as the whimsical child turned into an uncertain and taciturn young man who matured too rapidly. Louisa fretted about him. "He has a remarkably discriminating mind," she had reported to John Adams on his grandson's fourteenth birthday, but added, "It is blinded by an invincible obstinacy of character." To Charles Francis, Louisa was very frank. "You have lived long enough with your father," she told him, "to be aware of the great importance of regularity and method," and she urged him to build these good habits around his work. She said, candidly, that if he failed, it would be a particular grief, since "you have been the one of my children who has lived the most with me and I should be accused as the principal cause of your ill success." She knew too well that the guilt of watching a son fail would be more than John Quincy could bear alone.

The Secretary of State was especially vexed when it appeared that none of his children exhibited the love for literature he and his father shared. For instance, in 1821, when Charles Francis undertook to enter Harvard at age fourteen, John Quincy learned that his son had been given a conditional admission because he was deficient in Latin. This made the father rush to see for himself: "I could not avoid the suspicion that he had been unfairly treated." There followed a half-comical encounter between John Quincy and his alma mater: John Quincy believed an old foe in academic politics was avenging himself by mishandling Charles Francis, while President John T. Kirkland defended the college. Finally, Charles Francis was given a second exam at which John Quincy insisted upon being present. Even with this tension, Charles Francis did well enough to have the limit removed, but the college had more bad news. John Quincy was notified that George stood only thirtieth in his class and that John was a puny forty-fifth. When the Secretary of State called upon President Kirkland to investigate this, the administrator was too busy to be disturbed.

Harvard thereafter seemed an endless reminder to John Quincy of his failure to hand on to his sons the stern stuff of which scholarship is made. "I had hoped that at least one of my Sons would have been ambitious to excel." Instead, "the blast of mediocrity" was demeaning the family name. Feeling helpless and angry, John Quincy ordered severe reprisal—John and Charles Francis were not to be allowed to visit Washington during Christmas vacation in 1821, a terrible deprivation. George could come home but only for remedial attention from his father. Meanwhile, the dark suspicions that Harvard was being unfair to Adams boys persisted until President Kirkland spoke ominously, if vaguely, of other misdoings. This was sufficient to make even John Quincy retreat meekly. There was nothing he could do thereafter but implore Kirkland to help the three Adams lads improve.

As Christmas approached that year, Louisa tried to soften her husband's heart while remonstrating with her sons for betraying their father. Nothing availed; John Quincy stood firmly for punishment. He now wrote so much about his own feelings as to remove any doubt that the boys were in disgrace because of the "mortification" they had brought their father, a humiliation which John Quincy harshly told one son "was mingled with disgust." Were they to come to Washington with such miserable grades, "I would feel nothing but sorrow and shame in your presence." The stigma could be erased only if each boy dramatically raised his academic standing, John Quincy at first demanding they rank in the upper six, but then relented and stipulated a place among the top ten students. What his sons must attain, said the father, was rank comparable "with that which your grandfather and your father held." When that happened, he promised to embrace each not only as a grateful child, "but as my benefactor."

Then came news that Charles Francis was fifty-first in a class of fifty-nine. John Quincy was aghast, spoke of blushing for shame, and told the youth, "It is in your power to fill my bosom with anguish, or with delight," but only by "a standing which will not fix a stigma upon your name." Whether the father's sternness inspired the boys or whether they wanted to miss no more holiday seasons in Washington, their class rankings improved considerably after 1821 and left John Quincy somewhat mollified. Both he and Louisa took a special interest in seeing that their children developed social presence. John Quincy complained how "my dear Mother's constant lesson in childhood" had been "that children in company should be seen and not heard," treatment which he believed had left him incapable of practicing the art of conversation. Thus, when his sons joined their parents in Washington, they were taken to White House affairs, to balls and dinners, and, of course, to church.

There was still more devastating news from Harvard. In May 1823, on the eve of John's commencement, the Adamses were notified that a large number of seniors were being expelled for rebellion. The faculty had refused to allow one student who was very popular with his fellows to have a proper place in the graduation exercises, and there had been riots, disruption of chapel services, and even an uproar when the college president tried to speak. The college announced that the hotheaded John Adams was a leader of all this, so that he and over fifty other seniors were dismissed. John's fate made his father give up hope for the intellectual development of his sons. He ignored his father's plea that his errant namesake be received by John Quincy "tenderly and forgive him kindly."

Not even Louisa could describe to her sons what this expulsion from Harvard had meant to their father, while she herself was worried about what future troubles her wayward boys would find. "My soul shudders with horror." Meanwhile, John Quincy swallowed the bitter medicine of begging President Kirkland to rescind the action, urging that the miscreants be taught humility but that they not be humiliated. The college authorities were immoveable, however, and finally, when John Quincy refused to be appeased, told him that there were many unsavory details of the rebellion which had not been revealed. At this quiet threat, the Secretary of State retreated. Not only did young John Adams not receive his degree—it was awarded posthumously fifty years later through the good offices of his brother Charles Francis—but the young man was forced to suffer the silence of his unhappy father for many weeks before John Quincy finally sent him a curt note: "I wish to spare you and myself the pain of expressing my feelings on this occasion." John was ordered to Washington to run errands for his father.

Amid the disappointment over their second son, John Quincy and Louisa had fresh worries about their youngest, who began drooping under feelings of doubt and guilt. He acknowledged his "depraved habits," and proposed leaving college a self-confessed failure. Both John Quincy and Louisa tried to bolster his nerve and morale, the father thundering against "dissipation and licentiousness," horrified that another son was running away from stern discipline. The temptations Charles Francis feared were everywhere, John Quincy assured him. The boy should stand firm, keep his morals pure, and seek to use his talents if not with "lustre," then at least in no way to "discredit your name and parentage." Louisa, however, went right to the center of her son's problem. She told him quietly that he would succeed admirably if he could only conquer his "childish fear of ridicule." She said she would even use her "very little influence" with John Quincy to permit Charles Francis to drop out of Harvard for a year and take up private study, if that would renew his shattered confidence. The step proved unnecessary, for he soon regained command of himself, following long talks with his grandfather at Quincy.

Louisa's sons knew they had in their mother a friend at court, so that they sent her letters surreptituously. She begged them to understand how their father's expectations of them came from an overpowering love, while she told John Adams, "it grieves me to the heart when his children cause him uneasiness. A better father could not exist." Her grown children were one of Louisa's severest trials, for she realized that she would have to help them with the burdens of an Adams heritage while mediating and softening their father's harsh directives and punishments. Louisa greatly admired John Quincy as a public man as she watched his triumphs as secretary of state. She told John Adams that she only hoped that her children would "walk in his steps." If this happened, "then indeed I shall have room to be proud with such a husband and such sons."

Instead, Louisa found herself increasingly sharing her husband's chagrin. Their children's problems were not limited to youthful difficulties at Harvard. Indeed, there were temptations at home in the person of Mary Hellen, an orphaned daughter of Louisa's eldest sister who had become after 1817 an adoptive daughter in the family circle of John Quincy. Her father had commended her to the Adamses as "one of the sweetest little angels." Actually, she proved to be a vixen, whom Louisa later informed, "you are in the habit of behaving shamefully." Mary captured the hearts of all three Adams boys in turn, beginning with Charles Francis, who was her age. Mary, Charles Francis acknowledged, taught him how much physical desire influenced a man's relation with a woman. After his "difficulty" with Mary, he concluded that platonic ideas about love were nonsense, and vowed that someday he would tell the story of this vexing creature. Louisa observed all this and was much troubled. "I have very little faith in the chastity of the person," she noted in 1821, watching Mary turn from Charles Francis to George, who was soon another victim. The two became engaged, to his parents' dismay. Two years later, Mary shifted her interest to John, whose sardonic ways intrigued and even subdued her. George's shattered romance added to his enormous load of self-doubt, while Charles Francis muttered, "I have as much right to consider myself as having been very ill treated by her as George has."

This unhappiness reminded John Quincy of the occasion, long before, when he had taken George and John for a garden walk. "I took George by one hand and John by the other. . . . Come said I, great *burden* on this side, and *little burden* on that." Twenty years later it was difficult for John Quincy to tell which son was the heavier load. For a time John remained the greatest test of his father's inspiration as a parent. Since a legal career was out of the question for his son, John Quincy decided that perhaps John could be useful by managing some family investment, a plan that would keep the son under the parental eye. This led John Quincy into one of the great follies in Adams history, his purchase in 1823 of the Columbian Mills which then perched along Rock Creek in Washington. It might not have happened if he had not possessed astonishing patience and kindness for Louisa's hapless Johnson relatives. A cousin of Louisa's, George Johnson, begged for aid from John Quincy, thereby beginning a catastrophe.

George Johnson owned a milling business on which he had a $20,000 mortgage he could not handle. To stave off foreclosure, he offered John Quincy the mills if he would pay the debt and keep George in charge. Consulting no one, but professing to see a great business future for his sons, John Quincy Adams became owner of the Columbian Mills. He quickly discovered that the bank debt was only part of the difficulty. Vast sums were needed to restore the business to reasonable capacity. The new proprietor put all his available resources into the mills, which meant mortgaging his Washington residence. Kept for a time as manager, George Johnson promptly displayed how he had ruined the mills in the first place. John Quincy commented with unusual restraint: "These are circumstances of no encouraging tendency." Once the mills began to function, the market for flour dropped. "I will not despair," said the wretched new owner.

Soon cousin George had to be fired, while Louisa looked on tight-lipped. The family's finances were "in such bad order," she told her son John, that it would take "a large portion of grist to bring us up." It was John who now received this great responsibility. He needed to demonstrate not only his own ability but to help his father prove that the investment had not been folly. So it was that young John Adams was handed a task for which he had little talent, and which would have taken a business genius to manage profitably. An experienced miller would have been appalled at the bad location and the outdated equipment of the mill. John Quincy, however, was convinced that hard work and prayer would overcome almost anything, and his son was often reminded of this. John readily agreed to try the business. Disgraced in Boston, he had no taste for seeking a career in Massachusetts. The flour business was his only hope.

The new manager of the mill went to work while his father looked over his shoulder. Lest the boy's new duties cause him physical difficulty, John Quincy insisted that the son get exercise by accompanying him on horseback rides. But John was now beginning to have more than he could handle, and he begged to be let alone. "I have already broken down him and two horses," said John Quincy.

Such were the dubious satisfactions the Secretary of State could draw from the faltering careers of his offspring. Much disappointed, he allowed himself to enter the campaign for the presidency when James Monroe retired in 1825. As the election of 1824 drew near, John Quincy began talking less candidly about himself in his diary. An Adams was never supposed to seek public office. Instead, he was to be ready if the American people declared their need for him. Entries in John Quincy's diary for the years after 1821 have hints enough to show that he was painfully conscious of his hunger for the highest office and that he was being drawn into distasteful strategy to prevent William H. Crawford, Henry Clay, John C. Calhoun, or Andrew Jackson from taking the prize.

When the voting returns in 1824 showed no majority in the Electoral College for any one of the contenders, the issue went before the House of Representatives. There, on 9 February 1825, John Quincy Adams was chosen president after a demeaning negotiation among politicians. He remained forever silent about what price he was required to pay against the Adams family's aloofness from sordid partisanship. The triumph was tarnished from the start by talk of corrupt bargains the new President had supposedly entered into. This moment was one of the most painful in the Adams story. Once again one of the family had shown the power of ambition and partisanship, those personal and social weaknesses which Adamses were supposed to exhort against and transcend. The taint could not be erased. John Quincy Adams as president was as noble a failure as he had been a success in the secretary of state's office.

His election as president drew from John Adams predictably mixed emotions. While there was much gratification, John Quincy received a warning from his father: "This is not an event to excite vanity." But neither father nor son could have foreseen the parallels in their experiences as the nation's chief executive—that neither would receive a second term; that both would be thwarted in their hopes for lofty administrative accomplishment; and most of all, that while in office they would suffer as much from domestic affliction as from political disappointment. The new President gave himself wholly to his dream of statesmanship, remembering how his father had admonished him eight years earlier, when he entered Monroe's cabinet, "you must risque all." Indeed he did—and lost, for John Quincy sought to persuade citizens to increase federal power: to surrender themselves to a wise government which used its powers to create internal improvements, public education, and scientific enquiry for the benefit of all. Neither the times nor the Congress encouraged such interpretation of federal power in the Republic. The Adams administration was stymied.

Absorbed by his difficulties as president and his domestic worries within the White House, John Quincy took no great alarm at news of his father's rapid weakening, and thus he missed John Adams' final moments, as he had his mother's. Arriving at the Old House eight days after John's death, he faced the task of lifting the family from the lassitude which had developed during his father's decline. The prospect was so dismaying that John Quincy announced it was time "to prepare for the church-yard myself." Meanwhile, however, duty must be faced, and the President wandered about the mansion and the familiar Quincy scenes, trying to realize that these duties must be carried without comfort from his parents. However, it was only when he listened to memorial sermons which mentioned "my almost adored mother" that John Quincy's emotions gave way. He reported being moved "more than I was aware it was in the power of human speech to do." Nabby and Abigail's names overwhelmed him, and he gloried once more at "having been the brother and son of such unsurpassed excellence upon earth." A year later, he arranged a separate tomb for his parents, moving their coffins from the family vault in the cemetery to the newly constructed church across the road. The building was erected through funds bequeathed by John Adams. It replaced the old structure where John and Abigail had worshiped. Now they rested in a crypt beneath the entrance.

As head of the Adams family, John Quincy faced problems which severely tested his devotion to his household and to the principle of duty. Even a grieving Louisa, who was devoted to the memory of her father-in-law, was alarmed when she read John Adams' will. She saw at once that it put "the whole burden of affairs" on her already beleaguered husband. As executor and trustee, John Quincy was responsible for having the money at hand to pay the fourteen bequests left by his father, and to distribute benefits held in trust. Under John Adams' will, John Quincy had the chance to keep the Old House and surrounding farm, provided he paid $12,000 into a trust as Thomas' fair share. All books and papers were to be his, except that half their value was to be paid to Thomas' account. The vast assortment of land parcels around Quincy which belonged to John were to be auctioned, along with other items, and the income to be divided among the fourteen heirs.

Impoverished by his unfortunate purchase of the flour mill in Washington and from earlier selfless steps to aid his parents, John Quincy

was nevertheless determined to keep the Old House and to purchase
as much of the Adams lands as he could. He already owned the area
around Penn's Hill and the cherished Mount Wollaston farm with its
375 acres. On 19 September 1826, he bought at auction numerous
family plots of marsh and woodland for a total of nearly $10,000, greatly
enlarging the debt he owed to the estate of his parents. The step gave
him claim to eighty percent of the ancestral acreage, "which," said John
Quincy, "I hope will afford an asylum to my last days."

Consequently, when it was finally determined that John Adams' estate totaled $44,709.47, of which, after expenses, $42,000.00 remained
for distribution to fourteen heirs, most of this amount came from John
Quincy's pocket. He had taken the land, in effect putting him into debt
to his fellow heirs. To meet this obligation, the President of the United
States had to borrow money, which only added to his already heavy
debt in Massachusetts and Washington. This new burden, one the family would carry for many years, was assumed over Louisa's protests.
Was it not enough, she wondered, that her husband must now have to
care for his brother's family, "relatives who for years have been jealous
both of your talents, your station, and your fortune?" Must he also
proceed to "waste your property and burden yourself with a large, un-
profitable landed estate which has nearly ruined the last possessor,
merely because it belonged to him?" The only concession Louisa recommended to family sentiment was retaining the Old House. John
Quincy waved aside her advice.

Inside a year, Louisa's shrewd assessment had been confirmed. With
his brother still entrenched in the Old House, John Quincy managed
to rent the surrounding farm. However, tenants who paid only $435
annual rent clamored for repairs and improvements, for the place had
deteriorated sadly in John's final days. Ruefully, John Quincy began to
see what family sentiment was doing to him. The Old House had cost
him $14,000 in borrowed money, for which he faced $850 annual interest, nearly twice his income from property. "This is no bargain to
boast of," the President wryly acknowledged. His own largest debtor
was his brother, Thomas, from whom no payment could be expected.

After Thomas had seen his father, like his mother, die while he was
still a sorrow to them, he determined once more to strike out on his
own. John Quincy offered to allow him and his family to remain in the
Old House until he completed his term as president, and meanwhile
he arranged for what he called their "comfortable subsistence." In November 1826, Thomas told his brother that, while he knew John Quincy

might disapprove, he wished to move into Boston, there to practice law
or pursue business. In a rambling letter, Thomas talked pathetically,
seeming to realize how difficult this last grasp for freedom would be.
He and his family had some regular income from money bequeathed
in trust by John, and there was help from his brother. But this was
surely not enough, he lamented, to do what he most wished—to start a
business in New Hampshire.

Even if he could not go to another state, Thomas said, he must somehow escape Quincy, "where no good fortune has ever attended me."
So Boston it would be, and while Thomas had no idea of how he would
manage, he described the immeasurable relief "from the agitation of a
fevered mind" which accompanied the decision. Perhaps Thomas
thought this announcement might produce more money from John
Quincy, whom some family members erroneously considered a man of
endless means. However, the elder brother had neither the money nor
the faith to support a fresh beginning for Thomas. His plans collaps-
ing; the younger brother had to beg a renewed lease on the Old House.
After that, he sank once more into alcoholic helplessness, creating such
an uproar in the family mansion that Louisa refused to visit. There was
scarcely time to think of Thomas' plight while John Quincy and Louisa
lived in the White House. Washington held enough problems. Their
three sons were often with them, along with Mary Hellen and others
of Louisa's Johnson relatives, many of whom were quite willing to
sponge off the Adamses. In these surroundings, President Adams tried
futilely to lead the nation beyond provincialism and partisanship, while
his sons contributed the worst part of his ordeal.

Troublesome though John's presence in Washington was to John
Quincy, it was still a modest trial compared to that brought by George.
The eldest son was usually in Boston where he was charged with overseeing property and making a great man of himself. There he received
almost daily exhortation from both father and mother, with John Quincy
stressing particularly how "it is indispensable that you should keep a
continual watch upon yourself." He also told George he must be ever
conscious that his parents relied on him—did he not know, John Quincy
asked, "how much of the comfort of my future life depends upon your
conduct?" More than material wellbeing was on the father's mind, for
a son's virtue was his father's virtue.

George was warned by his father of what happened to men who led
helpless servant girls astray; he was told to develop a religious faith;
and he was advised to read John Adams' letters written at George's

age—"and then tell me whether you are his grandson." Most often heard were fervent pleas for the son's temperance. "I have been *horror-struck* at your *danger,*" said the father, "*danger* that distresses me even to agony." It was not so much George's failure to succeed in law and business which concerned John Quincy. It was his son's sinful ways. As the father's last hope to divert George from the fate of his Uncle Thomas, he begged the young man to read the Bible: "Burst the bonds of that wretched infidelity which is the natural offspring of Licentious life and which has been stealing upon you with your vices." Helpless and thwarted at having to stand afar while his boy fell into disgrace, the President grew almost frantic as he charged "licentious sensual indulgence" with being the cause of "weakness and imbecility of intellect." Indeed, he said, "dreadful must be the perversion of mind which rejects the multiplied blessings" of being an Adams.

Despite Charles Francis Adams' later efforts to obliterate George's tragedy from family records, enough remains to glimpse the unhappy soul of this eldest son. For some reason, deliberate or otherwise, a fragment of an autobiographical piece in George's hand, begun in 1825, has survived, along with a few pages of a diary for August of that year. The autobiography reviews George's life before entering Harvard in 1818 in a manner which suggests that the young man knew himself better than his parents realized. He seemed to perceive that he lacked the virtue most exalted in the family—the stern command of himself necessary to move through the treacherous world. George blamed this on the lack of discipline during his parents' long absence in Russia. From this, he claimed, stemmed his penchant for fanciful literature, his craving for "narrations of crime, tales of terrible depravity, mysterious horror, and supernatural power," and his fascination with stories of suicide. In short, his parents were at fault since they had left him behind during his formative years, or so George seemed to think. He had, of course, often heard his mother's guilty cry that her sons might have improved had she not been taken from them in 1809. When he entered Harvard, George recalled—at the point the autobiography breaks off—he possessed a wild imagination and no mental discipline.

The diary fragment for August 1825 seems to have been written during one of the periods when George casually considered reform. These 1825 resolves for self-command and intellectual enrichment lasted a few days, and were soon replaced by diary entries explaining why tardy rising and long afternoon naps were necessary. Then, in late August, the journal ceased, with George adding a postscript at the year's

Abigail Adams in 1800, by Gilbert Stuart. This copy by Jane Stuart belongs to the Old House. The portrait shows how much the First Lady had been affected by the illnesses and family worries which often kept her in Quincy away from the President, though he yearned for her presence. *Courtesy of the National Park Service, Adams National Historic Site.*

Left, George Washington Adams in 1823, by Charles Bird King. At this time the family still hoped for the young man's success. *Right*, John Adams 2d in 1823, by Charles Bird King. John was then a cause for family shame, having been banished from Harvard. *Both, courtesy of Mrs. Waldo C. M. Johnston. Photos by Mary Anne Stets.*

Mary Catherine Hellen, probably around the time of her marriage to John Adams 2d in 1828. The daughter of Louisa Catherine's older sister, she entered the Adams family as an orphan in 1817 and remained, until the close of her life, a sad and troubled woman. *Courtesy of Mrs. Waldo C. M. Johnston.*

Charles Francis Adams in 1827, by Charles Bird King, a portrait never highly regarded by the family. It hangs now in the Old House. *Courtesy of the National Park Service, Adams National Historic Site.*

Abigail Brown Brooks, "Abby," about 1829, near her marriage to Charles Francis Adams. Probably William Edward West or Charles Bird King is the artist. The portrait belongs to the Old House. *Courtesy of the National Park Service, Adams National Historic Site.*

John Quincy Adams in
1818. A bust by Pietro Car-
delli which Louisa Catherine considered a superb resemblance. It is in the Old
House. *Courtesy of the National Park Service, Adams
National Historic Site.*

First known photographic likeness of the Old House, a daguerreotype made
near the time John Quincy Adams died in 1848. Original is at the Old House.
Courtesy of the National Park Service, Adams National Historic Site.

end to say that keeping a diary required too much time—the sort of easy evasion that so enraged John Quincy. He knew that his son's social life at night required him to take naps of three and four hours in daytime, losing those precious moments which could go to the family practice of keeping a diary.

George conceded that journal entries had been useful "in preserving the purity of my heart from dangerous decline," but he also confessed that he had lost the battle to keep pure—and that the year 1825 had seen in him "a change far from commendable, and far from happy." His last diary words predicted that 1826 would bring a "crisis" in his life, one involving a battle against "vices the germ of which has been planted by past irresolution and has recently alarmed me by its gradual expansion." He sounded much like his father as he concluded 1825 with a prayer for aid in escaping "from the domination of vicious and corrupt passions," and as he confessed to being "melancholy."

The diary had hinted at what was ahead. At one point George spoke of the radiant loveliness he found among young women of the lower classes. As his romance with Mary Hellen, his cousin in Washington, dissolved, George took as his mistress a servant girl in the house where he boarded. With her, evidently, he could believe himself in command of life. He continued to drink and gamble, yearning for an identity he could claim as his own. Although George's letters to his parents were usually windy, evasive documents that sought to placate them, just as his Uncle Charles had done in similar circumstances long before, he did throw light upon his dilemma in a revealing comment. To his mother he claimed he could never be both himself and anyone acceptable to the family. "I must form an artificial character or be forever nobody," he said in 1825, adding sophomorically that all men were marionettes in a show box, "and I a petty puppet like the rest." George begged Louisa to show this letter to no one, meaning, of course, his father.

John Quincy bore down upon his son in 1826, perhaps sensing that time available for his reclamation was short. George responded uneasily by accepting a seat in the lower house of the Massachusetts legislature where his poor habits, laziness, and faulty judgment betrayed him. He failed, of course, in the face of renewed admonitions from his father. By summer's end in 1826 George was seriously depressed and withdrawn, undone especially by the reproofs which fresh disappointment and frustration now brought from John Quincy. Arriving in Massachusetts, the father demanded that the eldest son accompany him in the exhausting work of surveying the new plots of land purchased from the John Adams estate. When George faltered under the strain and wet weather, an ordeal which only made John Quincy labor the more, the father taunted his son, embarrassing him before the strangers who made up the survey team. This was the final blow, and George collapsed in the belief, as his alarmed mother put it, that "he is unfit for the society or the duties for which other men are born."

Stirred by George's alarming condition, John Quincy reversed his tactics in September 1826, staying with his son constantly, even sleeping with him. He showed a mixture of kindness and fright which finally prompted the young man to confess his misdeeds. John Quincy cautiously took hope, praying that the son be restored sufficiently from his evils to become "a useful assistant to me and a successor to my endeavors for accomplishing the benevolent intentions of my father." At that point, John Quincy decided to set the proper family example by seeking membership in the family's Quincy church, which, like so many New England parishes, was slipping into Unitarianism. The President conceded that he should have done this thirty years sooner, instead of allowing "tumult of the world, false shame, a distrust of my own worthiness," to persuade him to postpone the step. In October 1826, John Quincy became a church member, but he had no success in urging George to do likewise. The son had been told that church attendance induced meditation and self-scrutiny, modesty and benevolence, while it quieted the passions.

Apprehensively, John Quincy returned to Washington, and wrote to remind George of his latest promises. Louisa sent messages hoping that George had achieved "one great conquest over yourself" so that he might yet master "those passions which take the most formidable hold of our nature." To carry on as he had in folly and irresolution "leads to imbecility if not to ruin." Yet within a few months the parents had new worries as reports reached Washington that George's law office was a center for drunkenness and fornication. The relapsed son neglected to answer his father's frantic letters for six months. John Quincy was left in frustration, and since public life offered little comfort at the time, he lavished his stern attention on his second son John, insisting anew that they take together the exercise the father considered so essential for right living. The presidency had not daunted his respect for the strenuous life. Of John he reported: "He says that I ride so hard that it turns him inside out."

It developed, however, that John's mind was on matters other than exercise and milling. He had set out to steal his brother George's fiancée, and after a stormy engagement, the marriage between him and Mary Hellen occurred despite strong opposition from the groom's parents. The couple's wedding on 25 February 1828 in the Episcopal church near the White House so upset John Quincy that he forgot about the ceremony and had to be reminded to attend. Mary became pregnant at once, a state she considered invalidism, to Louisa's disgust. She was similarly annoyed by the couple's decision to live in the White House with its numerous servants. The situation depressed both John Quincy and Louisa, especially when the hapless John became a father on 2 December 1828. The baby, Mary Louisa, quickly became the charge of her grandmother when the mother seemed too weak and irritable for such duties.

This was only one unhappy episode in the terrible ordeal of the presidential term as far as Louisa was concerned. The White House years taxed her perhaps more heavily than had the fearful days in St. Petersburg. The First Lady had not been able to lose herself in political battle as her husband had sought to do. Instead, she was often left alone to fight off a serious depression, worsened by experiencing menopause, a time, she said bitterly, when "love ceases." She began to write incessantly, seized by what she called a "mania" for scribbling. In a comparatively cheerful moment, she put down her astute observations of her family. These character sketches are in a story called "The Metropolitan Kaleidescope" which pretended to be about a prominent English family named Sharpley, but which was actually an assault upon Washington life. John Quincy was Lord Sharpley, a man of vast talent and attainment, and "a fond father, a negligent but half indulgent husband." He was "full of good qualities but ambition had been the first object of his soul," wrote Louisa.

Of herself as Lady Sharpley, Louisa said: "She was the oddest compound of strong affections and cold dislikes; of discretion and caprice; of pride and gentleness, of playfulness and hauteur . . . irritable one moment, laughing the next," a person who had never cultivated her opportunities. Louisa described herself as having allowed a warm heart to make her suffer repeated disappointment, leaving her sour at the age of fifty, trapped in a frigid environment when she actually needed affection. Mary Hellen was first named Miss Flippant, but Louisa altered this to Miss Manners; as a niece of Lady Sharpley she was pictured as cold, indolent, and craving the unattainable. George was called a poet and enthusiast who was at heart a simple child. John was said to mistake "spleen for wit and sarcasm for playfulness" and to be "often over-bearing and haughty" with a mind "tainted and irritated by false principles." At nineteen, Charles Francis remained a mystery to Louisa. She described him as the third Sharpley son: "his passions like his father's were mighty but they seldom found vent, but he required all his father's skill to keep them within due limits."

This striking piece is the most clearly autobiographical writing Louisa did, outside of her journal and memoirs. But she put many of her views in the poems and plays she wrote while she was First Lady. A drama, entitled "Juvenile Indiscretions of Grand Papa," ridiculed the weakness and foolish amorousness of men, and extolled the wisdom, shrewdness, patience, and forgiving nature of women. In another play, "Suspicion or Persecuted Innocence," the women display only superior qualities, while the men are weak and emotional. The heroine, whom Louisa modeled after herself, announces: "Men ever dread the weakness of our softer sex," until "in the hour of peril" woman shows her true power and nobility. None of these writings was printed, although in 1826 Louisa was so enthusiastic about being an author that she secretly sent some poems to George in Boston, asking that he get them published for her, using only the signature "Lo."

Shortly before she began writing, Louisa and John Quincy marked their twenty-fifth wedding anniversary; Louisa felt that she was unchanged from the Johnson girl of long ago—"the same romantic, enthusiastic, foolish animal so unfit for real life." Her marriage she deemed a blend of bad and good, conceding rather coldly to John Quincy that "all in all we have probably done as well as our neighbors." There were times, however, when she could be so pleased with her husband that once she said, "I actually threw my arms round his neck and shed tears of pleasure." Louisa liked to refer to her "warm and affectionate" nature, to her "deep and powerful" passions, and thus to marvel that she had nevertheless remained chaste: "I wish I could say as much as regards errors of temper." Especially in her White House days, Louisa found herself besieged by "strange, exaggerated ideas" which she could find no one willing to try to understand, and this in turn made her talk the more about woman's dismal fate.

With her husband too busy to notice her and her children absent or indifferent, Louisa said she knew why so many women went insane. It was because they were ignored in later life by husbands who cared little whether they lived or died. Women, she insisted repeatedly, were bundles of "passions and affections," and could not endure isolation. In this mood she sent a painful confession and admonition to Charles Francis. "As it regards women," she wrote, the Adams men were "one and all peculiarly harsh and severe in their characters." They had no sympathy or loving warmth, so that women, who by their nature faced "almost unmitigated anxiety and suffering," could expect only painful relations with men in the Adams family. Louisa begged her son to realize that "in the intercourse between man and woman it is astonishing how far even a little tenderness of manner and appearance of affection will [go] towards promoting her happiness."

The White House years nearly wrecked Louisa and John Quincy's marriage. By the summer of 1827 the couple were hardly exchanging pleasantries. Louisa remained in Washington, while her husband retreated to Quincy, and her letters to various Old House occupants, while otherwise solicitious—perhaps even pointedly so—did not even mention greetings to John Quincy. When the President had returned to Washington, he fled the White House to wander in Rock Creek cemetery, where he worried about his distraught wife. He wrote in his diary enigmatically, "The heart knows its own bitterness, and a stranger intermeddleth not with its joys." Meantime, Louisa was nearly overcome by self-pity; she claimed her life held only unutterable suffering and that she would welcome her coffin, "that little house where sorrow and treachery are no more." As her melancholy deepened and she talked of craving death, physicians examined her, informing John Quincy that his wife's disorders were "not mortal" but were a recurring nervous irritability resulting from disease "in the right ovarum." To her astonishment, a Boston doctor assured Louisa that she would live to be a very old woman.

By this time, late in his administration, John Quincy's ordeal seemed almost unbearable. At home he was consumed by worry for his wife and sons, to say nothing of Louisa's relatives, who often were like locusts besieging the White House. Then, too, there was his own brother's plight in Quincy. And now John Quincy's public career seemed about to close in failure. During 1828, the last year of his presidential term, the signs pointed unmistakably to a defeat in the election. His opposition had succeeded in assailing, even ridiculing, his lofty proposals for a strong national Union through sale of public lands, protective tariffs, internal improvements, uniform bankruptcy laws, sound banking for dynamic business and industry, a national university, a naval academy, and an observatory—to mention only some of the President's farseeing ideas. Unpopular though these were, Adams' foes made even more capital from arguing that he had won his office unfairly and thus was scornful of the ordinary citizen.

Faced by overpowering enmity in Congress, John Quincy could only watch as his dream for a sophisticated material and intellectual basis for the nation was shoved aside by a new faction devoted to Andrew Jackson and to a different view of progress. The scene brought to mind old John Adams' prediction that his son must be prepared for defeat by public misunderstanding and ingratitude. But President John Quincy Adams was never quite so willing as President John Adams had been to acknowledge that such a response was the most which stern virtue could anticipate from the world. The younger man clung to higher expectations because of a crucial difference between his nature and that of his father. John Adams genuinely rejoiced in both the prospect and the actuality of retired life in Quincy, whereas John Quincy had displayed as early as 1808 an impatience with existence away from politics. Now, twenty years later, facing defeat and crushed by private anxiety and debt, President Adams had a new worry: "my time will be all upon my hands."

Even in 1820 he had begun to dread retirement and the "lifeless languor of indifference" which came over him when he was left to his own devices. What he needed constantly, he acknowledged, was *"an object of pursuit."* With such stimulation slipping away by 1827, and his family a tangle of disappointment, the President became as depressed as Louisa. He lost his appetite, his clothes merely hung upon him, and he was weak and indifferent to everything about him: "My own career is closed," he told his diary, "my hopes such as are left me are centered upon my children." His thoughts strayed back nearly forty years before, when, as a melancholy young lawyer, he read a poem which had reminded him that the only real blessings were health, virtue, and an eye with which to see nature. And, too, he tried to take solace from Charles Francis, who had been repeating grandfather John's teaching that serving the public was "galling" and that "all generous minds" should abominate American politics. Not very convincingly, John Quincy conceded the point. He must, he said, leave the political sphere in "the care of itself." Meanwhile, he plodded through the last months in the White House, trying, he confessed, to disguise his despair with a forced cheerfulness. He felt the rest of his family was indifferent to his plight, one of the few murmurings of self-pity he ever allowed himself in his diary. His existence, he feared, would soon cause him to "bust every ligament of self-control."

The President hardly noticed that as he grew more melancholic, Louisa was being restored in health and spirit. Soon, there was a reconciliation, arising from Louisa's delighted realization that her husband was virtually certain to be defeated in his campaign for reelection. For Louisa this offered two consolations. First, it precipitated a family crisis, and Louisa rose splendidly to any opportunity to comfort and to cheer. Second, John Quincy's defeat meant that she could be released from her captivity in public life. She was eager to retire, she said, and to embrace at last "domestic happiness in the midst of my children worth a thousand times more than all ambition can confer."

When Andrew Jackson was victorious in 1828, Louisa responded admirably, exhorting courage to all the family and predicting better days. She was not even disturbed by the uncertainty of where they would live once they left the executive mansion in March. John and Mary were building a house with some money Mary had inherited from her father, but that was unfinished. Cheerfully, Louisa found a home in Washington where they all could live temporarily, and she proceeded in February 1829 to supervise removal to this haven, called Meridian Hill. It did not bother her in the least that she must now do without most of the nineteen servants she had commanded as first lady. Instead, she concentrated on keeping up the family spirits. "The great conquest we have to make is of ourselves," she wrote with keen insight to Charles Francis. "Our tastes, our tempers, our habits vary so much from those of the herd that we can never be beloved or admired, but we may and must be respected—unless we forget the respect we owe ourselves."

Now content at last to be an Adams, Louisa anticipated a new career as family matron. Her altered outlook was evidently contagious, for Louisa reported that all the rest were "bearing their part in the intercourse of domestic comfort." She relished returning to "the duties of good house-wifery," while she was especially pleased with the astonishing attention and good temper John Quincy began to show toward her. Louisa looked forward to returning to Quincy later in 1829, where she and her husband would for the first time take command of the Old House in their own right. There she hoped to sustain and cheer John Quincy in his dreaded retirement, meanwhile keeping him mindful that she had an answer to the paramount question asked by every generation of Adamses: "What is this greatness so boasted?" Louisa's reply: "A vision from which people awake without pleasure and with all the dull realities of life . . . glaring in the view."

To establish the haven at Quincy, Louisa first had to remove Thomas Adams and his brood, a task John Quincy could not bring himself to face. Louisa sent orders to Charles Francis: Uncle Thomas must be elsewhere by April 1829. No matter how they might complain, he and his family would have to leave. As Louisa said: "Nothing shall induce me to let them reside with me. I must not multiply evils on myself." Louisa was considerably upset with her husband for taking renewed interest in his dissolute brother. Earlier, when he was depressed, John Quincy had startled everyone by announcing that he found more pleasure in the company of Thomas "than in that of any other person." At one point he even talked of building a new house for himself in Quincy and of leaving Thomas in the old family mansion. Charles Francis, however, bestirred himself—and Uncle Thomas—so that the Old House stood vacated and unfurnished by mid-April 1829. The removal had not been pleasant, but in a flurry of activity Thomas had established his family in a house dating from 1641 which sat a little over a mile away, beyond the family church.

The prospect of being wrenched from his family home and associations strongly affected Thomas. He began to read family letters and talk of his ancestry and weep. He told his son and namesake, who had a chance in life since John Quincy had arranged for his admission to West Point, that he was deeply moved by thoughts of Adamses extending back 200 years, and that he, "alas! perhaps a degenerate son," wanted to be remembered as grateful to their memory. Thomas was much dismayed by John Quincy's defeat, but not so much as was the President's son John. Like Uncle Thomas, he too stood to be much inconvenienced by this turn of events. John grew silent, grimly looking out through clouds of tobacco smoke, seemingly unaware that the flour mills were going to ruin under his command. "I do not know whether he will see his plans any clearer through the smoke," his brother Charles Francis observed, "but it is somewhat time that he should."

In contrast, John Quincy's depression continued to dissipate; his appetite returned; he gained weight. Louisa was astounded: "Your father really and solemnly and without exaggeration seems to enjoy the idea of shaking off those trammels which have bowed him to the dust." Actually it was not so surprising. John Quincy had found a new diversion which kept him openly battling in the political arena. He found that he must defend his record as a senator in 1808 against revived attack by old foes in Boston. He went at it with zest, rejecting Charles Francis' advice that the issue was not worth the distress. John Quincy countered with the suggestion that his sons "mark the similarity of my fortunes with those of my father."

Then John Quincy revealed why he relished this new fight in a statement which explained the next twenty years of his public career: "I was born for a controversial world and cannot escape my destiny." This latest chapter, he said, "has taken from me almost all sensation of pain in my present distemper." Spurning the example of his father, for John Quincy feared calm waters would bring back his depression, he announced to George that "my life must be militant to its close . . . Repose is not happiness." He saw himself now as a noble eagle assailed by despicable little birds, and he gleefully used a line from Shakespeare, "sweep on, you fat and greasy citizens." Privately he continued to pray that he could be spared "indolence and despondency and indiscretion," while entering a retirement which he feared must be like "a nun taking the veil." Probably the most accurate expression of John Quincy Adams' state of mind came as he walked away from the White House for the last time: "I can yet scarcely realize my situation."

Private grief presently erased these public concerns. George Washington Adams had always been his parents' chief disappointment. His wayward behavior made John Quincy come perilously close to renouncing him. The father asserted in 1827 that their relationship would henceforth be that of attorney and client as George tried to manage family affairs, a task he botched outrageously. When John Quincy said he would not write to his son, George became depressed anew. He told his mother that he no longer dared to do anything lest he stir his father's censure. Louisa begged her husband to relent, warning him that George now claimed that, if his father knew the real extent of his helplessness, the son would either die or go insane. Charles Francis confided to his diary that "George has changed most immeasurably for the worse," and he joined his mother in efforts "to wean one we so sincerely love from his follies and save [him] from the mischief which so powerfully threatens."

John Quincy, of course, did not turn his face from his son, no matter how disappointed he was, and soon he was busily advising George anew on the importance of "unsullied temperance." Then came disclosure of large debts George had run up, hidden from his parents. By early 1828, the boy was unable even to pay his rent and faced creditors who were owed $3000. Crushed, John Quincy responded by gently telling George he would handle the bills by the subterfuge of buying the son's library but allowing him to retain the books. The parent was left with no choice but once more to beg the son to turn away from "licentiousness," from "lavishing" his money "upon—I cannot finish the sentence." Again giving George one more chance, John Quincy promised to pray that God might mercifully redeem the young man from ruin.

For the moment Louisa was less generous. "I shall write to him no more, for even mercy may be trifled with until it is lost," she said, but soon she took heart from the prospect of a visit George was ordered to pay to Washington in late April 1828. The stay encouraged both Louisa and John Quincy to believe that, if George could remain in their company at all times, he might yet be spared. When the improvement evaporated after George returned to Boston, the parents decided to move back to Massachusetts and take charge of him. They announced this would be their chief project when John Quincy's presidential term ended. George was directed to find a suitable house for him in the city, John Quincy fixing $10,000 as a maximum while Louisa looked on in disgust, knowing her husband already had more property than he had money to sustain. To early reports that George was behaving himself, Louisa replied that she had as much faith in the son as she had in a butterfly: "Love him I must. Respect him I may at some future day."

In this uneasy background, John Quincy and Louisa summoned George to Washington in April 1829 to help them move to Quincy, an order which shook the young man. George was "quivering under the fear of the merited reproaches which my father can though he will not give," Charles Francis observed, not knowing at the time that George had yet a new reason for dread. His parents' residence in Boston would surely disclose his liaison with his servant-mistress and the child he had fathered by her. As his parents' spirit improved, George cowered within himself. Even Charles Francis was now encouraging. He suggested that George's difficulties might all along have arisen from loneliness among inferiors and reminded his mother that the "ordinary run of society" was not enough to keep the interest of an Adams: "Our tone and habits of feeling at home are of a standard which makes the dullness of common conversation tedious and the life of other circles vapid." Charles said he still enjoyed George's company, despite his brother's sins, particularly when literary matters were discussed. Louisa tried to cheer George by predicting how he and his father would spend long hours together over poetry once George was safely in the family's midst.

As the time approached for him to escort his father and mother from Washington, George began avoiding Charles Francis, which caused the latter to muse: "I cannot help pitying the miserable weakness of his character while I regret it." After much wavering, George set out to fetch his parents, but left behind papers which Charles Francis later examined and pronounced as displaying signs of an incipient "Brain fever." George boarded the steamboat *Benjamin Franklin* at Providence on 29 April for passage down Long Island Sound to New York, where the practice was to change to land travel for the remaining distance to Washington. What followed, as the family assembled the facts later, suggested that George finally broke at the prospect of permanent residence with parents who combined great affection with even greater expectations. George was now twenty-eight years old, with little to show for it but dependency, a record of repeatedly disappointing his family, and the certainty that he was about to cause them more anguish. Having fled from his expectations so often, it seems that George made the final break by leaping into the Sound and drowning himself.

Suicide may not have been George's intent, for he evidently became wholly irrational as the *Benjamin Franklin* pushed toward New York. He spoke to other passengers of a severe headache. He mentioned that persons were trying to break into his room in Boston. He announced that birds were talking to him and that the vessel's engine kept prodding him with the words "let it be." After wandering frantically about the boat until 3:00 AM on the morning of 30 April, he asked the captain to put him ashore, claiming that the passengers were laughing at him, talking about him, and collaborating against him. The captain's attention was diverted for a moment, and then George's cloak and hat were next noticed at the edge of the upper deck. The distraught young man had vanished. His body washed ashore a month later.

Nearly unhinged by grief and guilt, John Quincy said that only God knew how George had died. He might have knowingly taken his life; he might have felt he could escape the boat and swim ashore; or he might have gone into the water in simple hysteria. At any rate George was gone, and John Quincy was inconsolable. When the terrible news reached Washington on 2 May, it was Louisa who displayed fortitude, showing again that in moments of the gravest crisis she had a strength which her husband lacked. The days that followed were anguished beyond description for John Quincy. "Oh! my unhappy Son! What a Paradise of earthly enjoyment I had figured to myself as awaiting thee and me," the father wrote in his diary. "It is withered forever."

Reviewing in "excruciating torture" how he might have dealt differently with George, John Quincy was confronted by his wife, whose calmness he watched in awe from his nearly crazed state. It was Louisa

who took command of the grieving household and asked John Quincy
to read to the family the consolation of the Episcopal Church's Order
for the Burial of the Dead. "I am the resurrection and the life, saith
the Lord; he that believeth in me, though he were dead, yet shall he
live. . . . For man walketh in a vain shadow, and disquieteth himself
in vain . . . teach us to number our days that we may apply our hearts
unto wisdom." After the final prayer, with its words, "we humbly be-
seech thee, O Father, to raise us from the death of sin unto the life of
righteousness," John Quincy was at last able to ask divine strength "to
bear thy holy will; and to bless thy name."

George's death brought John Quincy to Louisa in great need. "Reason is unseated," he said, as he spent entire days in his wife's consoling
presence, meanwhile pondering what he felt was the meaning of this
tragedy—to remind him "to controul the depravity of my nature." Saying, "I have nothing left to rely on but the mercy of God," John Quincy
managed to fix his thoughts anew on Massachusetts and retirement,
and vowed that in devotion to George's sorrowful memory he would
henceforth "commune only with God and my family." The effect of
the son's death had been shattering, and Louisa knew that her husband
was wrung, as she put it, "almost to madness." As a consequence, grief
had turned him into "a ministering angel always at my side," she told
Charles Francis. For the first time, Louisa could report that she had
received from him "the most soothing tenderness." The price had been
fearfully high, however, for George's death left both Louisa and John
Quincy feeling "unworthy" as husband, wife, and parents.

Eighty years later, their grandson Henry Adams looked back on this
tragedy with his special brand of mordant humor. Remarking that
"Uncle George's" literary efforts contained "the usual damfool com-
monplaces of the time" and that his "blackmail experience [the pressure he felt because of his illegitimate child] was what any other dam
fool might have had," Henry went on to say with characteristic face-
tiousness that George's "drowning himself showed a tragic quality far
above the Adams average." However, for Henry's father, Charles Francis, such levity was unthinkable, since he had to dispose of the difficulties created by his deceased brother. Indeed, by 1829 the problems
which John Quincy found too much to handle had become a harsh legacy for Charles Francis Adams.

✿ 8 ✿

Charles Francis

With the death of George and the weakness of John, the future of the Adams family in its third generation rested wholly with Charles Francis. The role was one which he early predicted for himself and which he accepted deliberately, even coolly. He was gravely affected by responsibility for the family, leaving him the one Adams who came close to attaining that command over self and the restraint in public service preached in the family gospel. Even as a youth, Charles Francis appraised his family with a candor which occasionally seemed disinterested. When he celebrated his twenty-first birthday, he had studied the Adams history, past and present, and observed that their affairs were "the most singular in the world" and that "hot water seems our element." He deplored what he called the "natural inclination of our family to keep the future in a mist." All of his relatives seemed to find it grand, he said, to have no definite plans.

Charles recognized particularly how his father believed that events would always overtake the Adamses, sweep them away from the humdrum of ordinary existence. This passivity so appalled young Charles Francis that he devoted his life to bringing order and autonomy to the family. Such goals were for him eminently reasonable means of assuring that the name of Adams "should not be said to deteriorate in the generation to which I belong." This became an overwhelming concern as he found himself increasingly alone with the responsibility. He began to refer to himself as the "last scion of the race" and "the only one who remains."

Of John Quincy's children, only Charles Francis had stood up, some-

161

times sternly and even cruelly, for his own identity against his father. George and John's failure and folly probably presented more of a lesson to Charles Francis than he needed. He had matured early, largely in the presence of old John Adams and in the shadow of Uncle Thomas' shame. The massive self-doubt which characterized the family and the equally typical scorn for the world drew him inward. Even as a young man, Charles Francis was restrained, contained, seeking release only in his diary and in an occasional burst of temper. For him, life meant anxiety and tension from the start.

John Quincy and Louisa had surprisingly indulged Charles Francis and his brothers as teenagers in such pleasures as cigars, wine, champagne, and oysters. Charles responded to this early frivolity. He had raged against the seriousness of college life, evidently pursued prostitutes in Cambridge, and kept a mistress in Washington long before he was of age. Then, in 1823 Charles experienced an illness, possibly emotional, which prostrated him. He recovered under his grandfather's care at Quincy. From this experience the authentic Charles Francis emerged, for he increasingly cultivated a kind of stoic resignation that left him determined to move through life cautiously and thoughtfully. "I came back a new man," Charles recalled. He also closely studied people, family members and others, and made of his diary a brilliant exercise in character analysis.

As a young man, Charles Francis had a remarkable understanding of his parents' strengths and weaknesses and of the family's burdens. What he could not fully perceive, however, was how earnestly he had to struggle to break free of that overpowering embrace with which John Quincy was smothering his two brothers. Where George and John took refuge in evasion and subterfuge in letters to their father, Charles faced differences squarely. He even fought John Quincy over the contention that early rising was a moral necessity. With charming frankness, he pointed out how the elder Adams managed to be up before dawn only at the price of afternoon naps and long snoozes in his chair after dinner. Much more bitter was their exchange in 1828, one Charles Francis vowed he could never forget. It involved the young man's request for a comfortable allowance after his marriage, made when John Quincy was severely depressed at the end of his presidency. He saw Charles Francis' expectation of money as one more evidence of his children's weakness and refused, making matters worse by observing that Charles was still a beggar, living on charity. The son was badly bruised and tried to withdraw from his father: "I neither can nor will consent to be treated like a child." Memory of the episode "burns like a rankling sore," he said months later, pained because he knew his father spoke the truth. Charles Francis was far from independent, even though Daniel Webster vouched for his legal qualifications.

Charles Francis wanted to marry Abigail Brown Brooks, whom everyone called Abby. She was the youngest daughter among the numerous children of Peter Chardon Brooks, who probably deserved his title as the richest person in Boston; his wealth was derived largely from insuring ships and from banking. Abby would bring more than money to a marriage, however. Her family line included Boylstons, Cottons, Gardners, and Saltonstalls on her father's side, while her mother was Anna Gorham, daughter of Nathaniel Gorham, first president of the Continental Congress, and had Calls, Coffins, Gardners, and Stimsons in her ancestry. Abby Brooks was born 25 April 1808, nearly a year after Charles Francis. She was a person of lively spirits and some beauty. Perhaps because of her appealing quality of insecurity, Charles found Abby irresistible, and she charmed both Louisa and John Quincy, who approved the match, although Louisa found her "quite an oddity." Charles Francis appears to have rejoiced in Abby for herself and not because of her wealth. "This young lady's character is so singular that I am much struck with it," the nineteen-year-old confessed in his journal.

Charles Francis had met Abby in Washington. Their families were acquainted—a legal commission from Peter C. Brooks had sent John Quincy in 1809 to Washington, where he stayed seeking a diplomatic mission. Abby came down from Boston during the 1826 social season to visit her sister, Charlotte, who was the wife of Edward Everett, then a congressman from Massachusetts. Abby's doting father sent notes to her from home, written in a style more homespun than the Adams correspondence, as when he told her, "It has froze like a dog for two days past." Early in 1827, Charles Francis and Abby fell in love. Within a month he had proposed to her. Abby was drawn to him but hesitated out of deference to her father's wishes. By mid-February, the two families had adjusted to the romance, but Peter Brooks still had to decide if he should surrender Abby, admittedly his favorite child, to a young man he had never met.

For a month there was uncertainty and confusion. Not realizing how deeply in love the couple already were, the Brookses sent words of caution, counseling delay. Both John Quincy and Louisa rushed to help their son, the President of the United States telling Abby's father that, though young, Charles' character was "sedate and considerate." He was a studious, honorable, reserved young man, with regular habits, all high marks, of course in the Adams grade book. John Quincy said he was comfortable with early marriage for a disposition so self-controlled and above "ardent passions."

It was probably Louisa's letter, however, pleading that the Brookses reconsider, a document which does not survive, that must have turned the tide, for on 15 March 1827, Peter Brooks replied to the First Lady, cheerfully consenting to an engagement. He simply had not realized "how fast things had proceeded with the young folks." Abby's father then added in a manner which must have touched Louisa, "Having been young myself, I ought to have known better, and I will try to make it up to our young couple another time." Peter Brooks was a generous person and a man of his word. He lived for twenty more years, during which he found many ways to make life pleasant for Charles Francis and Abby. Louisa paid him the highest tribute in her power, saying that he was very much like her own father.

Even with Brooks family hesitations about a quick marriage and John Quincy's financial uncertainty, Charles Francis never regretted his choice. He continued to delight in the frankness and simplicity of Abby's manner, which he found "more engaging than the studied elegance of an accomplished belle." Certainly there was nothing threatening or mystifying in Abby, for it soon was apparent that she depended heavily on her husband for reassurance, as she had on her father. She required such support in ever larger amounts all her life. Abby Brooks in her own way found membership in the Adams family as difficult as Louisa had.

For over two years Charles Francis and Abby were engaged, during which time they were often apart, which meant frequent exchanges of letters. He urged Abby to be spontaneous and unstudied in her writing, as he wished he could be. He promised to try to be affectionate, saying that his restraint came from "the example of a taciturn family." Both succeeded, for their written exchanges were from the beginning ardent and comfortable, even on the occasions when Charles Francis had to reassure Abby of his love. The Charles Francis who appears in these letters is found nowhere else, while Abby's words were surely enough to melt even an Adams heart. "I sometimes have blushed to think how much I have said to you," she wrote barely a month after their engagement began, "but why should I be ashamed to confess that I love you better than all the world beside." Even at a distance "you have complete power over me."

Charles was envious of this simple outpouring and told her of his "utter inability of language to express my feelings," a flaw he attributed to the inhibiting Adams atmosphere. In his family, he said, praise and endearment were put at so high a price that it was left for action to display them, since words would only increase "the vanity and pride so natural to us all." On occasion, however, the young man could make his case nicely, once telling Abby: "I do not think you perfect, but considering every thing, I would not exchange you for the most perfect thing that ever lived." He called her "my dearest, sweetest, little Abi," and announced that he was grateful to "my lucky stars which led me to you." However, Charles Francis seemed less fortunate in his financial affairs, for upon reaching age twenty-one in 1828 he was still an impecunious young lawyer whose income was an allowance from his father of $1000 a year, paid in exchange for looking after the family property. These facts depressed him, for he knew his marriage would be long delayed if it must await his independence. He proposed to waive this consideration and marry soon, or else end the courtship.

Abby's father once more relented and promised to provide an income and a house for Charles Francis and his bride which would make the pair comfortable indeed. Marriage suddenly was possible. The Brooks generosity left Charles Francis feeling obliged to assure his diary that "I love Abby dearly for herself alone, and I do not wish to marry her family." He and Abby were wed in the Brooks mansion in Medford on 3 September 1829. It was an evening occasion which was missed by Louisa, who felt it was too soon after George's death to leave Washington. John Quincy was present and behaved handsomely, being very gallant to Abby during the lively party afterwards. The bride's name especially moved the ex-President "because it was that of my Mother and my Sister." He gave her a cameo ring showing two hands joined. To Charles Francis John Quincy presented Stuart's portrait of John Adams painted shortly before the patriarch's death.

The marriage was evidently happy from the start. Charles Francis, who fancied himself experienced in the physical aspect of love, confessed he was delighted in this regard with his wife. He said he "had formed moderate expectations only and they have been entirely exceeded." Abby had given him "moments of felicity which could not be improved." Unfortunately no record exists of Abby's thoughts about their honeymoon. The day after the happy event, Charles resumed his responsibilities in Boston, the couple having moved into a home bought for them by Mr. Brooks on Hancock Street.

Charles Francis was scarcely the typical newlywed. By 1829 he was responsible for clearing up several family difficulties. The first of these he faced at the time he and Abby were trying to plan their marriage. It fell to Charles Francis to tidy up after George's sudden death, a task which soon invoked his remarkable capacity to take command. Discovering that his worst suspicions were justified and that his brother's affairs were in a scandalous state, Charles Francis set out to remedy them as much as he could while keeping from his parents most of the sordid story. He felt the record of his brother's life should be obliterated, so Charles consigned most of George's papers to the flames, and in the process turned the errant brother into a pathetic but safely vague memory in family legend. The spirit which motivated him to read and then destroy almost all of George's manuscripts also made him assert that, while George's "fate was melancholy," it was nonetheless timely. "He would have lived probably to give much misery to his friends and more to himself," he wrote, and George died while the family might still cherish him and lament his end.

Charles tried to make respectable what little he felt could be allowed to survive of George's story. He had several motives for this, including the guilty notion that he himself might have interfered to keep George from the dreaded trip to Washington. Consequently, the younger brother was obviously relieved to find the papers disclosed no "alienation of mind" in George before he embarked. Charles Francis also knew that his father had "almost lived" for George's future, so that no record of sin and folly must be left to increase John Quincy's terrible grief. Finally, he was himself struck by the irony of his brother's life, where great promise and every encouragement had produced no more than a choice between "death or great trouble of mind." Charles Francis drew a lesson an Adams scarcely needed: "it makes me feel more deeply the singular weakness of humanity."

Reading these papers led Charles Francis to wonder about his own ability to deal with the world without betraying family expectations: "I have only steadiness of character without the boldness of enterprise essential to success," he observed in September 1829 after studying George's journals. Certainly he had none of his late brother's imaginative flair and promised brilliance. But what were they without steadiness and solemn determination? Charles Francis found himself unwittingly repeating much of the old Adams litany as he tossed George's pitiful aspirations into the fire: "Unity of purpose in life seems to be the great secret of success. My brother's records show much and bright talent, but constantly diverted by the seductions of pleasure, and even the trifles of life." For Charles Francis this meant "all that remains to me is to benefit as much by his good purposes as I can."

There was more to George's story, as Charles discovered when he turned to close the deceased's sordid personal affairs. George's wild spending to escape his tormented self-doubt left many obligations, particularly debts to John Quincy. However, the most unhappy task came on 13 May 1829 when Charles Francis opened another of George's trunks and found there a letter addressed to himself. George had written it in the event he died, asking Charles to look after Eliza Dolph, the mother of his child. She had been employed at the home of Dr. Thomas Welsh, who was married to a niece of Abigail's and at whose residence the Adams men had frequently boarded when in Boston.

Keeping his parents in the dark, Charles Francis followed his brother's request, determined that the family would yield no more than required. Charles paid Eliza's medical expenses, and the young lady soon found another job; she and her child disappeared from the Adams story. But there was talk of blackmail from persons whom George had engaged to help Eliza during and after the pregnancy, and who now promised to create a scandal if they were not rewarded for their trouble. They claimed that George had made financial pledges to them before his death. The matter was so serious that Charles Francis reluctantly told his father the story, which left the broken parent filled, Charles said, "with infinite regret and mortification." While his father watched from afar, Charles Francis then boldly repudiated the claims, telling the man who threatened to inform the public: "You are welcome to all the benefits a disclosure will give you."

Matters grew much worse before they were over. A pamphlet was published by the chief claimant, Miles Farmer, who reported that George Adams had forced upon him "the whole train of evils, which flesh is heir to." This furor, displayed before all Boston, led to a lawsuit arising when two of George's associates assailed each other after getting nowhere with the Adamses.

There was little of this in family correspondence. Louisa professed to know nothing of George's misdeeds. She grumbled at what unpleasant allusions she did hear, and even chided Charles Francis for coldness toward George's memory. Somewhat irritated, he told his mother she did not know the whole story and reminded her that "no member of the family has been like me dragged into Court" and obliged to listen to references which he was "unable to defend." He suggested no more be said about George's past. "I have been the only sufferer of late," he said, hoping that his brother's frailties could be left in the tomb.

George was finally buried in the family mausoleum in Quincy. His body had washed ashore on 10 June 1829, discovered by a lad on City Island about ten miles from New York City. John Quincy learned this while he was en route to Quincy, and stopped to hold a brief funeral. Hoping perhaps that there might be an error, John Quincy tried to pry the coffin lid open; but he did not succeed and had to be content with reports that George's body had been unmarked by violence. When he saw his son's watch stopped at 3:40, his thought was, "Oh! where at that moment was the Soul of my Child?" Meantime, the discovery of George's body prostrated Louisa, who now implored John Quincy to bear with her as she finally expressed "the overwhelming grief" which she said had been mounting for four years as she had watched George, "with the most deadly apprehensions alas too fatally fulfilled."

John Quincy and Charles Francis eventually stood on a dark, rainy November day in 1829 to watch George's coffin being placed next to Nabby's in the family tomb. After a short prayer by the pastor of the Adams church, Charles Francis went home to write of George in his diary, "May Heaven deal mercifully with his Soul. His earthly remains lie with those of his Fathers." John Quincy recalled in his journal how much he had hoped and feared for George, how he had cherished those scant signs that George might succeed, and how the son had proved to be a kindly, even abstracted figure, one who eventually "could not endure the contemplation of its own infirmities." Then, sensing perhaps that his incessant demands upon George may have led to the boy's downfall, John Quincy uttered an abject prayer. "Let not my errors be visited upon my child!"

Father and son had undertaken this sad duty at the end of a summer they had spent mostly together in the Old House until Charles Francis' wedding. There had been something of a reconciliation after the exchange of harsh words a year earlier when the son had requested more financial aid. John Quincy had told Charles Francis that "I wish not to remember the reproaches which I received or gave last Summer and wish you to forget them too." Since Louisa wanted to wait a year before seeing Quincy again, the father and son shared much of the summer alone, sorting John Adams' letters and talking of presenting him to the world in a manner which would display his greatness. They considered at length John Quincy's wish to build a new residence of stone on the brow of the nearby hill overlooking the church and the cemetery. And, of course, there was the accumulation of boxes containing books brought long before from all over Europe, now to be opened, identified, and shelved.

Cared for by cousin Louisa Smith who, with John Adams' death, had now begun looking after a new generation of mansion occupants, the ex-President tried valiantly to enter into village life, visiting many neighbors for evening talks, and especially enjoying the Marston household where a game of chess usually awaited him. The parties for music and dancing were always predictable, he said—twelve men and twenty-five women making up "the usual Quincy society." When Charles Francis was not prodding him to look at books or manuscripts, his father stooped over his infant trees, trying, with mixed success, to coax them to sprout from seed. He also struggled to persuade Norman hens to propagate, losing this battle, too, although he came to feel a great affection for these chickens, each of which he named—"they never see me in the yard without running up to me to ask for their breakfast," he wrote Louisa. He was amused at how these "trivial" concerns occupied him, where before they had been "always annoying to me, because I have passed my life in ignorance of them." He even reported to her his struggle to refine a cask containing over 100 imperial gallons of Madeira wine, using the yolk and shells from a dozen eggs. "It seems like beginning the world anew," said he, keeping busy lest "the pangs of the past, and the terrors of what may be to come wring my soul almost to desperation."

Charles Francis watched his father narrowly: "he has a manner which I never before saw in him of quiet sadness," he observed, adding that his father's appearance was "really affecting." But he also noted that family pride had become "an absorbing passion" in the older man. Both Louisa and Charles Francis tried to direct this zeal into the long-promised biography of John Adams. As father and son worked to assemble John's papers, John Quincy prayed for divine aid so that he might show his father to future generations "as he was," and so contribute to the welfare of mankind. Shuffling and reading manuscripts, however, soon proved boring and even irksome to John Quincy. Despairing "of my own perseverance," he attempted to begin his father's biography, but this too went poorly. "I find myself sliding into digressions," he complained, and welcomed the diversion of planning a memorial stone to John and Abigail. After John Quincy completed the tablet it was placed

on the wall to the right of the pulpit in the family church and immediately before the Adams pew. From here the family could ponder the
words every Sabbath and visitors to the church may read them today:

> During an Union of more than Half a Century
> They survived, in Harmony of Sentiment, Principle and Affection
> The tempests of Civil Commotion;
> Meeting undaunted, and surmounting
> The Furors and Trials of that Revolution
> Which secured the Freedom of their Country;
> Improved the Condition of their Times;
> And brightened the prospects of Futurity
> To the Race of Man upon Earth.

John Quincy could not fail to include a glimpse of family doctrine: at
the bottom of the tablet, an exhortation to every "PILGRIM"

> From Lives thus spent thy earthly Duties learn;
> From Fancy's Dreams to active Virtue turn;
> Let Freedom, Friendship, Faith, thy Soul engage,
> And serve like them thy Country and thy Age.

When the tablet was in place, it was late autumn and John Quincy
could justify returning to Louisa and the distractions of Washington
life. But in the spring of 1830 came the duty once more of returning
to the Old House. This time John Quincy brought with him Louisa and
John's wife and daughter. He himself had undergone a change in disposition, which Charles Francis noticed at once. His father was dejected, bored, and foresaw no solace in the homely pursuits which had
amused him for a time the previous season. It was his old difficulty, an
inability to be disciplined in personal enterprise. Furthermore, the scenes
of home and John Adams' mementos now reminded John Quincy that
he, too, had retired in defeat and worried over a son's shame. Charles
Francis reported that the Old House was a miserable place that summer: Louisa in nervous prostration, Mary in a complaining slouch, and
John Quincy sitting, grimly dejected.

Then came an unexpected and electrifying development which
transformed the family's retirement. John Quincy was elected to a seat
in Congress. This event appalled his wife and son, while it delighted
him. Louisa fumed at returning to a life she detested and vowed to
remain in Quincy forever. Almost ecstatic at discovering that the American public had not abandoned him after all, John Quincy complained

that his wife had "no sympathy with my sensations." Yet he hastened to Washington, forgetting the pious determination of a year before to devote the remainder of his life to the family papers and books. Charles Francis quietly succeeded in persuading his mother to follow John Quincy. "My father," he observed, "is a singular man. He wants the profound wisdom which gives knowledge its highest lustre, he is not proof against the temporary seductions of popular distinction to resist which is the most solid evidence of greatness." Thereafter, it was Louisa who found the sweetest pleasure in summer flights from Washington to the Old House. Her health, always wretched in the capital, usually bloomed in Quincy.

As a congressman, John Quincy entered at once into the political furor which soon threatened the Union. With state rights enthusiasm and talk of nullification kindled by debates over public land policy, tariffs, internal improvements, national banking, and, finally, the right of citizens to petition Congress against slavery, he had another chance to encourage a strong and unified nation. As a fearless debater he was more successful in his cause than he had been as an executive in the White House. His fervor in this new endeavor almost erased the public and private sorrows he faced, leaving Charles Francis no choice but to assume the actual leadership of the family.

"My father has always loved to ride the whirlwind," the son shrewdly observed, "but he is not happy. His enjoyments are but one mode of intoxication." For himself, Charles Francis saw private life as his only reasonable course, for any significant public role might put him in conflict with his father. He began what he called "building up the reputations of a respectable man," by which he hoped to give the family character that feature "which perhaps it most wants, dignity." His task soon became more serious and lonely. His brother John died in 1834 after a steady decline. His last years were marked with mysterious but always debilitating physical disorders, along with depression and dependence on alcohol. Few papers remain from which John can speak for himself, but there is enough evidence to show a young man—a middle son—whose demeanor veered from boastfulness to sullen withdrawal. Clearly, he wanted to break from the family, yet he was never quite able to take his own stand. His brother Charles Francis perceived him as courageous only in talk.

John was much troubled when his father lost the presidency, and indeed became something of a recluse in the very community where he wished to stay. His attempts to manage the Columbian Mills, where the family hoped to make up its lost fortunes, kept him dependent on the family and made him subject to John Quincy's relentless scrutiny. When his parents were in Quincy from April to November, John sent no word about the mills, though John Quincy bombarded him with advice and solemn remarks. "You know how much I have suffered from the mills since I purchased them," the father wrote, although he tried to be realistic, calling them "my imaginary gold mine at Rock Creek." He did secure advice from Boston friends on management and market matters and shared these with the silent John.

John had much to be quiet about, as it happened. Despite good counsel, he had taken fright and sold his flour at 75¢ a barrel instead of waiting for the predicted rise, which would have brought him $2.50. This was bad enough, but John Quincy was more disgusted with John's whining. "Your complaints are distressing to me," he wrote, not so much for the losses but "because they mark an impatience under adversity." Knowing how easily depressed John became, Louisa sent cheering notes and even tried to find him a political job. Her words were probably offset when Charles Francis bluntly charged his brother with mismanagement.

Finally, in 1832, word began circulating that flour purchased from the Adams mill was spoiled. John Quincy threw up his hands in despair, and not only over the business. John's emotional and physical condition was now so serious that Louisa's brother-in-law, Nathaniel Frye, had to be given management of the mills. With his help, along with the reorganization of family resources by Charles Francis, the most pressing debts were paid. However, John Quincy was stuck with the Columbian Mills for the rest of his life. They were a fearful reminder of the disintegration of his second son, whose condition worsened through most of 1833 and 1834. Both parents watched in horror. John had never been easy to deal with. Now he kept to his room or crept around his home on Sixteenth Street near the White House. The construction of this residence had been paid for by Mary Hellen's inheritance and help from John Quincy. John was usually found in a state of undress, coming alive only when he got away from home on errands he called "marketing."

John Quincy spoke of his "strong presentiment" that his son was going "to a destiny towards which I cannot look with composure." It meant that decay was extending "with irreversible proclivity to my family." John must drop everything in Washington, the father urged, and take refuge in Quincy, where renewed life and hope beckoned. There, by taking over responsibility for the family's real estate snarl, John was assured he would find "a comfortable and independent existence." Accept the magic in "the seat of your forefathers," John Quincy insisted, as he begged the son to reside in the Old House. Only here, said the father, had he himself escaped "ruin," and here in Quincy was "the last resort for my children . . . when they meet with nothing but disappointment elsewhere."

After frightening dreams came to him about John, John Quincy talked at length with Charles Francis during the summer of 1834, conversations which centered upon the family's sad history. Charles Francis was deeply moved, and acknowledged that "perhaps there is no better subject for constant reflection to me, than the history of our family from the middle of the last century." Louisa meanwhile blamed her husband for some of the trouble. She told John Quincy that his "perfect nonchalance" in shifting responsibility for family well-being to his sons and demanding impossible results had destroyed the boys.

News came to the Old House on 18 October 1834, that John was near death in Washington. John Quincy tried at once to reach the capital in time to see his son, leaving Louisa behind, gripped by "almost hopeless anxiety." She added to her anguish by recalling once again how she had abandoned her two oldest sons in their youth to follow her husband. John died early in the morning of 23 October, a few hours after his father reached Washington to sit beside his unconscious son. Once more, the pathetic plight and death of a son brought John Quincy almost to distraction. He hovered over the corpse, kissing John's lifeless brow. When the news reached Quincy, rain was falling in torrents, and Louisa had to be given opiates in a scene Charles Francis said "will remain forever engraved on my Memory." Meanwhile, the physicians had no explanation for John's death. His father was left to pray: "My child, my child—in the bosom of your God, may never ending joys, joys unspeakable and full of glory, cancel all the sufferings of which your portion was so great here below."

John's remains were kept temporarily in Washington until they could be removed to Quincy in March 1835, at which time Charles Francis supervised his brother's interment in the family tomb. He had been less inclined than his parents to ignore John's failings, just as he had stared unflinchingly at George's weakness. In both brothers he saw a terrible lesson he knew he could not ignore, and it made him remember the time when "the scourge of intemperance" had afflicted his grandmother Abigail's brother. There was no arguing with the fact, he told himself, that "vices are hereditary in families." And when it came to strong drink, he said, "our family has been so severely scourged by this vice that every member of it is constantly on his trial." For himself, Charles Francis believed that only "divine goodness" had called him back from his own earlier bout with this weakness "which has weighed heavily upon our house." Although wishing peace to John's memory, he remarked, "I cannot regard the loss of either of my brothers as a calamity." Their deaths spared the family and themselves much misery, he said. Now the Adamses could be saved from "the harrowing anxieties of witnessing a remediless evil." John had simply determined to die rather than to face "moral ruin," Charles concluded.

Charles Francis concentrated even more now on family affairs, for the latest blow had taken most of what remained in John Quincy's capacity to deal with such matters. "My father is daily becoming more helpless in his private concerns," he observed. "*My* duties are fearfully heavy." Grim though the moment was, it marked a turning point in the family's affairs. Their rehabilitation was possible only because there was one son in the third generation who could withstand the destructive love lavished by John Quincy. Order and progress began to appear. Louisa begged her son: "Go on and do not suffer yourself to be intimidated or brow beaten as your poor brother was but pursue your course steadily and respectably." Every bit of this advice was needed after John's death, for Charles was obliged to pass through a most painful encounter between two generations in the family.

John Quincy's finances were in ruins when Charles Francis brought his grieving mother to Washington in November 1834. The father was shackled with two burdens, much land and much debt, a great deal of the latter arising from his late son's folly and extravagances. However, his most difficult and important obligations were still the quarterly payments which he had to make to the legatees under John Adams' will. So desperate was the situation that John Quincy had to send interest-bearing notes to the heirs in lieu of money. This situation was not new, but it had worsened. Louisa had reported in 1832 that John Adams' will had turned her husband into "a perpetual sponge" which the heirs "have no hesitation in squeezing." She described John Quincy as not knowing where to turn for a dollar as the tax collectors pounded at the door.

Long before the financial crisis of 1834, Charles Francis had studied both his father's plight and his character. Cautious and orderly in the extreme, the son was never able to unravel the mystery of his father's chaotic and extravagant ways. He certainly suffered as he watched John Quincy bring upon himself "laborious and harassing distresses." He did not, however, sympathize when his father only talked about his problems: "It is a little singular how thoroughly speculative he is in his views, and how vain it is to hope for practical results." The symptoms he saw included the Columbian Mills, the expensive style in which John Quincy and Louisa lived, the relentless demands entailed in John Adams' will, and the habit of acquiring and clinging to real estate, a practice that made Charles Francis say with feeling to his mother, "I have a great contempt for landed riches. They are not one step better than property on the Moon."

Since his youth, John Quincy had had difficulty managing in the material world. He not only possessed little knowledge of business affairs, but he actually prided himself in disdaining these grubby matters. Much of his life had been spent in flight from the routine, the humdrum, and the conventional. Ironically, while he himself had managed to run away from a legal career and other challenges, he became intolerant and angry with his sons when they tried to take the same course. Now in 1834, when his father's style had caught up with him, Charles Francis moved to his aid.

Even before he reached Washington after John's death, Charles Francis began issuing what must have seemed like orders to his father, but the son considered them essential if his father was to avoid financial disgrace. John Quincy had an immediate debt of about $15,000, but there were also the payments he had to make as John Adams' executor, as well as the fearful costs of family maintenance and taxes. Reduce your standard of living, Charles demanded first. Servants, carriages, magazines, and wines must go. The former President tried to be agreeable. "I shall walk to and from the Capitol as long as my legs will carry me," he promised, but the strain of worry and grief was forcing him into sleeplessness, depression, and impatience with financial demands. When Charles continued to challenge his father to be practical and self-disciplined they had an ugly quarrel. Beginning on 16 November 1834 father and son fought for three days. Charles Francis insisted that John Quincy must curb his extravagance and sell a quantity of property. This might yield the amount needed to pay the Adams heirs and to meet the large debts in the Washington area. It was not so much this draconian remedy that upset John Quincy as the fact that his son seemed to treat him as a near-incompetent. Wholly unaccustomed to such candor from another person, particularly a son, John Quincy counterattacked by reminding him that a child's place was to offer a parent unquestioning service. Nevertheless, Charles Francis remained immovable.

Louisa tried to make peace between the two. She pointed out to Charles Francis his father's depression over John's death. Forgive his harshness, she implored, and suggested that the son might himself seek to be more warm and understanding. Tenderness was needed now, Louisa said, watching John Quincy borne down "to the weakness of Childhood." One wonders if Louisa was tempted to remind Charles Francis of how much like his father he was when he was so cold and reasonable in his behavior. Now that her husband was a broken, stubborn man, Louisa could once again display the great courage which otherwise faded. It was determination which John Quincy needed to find, and Louisa begged Charles Francis to aid him.

Ultimately, Charles Francis returned to Boston with some concessions, but bruised by what they had cost in angry exchanges. John Quincy had finally agreed to borrow no more, while Charles Francis would seek to convert property into cash. The immediate catastrophe was averted, but thereafter it was only Charles Francis' quiet work which kept his parents solvent while John Quincy devoted himself to congressional projects. The son's verdict upon his father was succinct: "He is a singular instance of a man to whom prudence never was natural," while John Quincy urged his son to read the sobering words contained in the Ninetieth Psalm, where man is compared to the grass which grows for a moment, only to wither and be cut down. John Quincy disdainfully told his son to fret about the things of this world if he wished. Let him devote a life "to making money," hoarding up treasures, and then let him see if his was a happier existence "than that which has been the lot of your father and grandfather."

Following Louisa's urgent appeal, Charles Francis' reply was masterful and calm. Certainly a match now for his father in verbal combat, the son chose family gospel to refute him. He would not enter the perils of a wicked political world where the price of fame was happiness. Did his father not recall once yearning for a life of contemplation and study? This was to be his goal, he quietly observed, and achieving it required the greatest self-control and economy. Both were traits his father had never shown. He added, in a conciliatory tone, that he meant no rebuke for his father's career; he only wanted to serve his parent as "a calm as well as a disinterested advisor."

This was 1835, and it marked the start of new harmony and affection between father and son. Charles Francis had endured, shown fortitude, and secured his independence within the family. Thereafter, from an office which stood on Boston land acquired by John Adams in 1772, Charles Francis, and his sons after him, looked to the rents, investments, and improvements that needed daily attention. A stern disciplinarian, Charles Francis kept himself at the details of family business—he even did the household marketing early in the morning. But he was an Adams, not one of Abby's relatives to whom business was second nature. "When I look at the multiplicity of details resulting from the care of the papers, of the library, of the garden, of the farm, of my plantations, of my various estates, of my property, of my household and of my children," he said in 1858, "it seems to me almost bewildering. I try to move the whole little by little by acting upon each at a time." Affairs necessarily wrenched him frequently from his beloved library, making him resent his "Estates," even though they represented the new family well-being. "It is all care to me," he grumbled.

He succeeded nonetheless. Three years after he took over management of John Quincy's affairs, Charles Francis had improved the value of parental holdings by twenty percent. He put an end to the unfair taxes and other assessments the village of Quincy had used in taking advantage of his father and grandfather's attachment to the family properties. John Quincy Adams had been nearly "sucked dry" by the town, Charles contended. "These people still imagine that they can impose upon me as they did upon him," he said after John Quincy's death, "but they will find their mistake." Convinced that the value of property inherited from his father would quickly double in value under "a practical man," Charles set to work to do so, and in the process helped to alter the character of Quincy, a change he later lamented. Much of the town's shift to business and industry, especially the quarrying of granite, came from Charles Francis Adams' enterprise, which included in the late 1850s his labors as president of the Mount Wollaston Bank. The rural scenes which his forebears loved soon disappeared. Charles had made the Adamses more than solvent, but the cost was high. A railroad was built across family property near the Old House, for one thing; though Charles Francis quickly became accustomed to riding on it.

These financial activities, aided by the money that his father-in-law Peter Chardon Brooks extended to him and his wife, had increased Charles Francis' assets—much of them in land—to well over half a million dollars by the time of the Civil War. About all this Charles Francis was very silent, principally because he was never confident in his feelings of security. For too many decades, his forebears had struggled for money or squandered it. He remembered the follies in his family too vividly, so that Henry Adams spoke for all Charles Francis' children when he said in 1860, "Papa's a rum one in money affairs." Since Charles Francis knew what his stewardship for the family had achieved, he resented bitterly those among his fellow citizens who claimed he simply sat in the lap of money. "No living being but myself knows either how much or what I have done to save every body whose concerns I have had to take care of," he told his mother. "I therefore like to have the credit of it, instead of being supposed to have done nothing but float in wealth all the time."

His assumption of financial responsibility for the family also meant that he had to provide for dependent relatives, who seemed only to grow in numbers as time passed. They sorely taxed his limited store of patience, but he was dutiful to a fault. Thomas Boylston Adams and his family presented the most painful problem. There was no change in Thomas' behavior after he left the Old House in 1829, but John Quincy looked on in forgiving kindness. Louisa and Charles Francis, however, were less inclined to disregard what they called his whining, begging, and complaining. Watching his uncle deteriorate rapidly while town ruffians continued to tempt "the Judge," as he still was called, Charles Francis reflected: "This family is a thorn in our side. It is craving and ungrateful." Evidently Thomas' boys took on dandified ways and were lazy. This exasperated Charles Francis, who noticed how they forgot his father's "constant series of benefits." When the "Judge" went on a wild ride in a rented chaise, colliding with a wagon and destroying everything but himself (though he was hurled out and run over), Charles Francis refused to pay the bills. "I wonder what people will come for next," he mused.

On 12 March 1832, Thomas died, and Charles Francis was left to support his survivors with their share of the income from John Adams' estate. While John Quincy had little to say about "my dear and amiable brother," Charles Francis contemplated the wrecked life as he buried his uncle in the Adams tomb. "A man who paid a bitter penance for his follies," he thought, one who "left his children to share the same as his only legacy." Now began the task of trying to reason with Thomas' widow, Nancy, for whom, as Louisa observed, "enough is always a little more than she has." Charles Francis knew this well as he sought to placate the woman, a task "about as disagreeable as any I ever had to do with," he said. John Quincy gave the widow Adams the income from the two farms he had acquired by default when Thomas could not pay his huge debts to his brother. But this did not seem to please "our fretful widow." A disgusted Louisa said that her sister-in-law served to illustrate how "extreme selfishness destroys the kindly affections and turns us into brutes." John Quincy made no comment, not even when he attended the sale of his brother's library and discovered that some of his own books were being offered.

Besides enduring Thomas' widow, Charles Francis had to keep an eye on the orphaned sons. There was John Quincy Adams II, who, to the family's disgust, displayed "an invincible repugnance and incapacity for learning." A military career was arranged for him. Another son, Joseph, took up his father's habit of accumulating large debts, which he then tried to pay by writing checks on John Quincy. Another was Isaac Hull Adams, for whom a place as a student in West Point was found—and soon lost when he disgraced himself by twice leaving his post of duty. Only Thomas Boylston Adams II seemed able and diligent, so that John Quincy and Charles Francis fastened much hope on him, until his premature death in 1837. Thereafter, Charles Francis had these ill-fated relatives in his perpetual care. In 1845 Thomas' widow died after enormous suffering from cancer. Charles Francis and his family worked tirelessly to relieve the patient and to cheer her children. None of the latter had any descendants, and only two survived, unmarried, into old age.

Charles had more troubles than just those of the Thomas Adams family. His letters and diary describe his efforts to cope with other kinsmen, most of them impecunious. The descendants of his Aunt Nabby Smith harassed him until he was forced to forbid her wastrel son William, who had caused John Quincy much grief in St. Petersburg, to appear in Quincy. Visits from Nabby's daughter, Caroline de Windt, were dreaded because of the emotional disorders she and her children shared. Then there were the difficulties presented by his mother's sisters and brother. The latter, Thomas Baker Johnson, made Charles Francis his trustee for a modest estate acquired in New Orleans in spite of Johnson's eccentricities. Upon his death, Charles Francis administered these funds in behalf of Louisa's many sisters, all of whom came, sooner or later, to depend upon him for their maintenance. The last one died in 1877.

An even more pathetic responsibility was Mary Hellen Adams, the widow of his brother John. Once a dangerous flirt who had tormented each of the brothers, she seemed dispirited after her marriage, and two pregnancies had proved difficult. Then John died, and four years later her daughter, Fanny, fell sick in Quincy, dying on 20 November 1839 in the bed in which she had been born, after an anguishing illness. With winter setting in around the Old House, the occupants were devastated. After kissing his grandaughter's still warm cheek. John Quincy went to his room, as he later put it, in "stupefication." Once more Charles Francis had to guide the family in crisis. He arranged for an Episcopal service to be read in the Old House, and he persuaded John Quincy to join him in following the little coffin to the family vault, where Fanny's remains were laid upon her father's casket.

Thus, after barely a decade of marriage, Mary Hellen Adams had become a brooding, prematurely aged woman. She lived with her parents-in-law, entering rarely into society in Quincy, a place she always disliked. For many years after 1839, Mary—who was known to the Adams family as "Mrs. John"—managed John Quincy and Louisa's household under Charles Francis' direction. Then, in 1853, soon after Louisa's death, Mary's remaining daughter, Mary Louisa, married William Clarkson Johnson of Utica, New York, who, as a grandson of the first Charles Adams, had visited the Old House. It was there he had come to know his third cousin, Mary Louisa. She was his second wife. Mrs. John tried to make a new life for herself around her daughter's growing family. Charles Francis did what he could to keep her cheerful and each quarter supplied the $348 which represented her income under the wills of John Quincy and Louisa. He even teased her about her endless complaints: "You have somewhat of your family characteristics about you and like my mother, though much ailing, keep on recovering to a good old age."

Then, on 16 July 1859, Mary's world collapsed when her daughter died of "brainfever" in Utica, leaving her, as Charles Francis put it, "desolate indeed." Soon afterwards her son-in-law ordered her to leave him and her grandchildren. Charles Francis came to her rescue once more. "She is left on the wide world more solitary than anybody in my experience. I feel for her much," he said, "though my recollections of her are mostly painful." For many years, "Mrs. John" chose not to come to Quincy in response to his invitations. She lived in the Sixteenth Street house in Washington where he saw to her support and visited her. On such occasions he usually had to soothe her. One such conversation, he noted ruefully in 1860, "cost me more than two hours." That same year he had special problems with Mary's brother, Johnson Hellen, long a perpetual troublemaker for the Adamses.

Charles' efforts on behalf of the family endeared him the more, of course, to John Quincy and Louisa in their final years. Their greatest delight came in 1836 when he decided to build a home of his own on Adams land in Quincy. He declared that he had become hopelessly attached to the place, with its many family associations. When news of the decision reached John Quincy, he could not contain his elation, for to him all future family hopes rested upon this son and his sons. The old gentleman felt they must have a continued attachment to Quincy, even if their winters were spent in Boston. Charles Francis suggested a building site up the hill from the Old House, affording "instant communication with you," he told his father, who replied at once, proposing to give him whatever spot he wanted.

John Quincy called this bequest one of the few consolations left to him. He would die despondent, he said, "if I were under the conviction that no remaining drop of my father's blood transmitted through me, would survive me to cherish the spot where his and his fathers, and his child and my own children's bones are gathered in repose." Embroiled in Washington's political wars, John Quincy lifted his head in hope, drawn by this evidence of his son's admirable service and determination. He knew he had a sturdy son prepared to carry on the family tradition: "I leave the charge to you to redeem my good name in after times," he notified Charles Francis in May 1836.

His son did not fail him, for Charles Francis had become captivated by the Adams family story as revealed in letters exchanged by John and Abigail. "The history of my family is not a pleasant one to remember," Charles told his diary, as we have seen. "It is one of great triumphs in the world but of deep groans within, one of extraordinary brilliancy and deep corroding mortification." Eager as he was to gratify his parents and all they represented, he remained determined to be independent. In him was reborn the spirit of old John. "I would not have any of my children particularly distinguished, at the price of such a penalty upon the rest," Charles Francis said, cautioning himself to avoid the political adventures which had so captivated his father and grandfather. These campaigns for power and recognition, he believed, had been the cause of the great "trial and temptation" in the family and had produced such havoc. By choosing Quincy and family life, Charles Francis proclaimed his decision to reject the attractions of public service.

9

Satisfaction

Charles Francis' decision to make a place for himself in Quincy began a time when the Adams family found a large measure of satisfaction and relief from tribulation. After 1836 the family enjoyed for a season, the kind of life they had privately long desired. John Quincy gained national recognition and esteem, while Louisa found the repose and affection she craved. Charles Francis brought fresh attention to the name of Adams through publication of family letters and through his own achievements in the public and private spheres. There was special rejoicing over the success he and Abby experienced in producing a fourth generation. Here was promise of family renewal. From the moment of their marriage in 1829, the young couple was aware of their responsibility to produce children.

Doubtless carrying this great expectation, Abby became nearly paralyzed from anxiety. She went into a physical decline shortly after her marriage that mystified the doctors. When John had a second daughter, Charles Francis began to despair. Would fate eliminate the family name? If so, he thought it could be "with a very good end, for it would be better that it should cease than degenerate to become a proverb." Poor Abby sat dolefully through the christening of John's child in September 1830, mourning her barren state. Said Charles Francis, "I pity and sympathize with her. Married life has brought her as many thorns as it has pleasures." There were fresh consultations with physicians, but to no avail, and then, on New Year's Day in 1831, Charles Francis wrote in his diary, "Fears which once tormented me have been dissipated, anxieties have been soothed, and on the other hand, desires have been gratified and hopes exalted."

On 13 August 1831, a baby was born to Charles Francis and Abby, but there was no disguising their chagrin when this child turned out to be a girl. John Quincy confessed that, while he was relieved for Abby's sake, his heart was filled "with anxious hopes and fears." Still, he and Louisa took as much interest in this granddaughter as they did in the two born to John and Mary. The new infant, however, proved troublesome from the start, as if she sensed she had not received an unqualified welcome. She refused to nurse properly, which was awkward since Abby was producing great quantities of milk. Something had to be done to relieve her; and when John's oldest child returned from a visit to see the new cousin she rushed to tell her amused grandfather (as Louisa afterward delighted to report) that both the baby and a puppy had "sucked Aunt Charles' titties."

However, for Louisa, the latest granddaughter proved a source of concern. At first she thought the child resembled John Quincy; but after a time it was realized that she looked more like her late, unfortunate Uncle George. To complicate matters, on 23 October 1831, a beautiful Sabbath, Charles Francis and Abby put an end to mystery over the child's name by having her christened Louisa Catherine. Louisa was aghast, acknowledging how "superstitious prejudice" brought a "pang at my heart when I heard it," as she remembered the baby who died in Russia. Neither could she look at the baby without recalling George and his pathetic career: "God Almighty grant that her lot in this life may be cast more propitiously and that heaven may smile on her fate." In an uncanny way, Louisa sensed the unhappiness which lay ahead for her namesake. "She has superb eyes," wrote the grandmother, "but they indicate high temper and want sweetness."

Meanwhile, Charles Francis gave himself to parenthood with passionate seriousness, agonizing over the slightest distress of his first child. But he was soon distracted by different considerations, for Abby, once alarmed by her failure to conceive, quickly became pregnant again. On 22 September 1833 she presented the Adams family with the first male child of its fourth generation. John Quincy's response was joyful. "Judgement and Mercy!" he wrote in his diary, exulting that "a new world of humble hope is opened before me, which I had for some time not dared to anticipate amidst the misfortunes and sorrows which have befallen me in the last four years!" A few days later he was still overcome. "There is no Passion more deeply rooted in my bosom," he acknowledged, "than the longing for posterity worthily to support my own and my father's name."

Charles Francis was, of course, also elated, but his response was tempered by the sense of solemn practicality that had become his trademark. Now that the family was continued in a "male branch," he observed, he must not only manage its affairs with prudence, but must also labor constantly to guide properly "the course of the future successors." Within a few weeks, however, Charles Francis was accused by his mother of being shamefully prudent; Charles Francis had announced that the new child would be christened John Quincy Adams, but the ceremony would be a discreet private affair instead of a large gathering at the church. He reasoned this would be best because the elder John Quincy was at that moment engaged in a race for governor. Louisa was indignant that any grandchild of hers should be denied a proper baptism because of politics.

Evidently there was a heated exchange between Charles Francis and his mother, but Louisa carried the day, and the baby was publicly baptized in Boston on 27 October 1833 by his uncle, the Reverend Nathaniel Frothingham, husband of Abby's sister Ann. Later Louisa wrote Charles Francis that she was grieved "beyond reason" the Adams family should have allowed worldly matters to intrude on so sacred an occasion; nor could she believe that Charles Francis had been unwilling to proclaim "from the house tops" that his son bore so honorable a name. Her final sally must have stirred Charles Francis deeply, devoted as he was to the family gospel: "My son, I have nothing more to say in my justification than that he who humbly presents himself before the Majesty of his God, pure in heart and mind, will make man too secondary in object to dream of his existence." Louisa's statement summarized beautifully the everlasting gulf between her and the house of Adams, whose blood members invariably were tempted to value public opinion.

John Quincy was struck by the Reverend Frothingham's remarkable sermon for the christening—a most excellent message, he remarked. The sermon had taken a theme from First Corinthians; John Quincy recorded the text in his diary, vowing to take comfort and inspiration from it. It had been, also, one of his father's favorites: "But God hath chosen the foolish things of the world to confound the wise, and God hath chosen the weak things of the world to confound the things that are mighty." In this spirit John Quincy marked formally the new future for the family by presenting Charles Francis with the seal John Adams had used in signing the peace treaty with Great Britain in 1783. The seal displayed the Boylston family arms, which had entered the house of Adams through John's mother, Susanna Boylston Adams. It was to be an "admonitory memorial," said John Quincy, asking Charles Francis to see that the most promising male in the fourth generation would receive the seal and pass it in turn to "a descendant worthy of him to whom the Seal first belonged."

The second John Quincy Adams soon had a rival, for on 27 May 1835 another grandson was born, receiving his father's name. The christening of the second Charles Francis Adams was an occasion of delight rather than controversy, occurring on 31 October 1835, the centennial of old John Adams' birth. Four more children, over time, arrived,—Henry in 1838, Arthur in 1841, Mary in 1845, and, after his parents had passed their fortieth birthdays, Brooks in 1848. For this large brood Charles Francis and Abby created a home where anxiety and good humor, tenderness and discipline, learning and entertainment shared equal place. Charles Francis considered parenthood the most serious of all his duties, and here particularly he feared that he might degenerate as other Adamses had done in the past. To forestall this, he put into his diary "the landmarks by which I can understand myself." Little was omitted. There was even graphic account of his physical disorders (most of which the present-day reader might suspect were due to the anxiety which life as an Adams caused him. Not until he attained success as an author and statesman did Charles Francis' severe headaches and stomach distress abate). He wrote so often of his fears about dying while his children were still young that at times he seemed to be willing old age upon himself. Abby claimed that he relished "a positive determination to be old and useless."

A more serious matter, one which Abby and the children treated lightly but which terrified Charles Francis, was the early appearance of weakening memory, a disorder which, indeed, became ultimately incapacitating. Speaking during Fourth of July ceremonies in 1855, Charles Francis forgot his text completely. *"I broke down,"* he confessed, in an outburst in his diary. He saw this shortcoming as more than "a warning" of decline. It was evidence of his faltering as an Adams. "Crushed by the weight of two generations of distinction," he said, how could he hope for success? "Let me take it for granted that my life must be by the decree of providence a verdict of failure against myself." Insisting that "my health is failing," Charles Francis spoke of being brave and of finishing his duties before it was too late.

The family preferred to see such memory lapses as proof of his absentmindedness. A few months after the incident on the Fourth, Charles

Francis mislaid his umbrella while traveling, much to the amusement of his sons. "Poor old forgetful Pa," Abby wrote unsympathetically, telling him that "the boys shouted over the umbrella and calculated what that article must cost you a year." She reminded him how "you would scold the boys dreadfully if they were so careless." This episode illustrates how Charles Francis remained a warm human being within his domestic circle. The father who was teased about losing an umbrella also took his sons swimming, fishing, and for long walks. Such indulgence, he said, kept children "flexible and attached to the company of their parents." This was sufficient cause for Charles Francis to be more easy on himself: "I sometimes sacrifice time to them that might perhaps be more usefully spent."

Charles Francis brooded over his tendency toward "inertia," which had also haunted his brothers and all his forebears. "To counteract it in my children has been the mainspring of my policy toward them." Thus, he advocated all the conventional virtues with an intense affection. It grieved him to have his youngsters out of his sight—a "weakness," he called it which should probably be repressed. When one of them was ill, he became tense with fear. He tried to supervise personally the intellectual, physical, and religious growth of each child, and he himself rejoiced and sorrowed at each triumph or disappointment.

More than most parents, Charles Francis tried to live in the daily world of his offspring. He delighted in the long walks, the concerts, the games they played in the evening, reading aloud from Dickens and Scott, and the fishing trips. At the same time another part of him dreaded the thought of seeing these boys and girls eventually entering the evil world that lurked just beyond the family circle. Or, as Charles Francis stated it: "I hardly dare to look at my children with the hope that I can do for them what I ought in order to save them from the dangers which I barely escaped myself without shipwreck." His courage nearly failed him, he said, as he watched his progeny "rushing past into the vortex of life." He wondered if he had done all he could to inspire them to be honorable and independent. At times he fretted that he might have encouraged too much affection and too little respect from his children. "I see the difficulties that it occasions," he said, "and yet I should not alter my course if I had it to go over."

Charles Francis valued his wife as much as he did his children. Ten years of their marriage, he noted, had only made him treasure her more. He often said that Abby's enthusiasm for company prevented his becoming the complete recluse that John Quincy had predicted he would be. As they grew older, he continued to rejoice in the success of their marriage. He never compared his wife to his mother and grandmother Adams, those independent spirits. By contrast, Abby usually allowed him to think for her in all respects except for one—she did love society and was eager to entertain. Her vivacity in company was a trait many persons remarked upon. However, by 1836 the strain of three babies in quick succession together with the rigors of Adams family life had taken their toll. Abby was no longer lively around the house, and, indeed, was frequently troubled by depression. During this period, the children easily angered her, and at times she had spells of weeping.

Charles Francis was somewhat stoical in the face of his wife's tendency to wilt before what he considered trivial matters. Once, in urging Abby to fret less over largely imaginary difficulties, he touched on the central problem of her life. The intense Adams world left Abby fearful and uncertain about who or what she ought to be, a plight her husband had in mind when he said, "you are sometimes a little over-earnest to please others in matters that are of no moment and feel too much disappointed if you make any mistake." Still, this spirited, pampered girl upon whom her father had doted, was not always defeated by the ordeal of life in the Adams household. She was still willing to be coquettish about her womanly interests. As much as she had loved him at the time they were wed, Abby told her husband in 1841, "I love you ten thousand times more now, indeed I should make you blush did I tell what I thought and felt today." When Charles Francis responded with an ardor of his own, Abby—who was then visiting in Washington—was elated. "Thanks, thanks for your million kisses," she wrote. "I wish I could have them on the lips instead of paper." In return, she sent "forty thousand million kisses." Several years later, just before the birth of Brooks, she vowed to Charles Francis, "I love you far far beyond all things else, either in this world, or the world to come." In fact, she confessed, her love amounted "almost to worship which is wicked."

From Abby, Charles Francis also received much encouragement, not simply to leave his study and books for the company of others but to assume a public role, the step he was most reluctant to take. As her husband's writing and speeches attracted attention, Abby proudly rushed details to Louisa, although warning the older woman that Charles "would kill me" if he knew she were relaying such news. Nevertheless, Abby could often be candid with her husband. When he read aloud an early draft of his grandmother's biographical sketch, she was evidently bluntly critical, because Charles Francis reported in his diary that Abby "finds great fault with it, thinks it wordy and conceited and recommends its being wholly cut down and written over"—advice the author said he did not have time to heed.

Nevertheless, Abby was easily unnerved. Charles Francis often observed "that her element is quiet," which, of course, was something a devoted and sensitive mother of numerous small children rarely finds. Louisa, worried about her daughter-in-law, had gone so far as to suggest to her son that there ought to be no more babies for a time—"but mum is the word!" she warned. "I would not lecture." When a distraught Abby left off nursing her fourth child, Louisa praised this prudent step—"anxious milk never did good to any child." However, when Abby brought endless complaints and self-pity to her mother-in-law, Louisa sternly reminded her that her children were sent "by our heavenly Father" as a solemn duty. If the little ones momentarily took her from "frivolous pleasures," they would afford rich blessings later. Abby was often admonished by Louisa to seek "cheerfulness and matronly dignity."

By 1847, Abby was frequently suffering depression. On a visit to her brother in New York City, she surreptitiously consulted a well-known physician but came away distrustful and disgusted. The doctor had declared her distresses emotionally induced and, in what Abby acknowledged had been a kindly fashion, he had put questions to her which no doctor had ever dared ask before. Although the physician thought he could help Abby if she talked with him often, she fled. Soon afterwards, Abby was pregnant again and deeply melancholy as she approached her fortieth birthday. Louisa did what she could to cheer her; she told her how she herself had once pined vainly for more children: "Keep your spirits my dear daughter and think what a little plaything it will be to me to have another baby to occupy me." Abby was not so readily consoled, preferring to scold herself for succumbing to her ardent self, confessing to Charles Francis that "it gives me much pain to reflect that perhaps upon the whole, I never passed a more sinful year. And so old as I am too. God in his mercy forgive me. I shall not soon forgive myself."

Two months later, on 24 June 1848, Abby gave birth to her last child, Brooks. She had survived the "dark" she dreaded, and now occasionally began to show her old enthusiasm. Once, in 1851, she actually rebelled against Charles Francis' insistence that she ought to assume his restrained, even cold demeanor. His problem, said she, was that he did not understand women, for there were "a thousand things" that "affect a lady's mind and feelings" which men could not comprehend. She also survived the rearing of her large and active brood, although at times she evidently would have been driven to desperation had it not been for Charles Francis' support and presence. She never realized how often her comfort sustained an equally sensitive husband. Abby's children knew her traits well, of course, and tried merely to be amused when she still mothered them after they were older. One son marveled at his mother's "skill in making mountains out of molehills," and her insistence that grown sons, if left by themselves, would be unable "to go to bed and get up in the morning." Charles Francis and Abby both sent advice to the children upon all subjects, especially stressing that they longed for each son to have an honest love match as basis for his marriage. Money should be no factor.

This was relatively easy for Abby to say, since she herself had made such a marriage despite her wealthy background. Her children, of course, soon realized how important their grandfather's money was to them, since the Brooks family really paid for their winter residence in Boston, where they had to retreat from the chill of Quincy. They had acquired a house on Mount Vernon Street that served as the Boston seat for the Adamses from 1841 until Abby's death in 1889. With five children in 1841, the small structure on Hancock Street near the State House, which Peter Chardon Brooks had given Abby and her husband at the time of their wedding, was bursting, and, besides, Charles Francis wanted a larger study. Although the house at 57 Mount Vernon was far less glamorous than the homes of Abby's brothers and sisters, Charles Francis persisted in wanting to buy it, and accepted his father-in-law's offer of money.

The location was close to Charles Francis' Court Street office and even nearer to the books at the Boston Athenaeum, his favorite haunt. The rooms were large and potentially very attractive, particularly one which Charles Francis foresaw as an ideal study. The structure had been built in 1804, one of several planned by Jonathan Mason, who did so much to make Beacon Hill a graceful residential center. Charles Bulfinch had been the architect. Even so, Charles Francis conceded that his wife and others considered him "a little demented for selecting it when I might have chosen from houses a hundred times more showy." Dismissing the family's plea that he have his brother-in-law Edward Everett's place, a move Charles Francis considered wasteful and pretentious, he took $14,000 from his father-in-law and bought the Mount Vernon Street residence.

Here it was, near the top of Beacon Hill, that the fourth generation

of Adamses lived during the winter months. Here Charles Francis
looked to the physical and spiritual health of these children. His first
step was to bring city water into the building to replace that drawn
from cisterns. These were filled by runoff from the roofs, water which
Charles Francis described as "impregnated with smoke and ashes" because so many Bostonians had begun heating houses and businesses
with coal.

When Charles moved his brood into the house in early 1842, he also
made certain that it would become a place of moral purification as well.
Thus, each Sunday morning for an hour or two the study at 57 Mount
Vernon became a church. Charles Francis considered that the spiritual
instruction of each child was his most solemn obligation. He himself
remained devout in the vaguely Christian fashion of his father and
grandfather, attending two church services on Sunday. The children
were expected to trot along to worship at least once a Sunday, although
keeping the younger ones quiet often distracted their father from the
sermon, usually a worthwhile message from his brother-in-law, the
Reverend Nathaniel Langdon Frothingham. But Charles particularly
hoped to see moral values absorbed by his youngsters at home. There
they gathered, as soon as each could read, for Bible recitation and
memorization, prayers, and occasionally hymns.

To Charles Francis' delight, late in 1843 Abby joined the group. Her
mind "has lately been taking a serious turn," said her husband. Soon,
with Charles Francis beside her, Abby was kneeling in the presence of
the Reverend Frothingham, who had been invited to 57 Mount Vernon
to pray with them. The occasion, Charles Francis said, "moved me to
tears." He was similarly stirred when Abby and he were taken into
membership by Boston's First Church in Chauncey Place on 4 February
1844. He had contemplated this step for years, but feared that he could
not "act up to my pledge." Now, said he in purely Adams style, "Life is
to me for the future only a scene of duty." He hoped that the religious
awakening would strengthen Abby's "good disposition" and give "firmness to her determinations."

The experience left Abby more hopeful of gaining what she claimed
was her earthly goal, being "worthy" of Charles Francis. She said she
could never fully become so, but now she felt she was making some
progress. Her prayer to God was, she told her husband, "that I might
yet live to hear you say I was worthy of your love, and even respect."
Abby recognized that she associated Charles Francis too closely with
God, once telling him "you have truly been an angel to me," and that
"you will have your reward, and oh God, I shall have mine, may it be

better than I deserve." For a time after 1844, struggling with the problems created by so many children, Abby made Charles Francis her confessor, talking of shame, and pledging that each day she must do nothing to displease God "or my husband." Abby had the solemn responsibility of leading family worship when Charles Francis was away, but she was an easier mark than he, reporting once that, "as you were absent" the children had so eagerly begged to skip the devotional "that against my will I gave up."

With such emphasis upon reading and memorization, Charles Francis' scheme for spiritual enrichment must have smacked of schoolwork, for the children evidently found the devotional more dreadful than inspiring. He watched them closely, especially to see if these Sabbath experiences would help combat what he considered the greatest menace to his children, arrogance. "My endeavor is and must be to cultivate humility because in our family [that] appears to be most wanting," he told Louisa. Yet by the autumn of 1847, Charles Francis decided his attempt "to promote the religious culture of my children" was a failure. He found that they showed not the smallest sign of religious feeling. Two years later, he allowed the older children to go their way, while he and the younger ones turned to the new Sunday School exercises at the church. There, in time, Charles Francis taught classes and actually served briefly as superintendent of Quincy's First Church sabbath school.

Long afterwards, when all his children grew up without any outward mark of religious interest, Charles Francis announced: "It has always been doubtful to me whether the attempts of one generation to control the later ones . . . through a moral influence really do more good than harm in the long run." What Charles Francis appeared to overlook— and he missed very little worth noting about his own failings—was that the dismal fate of his brothers and uncles, combined with the high expectations of his family, had made him a person who constantly doubted himself and others. He continued to put extreme demands upon himself, his wife, and his children. Ironically, it was John Quincy who advised his youngest son as the latter began having children: "Your standard of morals is more elevated than belongs to the world in which we live and to the clay of which we are formed. . . . Let down a little your scale of Virtue till its last step at least shall touch the earth." And truly, Charles Francis was a much more successful parent than his father had been. He knew very well that independence is nourished by sympathetic encouragement; and he acted, for the most part, accordingly.

When Charles Francis and Abby's son Arthur died in 1846, not yet five years old, his parents were permanently affected. Charles Francis became even more concerned about the surviving youngsters. From the start, Arthur had been a great favorite, especially because of his extraordinary beauty. When diphtheria carried the child off after an agonizing struggle, the ordeal produced the only significant gap to appear in Charles Francis' diary. After a lapse from 4 to 9 February 1846, the distraught father wrote that "my poor beautiful boy, Arthur, expired at a quarter past six this day." On hearing the news, John Quincy wrote, "He has passed away, like a delightful vision."

Charles Francis' grief almost overpowered him, and Abby entirely disappears from the family papers for an extended time. A year later, the father still talked of yearning to be with Arthur, away from the "cares of mortality." It remained, he said, only for him to finish life "as honestly and usefully as I may." Thereafter, whenever a child or grandchild had trouble, especially illness, Charles Francis would hearken back to Arthur, whose death, he said, "has altered my character more than any other event or than all others put together." At length, he revealed why "I live over every hour of my former agony." Just before Arthur was stricken, Charles Francis had treated the lad "several times severe for the purpose of controlling what seemed to me to be too violent a will for his own good." This recollection evidently never left him, and he vowed to be gentler with his remaining children. Abby could not trust herself to speak about Arthur until, nearly twenty years later, she mentioned how she would gladly have died to save her son. So fierce were the memories, Abby said, that she and Charles Francis had never been able to talk about the circumstances of the death. They wept together silently on his birthday and the anniversary of his passing, she reported, a time when they recalled his dark eyes and golden hair.

Death came in a far more satisfying way to little Arthur's grandparents, John Quincy and Louisa, both of whom were now much consoled by the promising grandchildren at hand to carry forward the Adams name. The last years had brought considerable personal satisfaction to the old couple. With almost superhuman zeal, the former President had been battling in Congress for causes he cherished, and while the results seemed long in doubt, eventually he won praise not only from much of America, but also from his relatives. Such reward had seemed especially unlikely when John Quincy persisted for eight years after 1836 in what may have been the most valuable of his numerous services to America—his fight for the unlimited right of citizens to have their

petitions to Congress heard. In 1836 the House created the "gag resolution" in order to avoid hearing petitions against slavery. For a time,
John Quincy was almost alone in opposing this and was widely condemned for his stand. He came very near to being censured by the
House. When in December 1844 the "gag" was removed, John Quincy
earned admiration throughout most of the nation.

So remote did his success seem for a time that John Quincy inquired
of himself once more: why insist upon public service when all it led to
was failure? "I subject myself to so much toil and so much enmity with
so very little apparent fruit, that I sometimes ask myself whether I do
not mistake my own motives. The best actions of my life make me nothing but Enemies." Nevertheless, on he went, appearing at his seat in
the House undeterred by sickness, sorrow, and despair. Satisfaction came
in part from the curious consolation an Adams took in being "forsaken
by all mankind." "In such cases," said John Quincy, "a man can be
sustained only by an overruling consciousness of rectitude. To withstand multitudes is the only unerring test of decisive character."

Affairs kept John Quincy so blissfully busy that he had little time to
reflect whether he might be capable of standing on error. On his birthday in 1840 he did acknowledge how "the interests and affections of
this world" had so consumed his days that "self-admonition" had fallen
away. "The truth is," the old man said with some of the introspection
his son was making into an art, "I adhere to the world and to all its
vanities from an impulse altogether involuntary." He did, however, have
recurrences of the familiar depression, occasionally so severe that he
scarcely left his bed. These usually occurred when he could not avoid
wondering whether he had done enough in his life or if he had chosen
rightly. At these times he experienced what he called "a dejection of
mind" which almost made life a "lassitude," a "disease of debility," an
"imbecility which disqualifies for action." In these moments he felt: "I
have done nothing. I have no ability to do any thing that will live in
the memory of mankind. My life has been spent in vain and idle aspirations, and in ceaseless rejected prayers that something should be
the result of my existence beneficial to my own species."

At times, John Quincy looked wistfully at his cousins, Josiah and Ebe-
nezer Adams, as they, uneducated, pursued a simple agrarian life. He
especially envied their contented existence when he contemplated his
own lack of achievement. What had happened to the "spark of etherial
fire" which had entered John Adams' soul, he wondered, a spark which
had set John apart from the uninspired careers of his forebears? Even

after he passed his seventieth birthday, John Quincy dismissed himself as "a commonplace personage," far inferior to his father, one who was "about to disappear from the Stage leaving nothing behind me worthy of being remembered in after ages." Although he acknowledged being hounded relentlessly by his own ambition, he still felt his life illustrated how "the spark of etherial fire descends not from sire to son."

Then, shortly before his death, the tide in his public career turned. The Democratic Party which had ousted him from power in 1828 was itself defeated in 1840, leaving his allies, the Whigs, in charge of Washington. John Quincy and Louisa were suddenly feted as venerable royalty. Even more gratifying, however, John Quincy triumphed brilliantly before the Supreme Court in his defense of the African blacks from the Spanish ship *Amistad*. They had mutinied while being transported as slaves in 1839, and were captured by an American warship and brought to trial. In 1841, thanks to John Quincy's cogent presentation, the Supreme Court granted the blacks their freedom rather than ordering them returned to Spanish authorities. With this achievement John Quincy told Charles Francis that for the first time he felt he deserved the blood of the signer of Magna Carta that coursed in his veins. He was comfortable at last, he said, with the memory of that moment when, as a lad of sixteen, he had seen the original Magna Carta in the British Museum with the signature and seal of Saer de Quincy. It had been a sight which suddenly brought meaning to all the parental exhortations he had heard, and sustained him when he stood before the Court as the sole voice for freeing the black men.

Perhaps because he had needed to stand alone fighting for principle and winning against overpowering foes, as he had surely done in the *Amistad* issue and in the Gag Rule battle, John Quincy was vindicated on his own terms. Now he could be content, having successfully fought the malevolent world his parents had sent him forth to conquer. Fifty years of public life had been required, but there was now satisfaction, and John Quincy permitted this spirit to spill over into his family life. It meant new warmth and attention for Louisa, whom he began calling "Ma belle et bonne," a tribute which pleased the startled Louisa, who cautiously replied that "at sixty-five the title of *belle et bonne* is too brilliant for one as modest and insignificant as your affectionate wife." John Quincy would have none of such disclaimers: she would be *belle et bonne* "to me as long as I live myself, whatever your years may be." In another burst of tenderness, John Quincy saluted Louisa as "My beloved," and asked if she remembered when they had first read together the

Song of Solomon, especially the lines, "My beloved spake and said to me, rise up, my love, my fair one, and come away. . . ." With rhetorical gallantry he wondered if they were to lose "the quickening into life" as they sank into the "winter of mortality." John Quincy's answer must have touched Louisa. The spring in his soul, said her seventy-five-year-old husband, "will never grow old."

Louisa was not the only beneficiary of John Quincy's contentment with himself. He offered Charles Francis new support and took obvious pride in his only surviving son. Charles Francis, of course, continued to hear about his duty to God and country, but his father now seemed willing to acknowledge that his son had shown the special strength and wisdom an Adams craved to find in all his children. The busy old gentleman also took great pains in encouraging and instructing the grandchild who was his special love—John's surviving daughter, Mary Louisa—to whom he taught everything from French to logarithms.

To Abby, John Quincy sent in 1844 a ring bearing what he called his English signet, "with the cock and the motto 'Watch,' " while for Louisa there was a signet showing the lion carrying the cross, and for Mary Louisa another version of the "Watch" design. This latter signet, bearing Christ's plea for dutiful stewardship (Mark 13:34–37), had inspired the old President all his days. Was not every man given a duty? The divine charge was one John Quincy knew had been embraced by his own parents, just as he now begged those who followed him to remember: "And what I say unto you I say unto all, Watch." This scriptural injunction was an appropriate farewell from one generation of Adamses to another. Hold fast indeed to faith and duty for, as the family's favorite lesson from Mark foretold, "the sun shall be darkened, and the moon shall not give her light."

By 1845, there were warnings that the troubled spirit of John Quincy Adams would soon find rest. His palsy, a condition which also had afflicted his father, grew much worse—by 1839, he had been obliged to engage a barber to shave him thrice weekly. When this happened, the old man set to wondering what remained "when the faculties of body and soul are in the last stage of decay?" He survived a stroke on 20 November 1846, which toppled him as he was walking in Boston. Perhaps too vain to show his infirmity to Louisa, he demanded to be attended by a widowed nurse, who kept his wife at an amused distance. He did, however, take the precaution of having one more serious talk with his son, in which, as Charles Francis put it, "he reiterated his charge upon me in a manner scarcely necessary to make me fully comprehend my responsibilities." He also put his beloved diary in Charles Francis' hands, but "distinctly stating that it had never been written for extended publication, and it was not his wish that such publication should be made."

As soon as John Quincy was back in the capital, he insisted on appearing in the House of Representatives, whose warm welcome made Charles Francis, an observer, realize how much fonder of his father people were in Washington than in Massachusetts. There, a year later, on the floor of the House of Representatives, John Quincy Adams had a second stroke, and he died on 23 February 1848. He had still been remarkably active, despite increasing frailty, striving to find in duty fulfilled some source of continuing strength. As Louisa observed, half in pride, half in distress, "the House is his only remedy." Shortly before he was stricken, she had pleaded with him to stay at home; "he answered as usual that if he *did* he should *die!*" More than anyone Louisa knew that John Quincy could never have existed for long without excitement. She watched him suddenly grow very thin; there were marked signs of infirmity, even though his memory remained good. Yet in the hours before his final stroke, he seemed suddenly invigorated, attending church twice, teasing Louisa when she pressed him to take the liquor urged by the doctor, and then listening carefully as she read a favorite sermon by William Wilberforce about the use of time.

That was Sunday. On Monday he amazed his family by the briskness with which he marched off to the Capitol, although he was driven most of the way. At two that afternoon Louisa received "the shock which separated us forever in this world of suffering and trial." Hearing the news, Charles Francis started out that day by train, distracted somewhat from the gloom by a novel Abby had shoved into his hands, *Jane Eyre*. But even Charlotte Brontë's magic could not keep him from marveling at how his father had been "the great landmark of my life." When a message enroute confirmed that the old statesman was dead, Charles Francis observed: "the glory of the family is departed and I a solitary and unworthy scion remain overwhelmed with a sense of my responsibilities." Reaching Washington, Charles Francis found his mother once more responding to crisis with astonishing calm, even though, by some error, his father's remains were not brought to the house but rested in the Capitol. There the son hastened, drawing from the sight of his parent's corpse a redoubled sense of duty, something his father surely would have approved: "For the future I must walk alone and others must lean on me."

On 26 February, Louisa was too weak to attend the ceremonies, so Charles Francis was joined by his widowed sister-in-law Mary to watch the nation's leaders offer homage to John Quincy Adams in a House service which the son called "as great a pageant as was ever conducted in the United States." His thoughts strayed to the unrivaled example in public service his father and grandfather had set: "The only approach to it is in the lives of the elder and the younger Pitt." At the close of the ceremony, waiting to follow the coffin away from the Capitol, Charles Francis observed George Washington's statue "looking down upon the utmost pomp of man's last hour," with a hand appropriately pointing upward. The only unhappy note was at the site where John Quincy was temporarily interred. The Episcopalian rector, who was the family's clergyman in Washington, had wished to preside at this final service, but the congressional committee refused him because of local politics.

The great display of national sorrow, respect, and affection which had been shown in Washington was more subdued when John Quincy returned for the last time to Massachusetts on 10 March 1848. This was appropriate in the state which had so often denounced his stands upon unpopular causes as a young senator, as mature president, and as a venerable congressman. The old man might have chuckled when the rain poured down, and most of the Boston occasion had to be canceled. On the next day, in the more congenial setting of the Quincy church, a hymn was sung to some of John Quincy's verses, a splendid sermon was delivered by the family's friend and Unitarian pastor, the Reverend William P. Lunt—"I have seldom heard a finer production," Charles Francis noted—and at 5:00 PM the coffin was taken across the road and placed in the crowded family vault. Charles Francis wrote: "Who can tell the crushing feeling with which I looked at the end of that remarkable career." The most he permitted himself to say was that his emotions were "contradictory and various." He remembered especially that the funeral sermon had stressed faithfulness unto death. It was, said Charles Francis, "the noblest eulogy of him."

For four years John Quincy rested in the old burying ground with many relatives, his brother Thomas, his sister Nabby, his two sons, and others nearly as dear. When Louisa followed him in death, the two were placed in the crypt beneath the family church, where John and Abigail lay. Louisa had reached her own contentment; it had been a long time coming. Not until her husband had finally achieved some of the triumph he had sought did Louisa's satisfaction with life increase.

She began to struggle less, especially against her conviction that she was misunderstood. Having never stopped contending that society under-estimated female strength and talent, she started her memoirs in 1840 with the sarcastic title, "The Adventures of a Nobody." Louisa was eas-ily annoyed by women such as Mrs. George Bancroft who, she said, were always repeating what their husbands said. Such behavior only lent credence to the male assumption that wives were nonentities, and Louisa was proud to say, "When my husband married me he made a great mistake if he thought I only intended to play echo." She certainly never hesitated to speak out, even when her husband was seriously con-versing. Once, when the old couple entertained a Unitarian clergyman at dinner and the gentlemen were praising the power of reason, Louisa interrupted with a stunning inquiry about why laws were so necessary if reason was supreme.

Difficult though life had been with John Quincy—"I felt a desolate loneliness in the very midst of a family that I have too much idolized," she observed as late as 1839—Louisa had come to forgive her husband for his thoughtlessness and asperity. When the couple celebrated their Golden Wedding anniversary in 1847, she prayed that God would take the little time remaining to her husband to teach him "to struggle against the worldly passion which was in his soul." Not that she was free from this challenge; Louisa confessed to Charles Francis that "when my hus-band or my Son is attacked, my blood fires with uncontrouled anger, and reason loses her sway." Even so, Louisa tried to be more tolerant of John Quincy's passion to be a public hero. A few months before his first stroke she wrote a poem to him. She no longer scorned John Quincy's struggles against malice and despair—"Thy God shall shield thee from their deadly snares"—and her closing lines caught the yearn-ing that had tormented her husband's life:

> A bright renown shall wreath thy noble name
> And newborn worlds thy loving fame proclaim.

Meanwhile, she continued to be devoted to Charles Francis. "He is the only one of my children whom I never deserted," she said, still specu-lating about what might have happened had she not been torn from George and John in 1809. She sent her son many of what she called "Mothers old sermons," but now without so much of the old morbidity.

"The Madame," as everyone called Louisa, enjoyed good health dur-ing the last years of her husband's life. She was a grandmother who described herself eager to "romp" with the young family members, "to tell stories," "to quiz." "I long to play," she wrote her daughter-in-law, Abby, in 1842. Certainly Louisa took every opportunity to engage in her favorite pastime—never was there a more dedicated angler for smelt than Louisa Catherine Adams! And if deafness had reduced the lively style of her younger days, Louisa still enjoyed society, now appearing in dignified black—a color John Quincy loathed. The larger world Louisa had always disdained did not forget her, as when, on New Year's Day 1850, President Zachary Taylor, General Winfield Scott, and a host of senators and congressmen came to offer her their respects.

Although she was crippled by a stroke a year after John Quincy's death in 1848 and confined to Washington, Louisa welcomed many visits by her relatives from Quincy, who read to her and took her on carriage rides around the city. Nearby were her beloved sisters, and her son's widow, "Mrs. John." In this setting, on a May evening in 1851, Charles Francis had a long talk with his mother, of which "I shall treasure the remembrance among the most pleasing incidents in my life." Once so excitable, Louisa was now calm, speaking quietly of death and of her trust in God's mercy to forgive her sins. Charles Francis saw that his mother had achieved a serenity which "the philosophy of my father did not attain." John Quincy had fought anxiously "to live to the last." Charles Francis looked at Louisa, now nearly helpless as an invalid, and gained new admiration for her. "The cheerfulness of her unassuming Christianity" left him deeply stirred.

The mainstay of her life was more than ever her fervent, simple Christian outlook. It led her to denounce the baseness of man and especially those who shamelessly pursued "the bauble of an hour," reproaches found everywhere in her papers. Her indignation usually melted into an earnest spiritual plea. "Hear me O God, hear the cry of my desolation," she would pray. Louisa's beliefs never strayed far from the Episcopal Book of Common Prayer. Using it, she implored that her family might be saved "from the vices of our own hearts." It was this religious refuge which gave Louisa her final, triumphant peace.

On 14 May 1852, Charles Francis received the telegram he expected—his mother was dying. He set out on the next day for Washington, arriving in the early morning of 16 May to be greeted by a "small piece of black crepe attached to the bell handle." Louisa Catherine Adams had died peacefully at noon the previous day. The four years since John Quincy's death had given "a touching and tender close to the memory of her life," Charles Francis observed, recalling especially his talk with her the year before—"she hath left to me a sweet remembrance and an example for humble imitation." Her son's tribute would have moved Louisa deeply, since her aspiration in a life over-flowing with disappointments and sorrows had been to succeed as a loving, gentle mother and wife. Against fearful odds she had lived to fulfill her wish. In his own last years, Charles Francis cheered himself by recalling the "affection for my mother which was always the delight of my early days." She was, he said, "a true woman, frank, affectionate and faithful."

Who can guess what Mrs. John Quincy Adams' response might have been had she known that, at her death, both houses of Congress promptly adjourned in her memory, an unprecedented gesture to a woman. Louisa would probably have struggled between her scorn for the hollowness of public notice and her wish that women be properly esteemed. However, in the private world of the Adams family, there was nothing ambiguous about her distinction. While the awesome Abigail Adams was remembered for qualities of dominance and drive, Louisa remained for all her family what only the power of pure affection given and returned could inspire.

Louisa's death left Charles Francis very lonely. Distrusting the world, he could accept as genuine only the personal regard of his family and especially his mother. Now the one person whose kindness and affection Charles Francis said he never questioned was gone. "[H]er going leaves a blank which nothing can replace." Louisa's memory made him want to flee Washington at once; the city to him was "a great caravan of wild animals and the sooner I get out of it the better." He hastened back to Quincy, leaving his mother's coffin behind until he could arrange for his parents to be placed next to his grandparents. By 16 December 1852, the crypt beneath the town's Unitarian Church had been enlarged to receive the new occupants, who were interred there without ceremony, except that Charles Francis looked on with satisfaction. "The great duty was done," he reported to his diary. "There rest the bodies of those around whom as time rolls on, all the associations of this place will more and more cluster," a prophecy borne out over a century later by a stream of visitors from all over the world.

Charles Francis thought a great deal about his father; and twenty years later, he was able to offer an interpretation. After acknowledging how he still had "a feeling of delicacy toward my father," Charles Francis encouraged his son Charles to write about John Quincy. For advice, Charles Francis restated to his son his belief in character study. Do not concentrate upon the old man's public career, Charles Francis advised: "that is comparatively easy." The student of John Quincy must enter "the inner life of the man." Because this had never been done, the legend about him was constructed from superficial elements, leaving him merely "a bold, capable, ambitious, honest man," who was "cold, calculating, and violent." "It is impossible to imagine a poorer likeness to the reality," said Charles Francis. There was one quality, "the moral discipline of his life, which subdued the natural impulsiveness that belonged to him," which could properly explain him. Charles Francis made this observation in 1861 when he admitted he still could not read his parents' letters comfortably—"they move me too deeply." Now, however, the impatience and scorn which once inhibited Charles Francis' view of John Quincy no longer kept him from removing the "iron mask" he claimed made his parent a mystery. Behind it he discovered that John Quincy was not so different from himself—compelled to make moral discipline the beginning and end of existence. Charles Francis, at least, found it easier to master such a nature.

❧ 10 ❧

Another Generation

As Charles Francis returned to the Old House after overseeing the burial of his parents in the church crypt, his responsibilities as an Adams were, in important ways, just beginning. He was now, at forty-five, the patriarch of the family, properly seated each summer in the family's venerable Quincy mansion. Three years before her death, when her failing health prevented trips to Massachusetts, Louisa had insisted that her son take possession of the house. She was eager for him to be established in "the residence of your Ancestors." Charles Francis philosophized over his presence in the mansion: "I have come into it as a matter of duty and I trust that I may never do anything whilst in it to discredit those who occupied it before me." The Old House was in tumbledown condition and lacked a view of the sea, but it contained such overwhelming memories that Charles Francis and his brood quickly embraced it and the family position in Quincy.

For some years, Charles Francis had turned his attention largely to family tasks. Among these, the most interesting and often the most inspiring was reading and editing some of the manuscripts left by John and Abigail. His first project was to publish a group of his grandmother's letters. "She was a natural genius," Charles Francis marveled, discovering as he read through the manuscripts that "from her very clearly my father inherits his imaginative turn." In May 1840, he announced that "my volume of Letters of the old Lady is almost ready for the press." He thought of the project as "my first great adventure." It was also a sobering experience. The editor saw the disappointment John and Abigail had felt in their children. Probably in part because John

Quincy was still alive, he carefully omitted the letters which revealed his grandparents' despair, but Charles Francis could not forget this aspect of the family history. "Is not this discouraging?" he wrote in his diary. The book immediately sold astonishingly well, and by October 1840 sales were brisk enough to require a second edition. John Quincy himself was so moved by reading Abigail's letters that each one dissolved him in "a stream of tears."

This success brought Charles Francis some personal achievement in the world. The book, he said, meant "reputation that will endure." Immediately, he began plans for publishing a selection of the letters of John Adams, to the great delight of Louisa, whose memories of John were much warmer than of Abigail. "Families were created before Nations," Louisa reminded Charles Francis as she applauded his contribution to the noble sentiment of family veneration. Overcoming the mountain of paper left by John Adams, however, was a formidable task, as Charles Francis quickly discovered. "This is a prodigious work," he sighed, pushing on until August 1849, when a first volume was completed. The labor brought him the comforting assurance that "my Grandfather was a very great man, with a man's imperfections clinging about him." He admired particularly how John Adams never lost his delight in private life—such a contrast to John Quincy, dead now about a year, but who, Charles Francis said, had been "agitated by the restless worm." But the feature of John Adams that his grandson found the most gratifying was his refusal to accept the doctrine of moral perfectibility. John steadily insisted that man was still constituted as he had been "since the world began." With this Charles Francis announced his complete agreement.

When the first bound volume of his grandfather's letters was in his hand, Charles Francis was moved to say, "I felt as if I had not quite lived in vain." By 1852 he was sure that he had a cause, for he had peeped sufficiently into the later materials to recognize how remarkably revealing of men and events his grandfather's writings were. "I feel my anxiety increase inasmuch as I clearly see that his justification before posterity will in a great degree depend upon my presentation of his case." The job was now so compelling that Charles Francis was alarmed to find himself in church on Sunday mornings thinking of John Adams instead of the sermon. He wondered if he could "do my injured Grandfather the justice he deserves?" Toward that end, the public was given ten volumes of John Adams' papers by 1856.

Next Charles Francis thought to prepare his father's manuscripts for

publication, but he soon discovered that the contents were explosive—
"the persons to be touched are yet subject to suffer too much." So John
Quincy's diary was put aside for twenty years, while Charles Francis
devoted more time to understanding his children, whom he knew must
eventually carry on the traditions. Thanks to the old letters, Charles
Francis armed himself with a greater appreciation of how disappointing parenthood had been to earlier Adamses. It was a helpful perspective since Charles Francis' editorial career had been increasingly interrupted by difficulties with his three eldest children.

As with his grandfather, Charles Francis' first-born was a girl, little
Louisa Catherine Adams. Sister Lou, as everyone called her, never
overcame her anger and guilt at not having been a boy. She received
as much zealous attention as a first son might have, and she evidently
came to sense that she had all the talents needed for triumphs as a
male—except for the sex. Sister Lou's life became a long flight from
this accidental failure; and Charles Francis and Abby wondered, as John
Quincy and Louisa, and John and Abigail had done before them, what
they as parents might have done to save this child. Waste, only waste,
Charles Francis found himself saying of his first-born, over and over.

Sister Lou had proved so difficult to manage by the time she was
thirteen that her parents, in despair, decided to send her to the highly
regarded school for young ladies maintained by Catherine M. Sedgwick
in Lenox, Massachusetts. There Charles Francis hoped "that my
daughter may learn a lesson of moral discipline she does not obtain at
home. Here she is indulged and wayward." Off to Lenox Sister Lou
went in November 1844, delighted at the prospect of escape. Her father ruminated on how much of an Adams she was and how she must
learn to command her "haughty domineering spirit." Certainly the
young lady's tempestuous self-centered nature had not arisen from the
lack of love or attention in the house on Mount Vernon Street, yet she
grew so "perverse" that her parents feared to leave her at home. Charles
Francis sadly called the first member of the fourth generation
"thoughtless and self-willed."

Even with their daughter at the Sedgwick school, Charles Francis and
Abby were not reassured. They sent "exhorting" messages regularly to
Sister Lou—she preferred impatiently to call them "goodness" letters.
In turn, the young lady delighted in dismaying her parents and grandparents by tales of how she was the school's tomboy, how she was the
most daring sleigh rider of all. She informed everyone that when she
came home she would "astonish the natives." "I was intended for a

prodigy, and by degrees I am fulfilling the original idea." Miss Sedgwick, however, was determined that young Louisa should have something more substantial to show for the $62.50 which her parents paid as fees every quarter, although she acknowledged that her Adams pupil was a difficult case. Not only did the school have to report to her parents that the girl was careless and disorderly, but also that she suffered from "intellectual as well as physical indolence." Charles Francis must have flinched when Miss Sedgwick inquired if Sister Lou had ever been asked to do any hard work.

As time passed, the Sedgwick teachers discovered more about Sister Lou, making it easier to see why the girl had been sent to them. Louisa loudly asserted that all her classmates were beneath her, and Miss Sedgwick replied that not only was Miss Adams afflicted with the sin of pride, but she was the only person the school had encountered who was proud of her pride. Sister Lou, in turn, complained that she was being mistreated and misunderstood by the Lenox staff. After conceding to her parents that she had a difficult nature, the young pupil begged for some applause. "It is very fine for people that are amiable and good to be loved and praised, for they don't have to try, while those that try are disliked and blamed although I am sure if they all tried as hard as I do to do right, they deserve the praise that good ones get." The school, however, claimed that Sister Lou pleaded not to have her faults reported to her parents while at the same time she made no effort at improvement.

The causes for Sister Lou's struggle at home and at school are not difficult to find. From her own letters and the accounts of family members, she emerges as a highly talented person who was bored with what was expected of a woman. She yearned for the attention that would have gone to an eldest son in a family which so solemnly passed tradition, duty, and responsibility in the male line. Evidently Sister Lou went to extremes in seeking attention, making herself seem all the more unruly, egocentric, and unhappy. After her two years in Lenox, the family decided to have another try at helping the young lady by keeping her at home; but after a short time back in Boston, Sister Lou collapsed physically and emotionally. Her mother always remembered this episode with a "shudder," and spoke of her daughter's penchant for "dangerous intimacy." She and Charles Francis had had to break off at least one romance. Life with Sister Lou, Abby said, was simply "one trial and anxiety to us."

Thereafter, mother and daughter got on poorly until Sister Lou's marriage in 1854. The younger woman was high-spirited and aggressive, the elder was insecure and an admirer of convention. Neither understood the other; the girl felt the mother was small-minded and complaining, while Abby considered her daughter boorish. At times the strain at Mount Vernon Street became so severe that, Sister Lou confided to her father, life there "makes me so unhappy, that sometimes I feel as if I should go away and never come home." With her father she seemed to have a warm relationship, so that her letters to him carry a teasingly affectionate quality. Charles Francis tried especially to bring the girl out of depression, aimlessness, and selfishness by using such stimulants as concerts, books, and long walks. Hoping she might profit from a change of scene, he even encouraged trips to New York where Sister Lou could visit an uncle.

In New York and in Washington, where her exuberance both delighted and alarmed her grandmother Louisa Adams, Sister Lou found her sphere. She seemed to blossom when caught up in social activity—parties, dances, dinners—where, with her wit and polish, she could command attention. Abby was present at one such Washington triumph, and watched disapprovingly—"she has been to another party and is much noticed, which charms her. She likes it too well." In Boston and Quincy, Charles Francis tried to feed this hunger prudently. He helped his daughter establish a literary circle among the young ladies in Quincy, and he read aloud to them occasionally when the group met at the Old House. Sister Lou was also encouraged to be active in church, and for a time her energies went to Sunday school teaching. Momentarily, the experience so gratified her that in 1853 she embraced the faith and joined the old family congregation in Quincy. The appeal of this traditional role was brief, although Charles Francis worked hard to occupy his daughter with edifying activity. He took Abby and Sister Lou twice to hear Jenny Lind when she gave Boston concerts in 1850, although he grumbled a bit at the price—ten dollars a seat, and these not the best in the house, he noted. A grand piano was brought into the Mount Vernon Street residence in hopes it might attract Sister Lou's undeniable talent and energy.

Ironically, what finally seemed to occupy young Louisa, was the delight of leading many men a merry chase. The swains in Washington followed her to the train station when she left there after visits. She had become a brilliantly spirited creature, who, while no beauty, could use to advantage all she had heard and read in the rarefied Adams milieu. After 1850, Sister Lou seemed to live for her trips to New York

City, until finally, perhaps by impulse, she decided to claim a man for her own. Starting in 1851, Charles Kuhn begins to appear occasionally in her letters, and then, in December 1853, Charles Francis anxiously introduced this gentleman into his diary: "Mr. C. Kuhn came from New York and paid us a visit. His object was to apply for the hand of my daughter, Louisa." Having no basis to object, aside from mistrusting his daughter's motives, Charles Francis tried to be cheerful about assenting.

Louisa's young man came from a prominent family in Philadelphia; his mother represented the Lyle clan while his father, Hartman Kuhn, owned a splendid house on Chestnut Street. Born in 1821 and thus ten years older than Sister Lou, Charles Kuhn had received a lawyer's training before entering mercantile affairs in New York City. Both he and his father were graduates of the University of Pennsylvania, where his grandfather Adam Kuhn, a medical doctor, had been a faculty member for thirty years. However, between the Kuhns and the Adamses there would be no intimacy and little friendship, perhaps because Sister Lou always spoke scathingly of the Kuhns, complaining that they ignored her.

A large delegation of Kuhns, however, came to Boston for the wedding on 13 April 1854, an event which Charles Francis had tried to delay, but the couple would not hear of it. The ceremony took place in the study at Mount Vernon Street before seventy guests, after which the newlyweds departed for New York City where Sister Lou anticipated a life of flattering attentions and diverting excitement. With a troubled heart Charles Francis saw her go, musing over her "marked qualities both of good and evil," and praying, "God be merciful to her." To help the pair face life together, the Adamses had settled $1800 a year on them along with $5000 to furnish their new home.

The money was not enough to assure happiness, for the match was troubled from the start. After what must have been a disastrous honeymoon, Sister Lou spent much of the next four years in Massachusetts with her own family, and when she had to be with her husband in New York, she persuaded her father and mother to visit her often. What combination of circumstances made Sister Lou so wretched is now unclear. It was equally so at the time to Charles Francis, who traveled to New York a month after the wedding to aid the couple in spending his gift for furnishings. He arrived to find his daughter almost incapacitated by abdominal cramps. Charles recognized that the complaint was "not of ordinary character," for he knew something was amiss when

Sister Lou became highly emotional upon first greeting him in a crowded hotel dining room. As the months passed, she seemed to shun the company of her husband, preferring to spend her time with her best friend, Lucy Baxter, a young lady with southern antecedents. Evidently Sister Lou's recoil from marital intimacy was much more extreme than usually was experienced by well-brought-up young ladies of the mid-nineteenth century.

Although Charles Francis acknowledged that for "a really intelligent woman," the passage from the home of her childhood was always difficult, Sister Lou's behavior baffled him. Charles Kuhn evidently was part of the problem; he seems to have been distinctly Louisa's inferior in talent and her match in stubborn self-centeredness. Sister Lou grew so desperate that when her father could not remain in New York, she begged Abby to come to her, even though mother and daughter had rarely found each other's company congenial. Abby, startled, went at once, conferring with Charles Francis worriedly by mail. The parents finally agreed there was no alternative but to bring their daughter home, so that during the summer of 1854, Sister Lou was back in Quincy, a scene she had once yearned to escape. Her new husband remained in New York, and the girl recovered her former spirits with astonishing speed.

In the autumn of 1854, Sister Lou returned to New York for another try at her marriage. It was difficult going. By November, the pair was estranged for a time and Louisa took refuge with her companion Lucy Baxter. With the most skillful diplomacy, Charles Francis reunited the Kuhns, leaving Louisa basking in "dear Papa's" encouragement. "Your praise is the most valuable reward I can obtain," she told him. "I shall at least *try* to do what is right and best for us all." At one point, the Adamses sought to persuade Louisa to move permanently back to Quincy, but she refused: "I have seen too much of married children going home . . . to be willing to put any one in our family to the test."

Until 1858, when Louisa and Charles Kuhn turned to Europe for diversion, she tried to escape by being a young New York City socialite. This, with frequent trips to Massachusetts and visits from her parents, kept Louisa and Charles together, although she continued to mystify her family, which expected women to be more submissive or disciplined. Barriers to her happiness were obviously enormous and no matter how frenetically she tried to circumvent them they continued to block her. There are enough signs of identification with her father, of resentment that she was not a man, and of a need to dominate to encourage a view of Sister Lou as a woman who could not accept the conventional physical and social roles required of women by the times. Thus, she may have been one of the most pathetic of those family members for whom life in the Adams world was a failure. Possessing enormous talent without the gender to use it in those duties to which the family seemed called, Sister Lou drowned herself in foolish wasteful behavior. Her relatives watched in sorrow, never certain why she was so troubled.

Abby, of course, was convinced that her daughter needed babies to improve the marriage. Even here, though, there were ominous signs. Sister Lou tried to reject this part of the woman's role, exclaiming how she detested children, particularly those belonging to Charles Kuhn's relatives, toward whom she displayed an increased hostility. Not the nursery nor the bedroom, but the ballroom and the dining room, were the surroundings she preferred, where her wit and sharp tongue brought her to the center of conversation, particularly when the topic was politics. Here she easily outclassed her husband, and this made him angry and uncomfortable. Nevertheless, Sister Lou did become pregnant early in 1857.

Immediately, her parents begged her to come to Quincy for her confinement. They especially wanted the baby to be born in the Old House. Louisa refused, but did spend most of the summer of 1857 in Quincy, where Kuhn briefly joined the family, a time during which Charles Francis took a new interest in his son-in-law. Back in New York City, with Abby in anxious attendance, the first member of the fifth generation of Adamses was born on 11 October 1857, a little girl who, because of breathing difficulties, died soon after birth. Sister Lou was left for the moment "wild with agony." Within a few weeks, however, she was writing her cherished Papa about her angelic infant and "how much I should have loved her." Conceding that "perhaps I am not good enough" for motherhood, Sister Lou begged her father to believe that she had not been rebelling at being a parent. Then, abruptly, she predicted that losing the baby would prove "a blessing in time." Whatever may have been Louisa's deepest feelings about the child's death, the event brought the Kuhns' marriage to a turning point. They decided to abandon America to wander in Europe.

Except for an unhappy stay in the United States between 1860 and 1864, the rest of Sister Lou's life was spent in search of diverting pleasures, mostly in Italy. The Kuhns left for Europe in May 1858, to the dismay and deep disapproval of Charles Francis and Abby, who considered the exile an obvious evasion of domestic responsibilities which all right-minded adults should face. Their spirits were not much improved by the enormously long journal-letters which Sister Lou wrote as compulsively, it seemed, as her father filled his diary. These letters were her means for assuring everyone of how happy, petted, and triumphant she was amid the endless parties and balls which constituted the Kuhns' existence when they settled in Florence. Louisa seemed especially elated when she could describe those revels which lasted until 8:00 or 9:00 AM, leaving her to sleep all day until it was time to prepare for the next night's diversions.

Florence became a paradise for Sister Lou. Here were other dilettantes willing to admire her flashing temperament. Her letters disclosed a starved ego devouring a banquet which left a craving for more of such sustenance, while her moods ranged from ebullience to melancholia. These impressions made the Adamses even more uneasy. For a time they believed her when she said that her husband was detaining her abroad, but then, when Kuhn returned alone to America for a visit, her family realized they had been deceived. Abby was disgusted, claiming that what Louisa needed was a husband who could curb her. She burst out that should she ever see her daughter again, "I shall want to whip her as she deserves." When she had been thus found out, Sister Lou suddenly confessed that she shrank from what she knew would be Charles Francis' disapproving stare.

Despite Louisa's dread, Charles Francis was more concerned as a parent than he was stern. This is particularly evident in his attentions to his two oldest sons, where the family's hope for the future necessarily had to repose. The second John Quincy and Charles Francis each engaged in a different struggle to triumph as an Adams. They, at least, had the advantage of being male.

Or was it an advantage? In the instance of the eldest son, John, nature seemed to have been confused. The aggressive, restless, ambitious, demanding qualities which the era wished to see in men were given to Sister Lou, while John was awarded a yearning for domesticity, seclusion, children, and nature. He was seen as being indolent while his sister was considered relentless; where she was brilliant, he was limited in talent; John was charming while Sister Lou was abrasive. And from the beginning he was almost as troublesome a child as his sister.

While Charles Francis worried that the boy was such a slow learner, Abby was distressed by his behavior and his choice of friends. Louisa describes a charming tableau when her grandson was four years old, being eternally corrected by his mother. One day little John had had enough scolding; and, as his grandmother related it, he had said to Abby that " 'she had been little once, and said naughty things.' " Indeed, the lad announced, she had " 'put her elbows on the table,'—'all little children were naughty.' " The boy admitted he was, too—he did not know *why* he was; he " 'could not help it.' "

John may not have shown talent when books were put in his hands, but the fishing rod was another matter. By the time he was twelve his father had to admire the son's success as a sailor and angler. Once, at Woods Hole, young John caught twenty-three bluefish while his father hooked only three, their lines side by side. However, this was not enough and when Charles Francis sent his eldest son to Harvard in 1849, he said: "John shows tendencies which make me fear that he will not come out as I would have him." The Boston Latin School had not improved the boy whom his mother called "noisy and absurd." Harvard did little better, despite the reminders John received from his father about duty, family expectation, and how he was "now in the critical moment of his life." There was good cause for alarm. John got into trouble not only with the college but with the Cambridge police. He was fined for disorderly conduct, and endlessly admonished over neglected work, absences, and tardiness. His charm and possibly his name helped to offset this, and he managed to graduate in 1853.

By then, at twenty, the second John Quincy Adams was still a lively child, one who delighted in pretending he was his father when Charles Francis was absent from the table at mealtime. Once, John led the family to church services, announcing he had "acted" his father out perfectly in that he had "slept very sound." John's early path, like that of his father, led him into an unhappy legal career and to looking after the family's vast property holdings, as Charles Francis intended taking a larger interest now in literary and political affairs. During this time, John was obliged to live with his parents and even had to seek their permission to make trips with his friends, most of whom appalled Charles Francis and Abby. There must have been some escape for the young man in the amateur theatricals in which he proved successful. Serious responsibility made John uneasy, so that he often seemed to avoid or forget his tasks. His father observed sadly, "he is not of the class to fight the hardest battles of life," while Abby grumbled: "he really provokes me beyond measure." She told everyone how thankful she was that her son had so far escaped the depraved habits of his friends, young men from Boston whose names she was quite willing to announce.

While Abby and Charles Francis urged John to believe that he had untapped talents, they refused to acknowledge that one of these was the young man's enthusiasm for farm life. This was what he most wanted, especially if he could live in Quincy near the ocean. The citizens of the town liked this uncharacteristically genial Adams, and he soon found himself chosen for local offices. What little law he practiced was there, when he was not happily working with cows or trees. By 1860, John's letters abounded with affection for the Quincy scene and particularly for the Mount Wollaston farm. He frankly conceded he detested Boston life: "Boston is the most unutterably dull and stupid hole in this country," he told his father. Such a display of affection for Quincy eventually consoled the family, offsetting in their eyes many of John's shortcomings.

The first encouraging sign the family saw in John was his marriage to Fanny Crowninshield. His brother Charles, who had once hoped to woo her, described Fanny as "the wildest, funniest, and most fascinating girl I have ever met." Her greatest appeal, he said, was that "she's a high priced article and Lord, she wouldn't do or say anything vulgar if she tried to." John's brother Henry was also pleased at the prospect of the marriage. "He'll need an awful powerful team to keep him straight and make him work." Certainly, Henry observed, John's "life of man-on-the-town has lasted long enough." This view was pretty much the family's outlook. Unless John made a stabilizing marriage, "he has finished himself."

The second John Quincy Adams had become famed as a party-goer, frolicsome with the ladies. According to one story his brothers especially relished about him, John supposedly was doing his best to impress a young woman at a large dinner party. In order to take her in to the dining room without going through the usual crush when the meal was announced, he thought to escort her by way of the pantry. A stranger in the house, he grandly flung open what he took to be the proper door, only to discover it was a bathroom. What followed is unclear, except, as a gleeful brother described it, "The curiosity of the crowd was excited by a violent shutting of a door," and John fled the scene in mortification.

As it turned out, John hesitated about marriage as much as he did most of life's choices. In fact, it seems that he did not really wish to escape his bachelor's life, but found himself so associated with Fanny's name that he could not gracefully back out. By June 1860 he and Fanny were engaged, despite an irksome hesitation by her haughty mother, whom Abby indignantly called "ugly as sin." The more restrained

Charles Francis observed that the Crowninshields "are too exclusive for the habits of America." Abby was furious that anyone could think one of her sons was not qualified to marry any woman. "I feel as good as Queen Victoria," she wrote another son. "My father and mother were 'upper crust' and *you* may say what you please of the Adamses."

Fanny Cadwalader Crowninshield could claim one grandfather as a secretary of the navy and the other as a United States senator, while a great-grandfather, Jonathan Mason, had also served in the Senate. This was only the start. Through her mother, Harriet Sears Crowninshield, she could point to such forebears as John Winthrop, first governor of the Massachusetts Bay Colony, to Thomas Dudley, another early governor, and to such other illustrious names as Tyng, Browne, Williams, Sears, Willard, and Borland.

Although Fanny's father, George Caspar Crowninshield, had died at the age of forty, before he had time to make his mark, his father, Benjamin Williams Crowninshield, lived a long life, capped by somewhat lackluster service as naval secretary in the cabinets of President Madison and President Monroe. Benjamin Crowninshield appears to have been carried along by the reputation of his brother Jacob, whom President Jefferson sought to make secretary of the Navy in his second term. The nomination was approved by the Senate, and the record shows Jacob in the post, but he never actually served, preferring to keep his seat in the House of Representatives. He died suddenly in 1808 at thirty-eight. In politics and out, the family was associated with life on the sea. Benjamin and Jacob's rather bizarre brother George was America's pioneer yachtsman, builder of the successful racing boat *Cleopatra's Barge*. The family had a talent for making and losing and regaining fortunes in shipping and privateering. Under the first George C. Crowninshield, father to Benjamin and his brothers, they had become the most successful name in the maritime world of Salem, Massachusetts.

The founder of Fanny's paternal family was a physician, Johannes Kaspar Richter von Kronensheldt, who came to Massachusetts in 1684, fleeing Leipzig where he had slain an adversary in a duel. He married Elizabeth Allen, anglicized his name, and became highly respected. Other Crowninshield wives in succeeding generations came from the Williams, Derby, and Boardman families. Mary Boardman, Fanny's grandmother, was famed for being six feet tall. It was a family of audacity and prosperity with some public service. Crowninshields could also be expected from time to time to do the startling; there was even a scandal. Fanny's father had an uncle Richard, brother to Benjamin and Jacob, who had brought back from New York a wife whose reputation was never the best. They had a son, also named Richard, who committed one of the most sensational crimes of the century by murdering Captain Joseph White of Salem in a complicated plot which Bostonians whispered about for decades afterwards. Thirty years later, however, by the time Fanny became a wife, this chapter in the Crowninshield saga was left in silence—almost. In fact, Fanny's mother was heard to speculate whether an Adams was a worthy match for her daughter. Fanny had been one of the most sought-after belles in New England. The young lady was closely watched by her widowed and wealthy mother, who kept herself aloof, occupied mostly by her children.

John's family seemed to like Fanny more than he did at first. He made a desperate plea that the wedding be postponed, but the nuptials took place on schedule in the Crowninshield mansion on 29 April 1861, after which the couple went out to the Old House for a honeymoon. The magic worked. John was quickly tamed, his new "gentleness" became a matter of note, and Mr. and Mrs. John Quincy Adams II grew famous for their devotion to one another. Whenever John begged his wife to come dine with him in Boston if business dragged him from the country, Fanny was ecstatic. "Won't it be a *spree,*" she said once to her brother Caspar. "I am on my head with pleasure." To John, Fanny said that "even after being married three long years I feel as lonely and dull as ever when you go away and wonder if anybody loves their husband as I do you." Even her mother relented, at last, and visited Quincy, pronouncing the venerable Adams mansion "a nice old place after all."

John and Fanny produced in all six children, four sons—three of these by 1866—and two daughters. Again the old House was crowded by a full family. All this elated Charles Francis and Abby; the pleased grandfather assured John that he had "good and high qualities" after all, beneath his facade of "apathy and sluggishness." Charles Francis went so far as to predict that "fidelity to duty" would bring John a "useful and honorable position in Society."

Of Charles Francis Adams the second, there was never any uncertainty. He was remarkably opposite to his elder brother, for it seemed that Providence simply skipped from Sister Lou over John to Charles in distributing the more famous—or notorious—Adams qualities. This son, however, also brought Abby and Charles Francis more worry, for young Charles was determined that out of his ambition, his hunger for attention, his need to dominate, and his desire for material success, he would stand apart from the family. As irascible and abrasive as John was mild and congenial, Charles undertook to be an Adams entirely upon his own terms. He succeeded only in part, for which he blamed his father. As he became older, the second Charles Francis Adams flaunted a hostility toward his father which was unique in the story of the family's inner relationships.

Charles grew up resenting his fate as second son. It was clear to his parents how Charles improved when John happened to be away. If the older brother was at hand, Charles tried to irritate and excel him. Later, he was almost overcome when, at the famous Latin School in Boston, he ran across other boys who outshone him. He begged his father to be allowed to drop out and have a chance to "be even" again. Later, young Charles was miserable at the thought of Harvard and pleaded to be sent to Yale. If he and his brother Henry were in college together, "it will appear that he is very forward and I very backward which I could not stand." He wrote this at age sixteen, by which time he was angry that Henry, three years his junior, was being treated as his equal, or so he fancied. Meanwhile, when John was away from home at college and his father happened to be traveling, Charles could convulse the rest of the family at breakfast by elaborate pretenses of being in charge.

Naturally, Charles went to Harvard; any other school for an Adams was unthinkable. He tended to withdraw, even from his family, and he did reasonably well, winning a Bowdoin Prize in 1855. He had not told anyone of his intent. In 1858 he passed the bar examination, also without notifying his family, and Charles Francis was afraid that his son was attempting merely to race ahead without regard for excellence. Obliged to live at home after graduation, the custom for young bachelors of well-to-do families, Charles became wretched as he saw himself slipping into the Boston rut he so despised. "I wish I could get away from this limb of creation for a little while and find out whether there is anything else in creation, besides Beacon Street." He loathed the practice of law as much as had any other Adams since his great-grandfather's time. But most painful of all for him was his recognition that John was slated to lead in the family's new era. It was the eldest son's right.

Even as he grumbled, the young man berated himself for lacking the originality to find a new path. He predicted that for those in the family's fourth generation the danger was in becoming sheep following a bellwether. Yet how to avoid this Charles did not know, at least at the moment. When his father went to Washington and Congress, he stayed behind to help John and to pretend to practice law, all the while fuming that he was too old "to be ruled" and that he must "leave my father's house." To remain dependent and a functionary within a famous family meant, he said, that "the spikes and rivets" were drawn out of the new generation's making. However, despite all these brave words, he turned in some relief to joining his father in real estate speculation, finding here the kind of existence which lifted his spirits—investment, profit, expansion.

Rashness and greed soon betrayed him. By 1860 he had to confess to Charles Francis; "I am in too deep water." The experience of being rescued from debt made Charles even more furious with himself and with his lot so far. "My whole college life was embittered by dependence," he wrote to his father; "I was always in need of money and was obliged to go to you for every cent." Now, he proclaimed, he must be his own man as quickly as possible. When Charles Francis cautioned against such aggressiveness, the son brushed the advice aside. "It annoys me that you seem to worry about me, I am of age and can take care of myself and I really can't see why you should take such an interest in me and my doings."

Years afterward, when Charles sat down to write his autobiography, his memory of his father was bitter and resentful. He recalled none of the fishing trips, long walks, sight-seeing excursions, and reading which they had shared together, preferring to claim that Charles Francis was "hereditarily warped" and could not play the role of father. Charles even contended that his father forced him to try to make his own way promptly, thereby preventing him from taking glamorous excursions to Europe. He failed to recall the time in 1861 when Charles Francis had generously rescued him from financial disgrace. He certainly forgot his father's words: "what is the use of relations if it is not to befriend each other in difficulties . . . I have no other desire than that of aiding my children."

Henry Adams, the fourth child, was three years younger than Charles. It was, in fact, something of a miracle to his parents that this son was even alive. They never forgot how near to death Henry had come when he was three years old and gravely ill with scarlet fever. The crisis reminded his parents of the peculiar premonitions of evil which they had felt at the time of his birth, during a huge snowfall, on 16 February 1838. As an infant Henry grew "like a little pig," winning hearts, especially that of his grandfather Brooks, by the way he seemed to laugh with his eyes. Although Henry later insisted that he was christened in Boston by his Uncle Frothingham, the baptism actually was in the Quincy church. He received the name Henry Brooks, by which his parents chose to remember both the first Adams to settle in America and Abby's brother, whose recent death grieved her.

When Henry was stricken with scarlet fever in 1841, Charles Francis watched prayerfully at the bedside as he helped his son keep from choking to death. The rest of the family stayed at a distance. In due course, Charles Francis himself caught a milder form of the disease, but kept grimly to his role as chief nurse. Meantime, in Washington, Henry's grandparents were helpless with worry, Louisa saying that she had never seen John Quincy "give way to such excessive grief." For the first time he lost all interest in public affairs. The grandfather acknowledged that he could not even talk coherently under the "awful terror" about Henry, and that when mail from Boston arrived, "I was unable to read the letters." The crisis lasted nearly the entire month of December. Then Henry began improving, which allowed the family to move into the Mount Vernon Street house, a change postponed by his sickness.

Thereafter, the lad grew into a pleasing, studious, smallish person. His mother wrote in 1851 to Charles Francis, who was away, "Henry is Henry, when in the house mostly curled up in your big chair with a book and good natured, of course." In 1854, he had to take his turn at enduring Harvard College, where his literary interests made the ordeal bearable. Charles Francis was not greatly surprised when Henry, without fanfare, took a Bowdoin prize in composition and reigned as Class Orator in 1858.

However, in the meantime Henry did not fail to form the usual habits of Harvard men, being frequently admonished for smoking, making a commotion in the yard, lounging, and numerous absences from prayers and recitations. Nor at this time did he escape falling in love, intensely so, with Caroline Bigelow, daughter of Judge George Tyler Bigelow, who became Chief Justice of Massachusetts in 1860. There was a Quincy connection, for Caroline's mother was Anna Smith Miller, a neighbor of the Adamses. Henry pursued Carrie Bigelow for three years, not realizing that she was merely flirting with him. When he discovered this, he confessed to his brother, "It cost me the hardest heart-aches ever I had before I could sit quiet under the conviction that she is—what she is."

The memory of Carrie Bigelow made Henry talk like most wounded swains. He announced that "all women are fools and playthings until they've proved contrary." From the safe distance of Europe, to which he had fled in 1858 after discovering that his lady had thrown him over for someone else, the embittered Henry shouted, "By God, I grind my teeth even now to think how easily I let myself be led by that doll, who didn't even have the brain or the heart to exercise her power." He claimed that he was "cured" of any further interest in women. Recollections of Miss Bigelow's perfidy haunted him for years. His departure for Europe, ostensibly to study, was further encouraged by a hearty distaste for Boston, although his going roused grave misgivings in his parents, who feared that this step would make him unfit for the business of life. They ignored the fact that Henry's grandfather, John Quincy, had twice escaped to Europe to find himself.

For two years, Henry wandered on the continent. He rarely pretended to be studious, but he was always indignant when Charles Francis implored him to stop wasting time and money and to come home. Nothing, said his father, was more contemptible than a "gentleman loafer." Such letters from Charles Francis made Henry's plight only more painful. Like his brother Charles, with whom he often discussed the subject, Henry knew that an Adams must strike out for independence and usefulness. The prospect for him was no more certain or pleasing than it had been for many of his kinsmen.

"I am actually becoming afraid to look at the future, and feel only utterly weak about it," he wrote to Charles from Europe. "This is no new feeling, it only increases as the dangers come nearer." On the one hand, Henry talked of settling abroad—"I always had an inclination for the Epicurean philosophy"—while at the same time he announced in apparent seriousness that he planned a legal career in Saint Louis, where he hoped to make his mark. However, in all of this confusion Henry never lost his pleasure in writing. His admirable letters home about the European scene were published by the *Boston Courier*.

Eventually, with anguish and aspiration suppressed, Henry meekly announced that he would return to live with his parents while they were in Washington for Congress. Here he felt he might enjoy the scene as well as be useful. This decision came in July 1860 after he had spent a European winter and spring in what, at least by his description, was much revelry—cards, cigars, champagne, and German girls. One of his favorite stories was about emerging from a drunken stupor to find himself perched on a trunk in a lady's home, refusing to budge.

His cronies finally carried him to a hotel and put him to bed. After he awoke from this night of dissipation, the youthful Adams said: "I've a damned headache today and almost swear I never'll drink any more as long as I live." With similar resolutions, Henry now reappeared in America to try a new life helping Charles Francis who, as a congressman, was seriously engaged at last as a statesman.

⚜ 11 ⚜

London

In spite of his success as an author and editor, Charles Francis could not escape the feeling that there was something more he needed to do in order to see himself a worthy Adams. On the Fourth of July, 1843, he looked carefully at a subject which appeared frequently in his diary—his ambition. That part of himself, he concluded, was not the ordinary selfish impulse, but was derived from "my peculiar situation as the third of a distinguished line." This meant that he faced two alternatives. He could submit to his natural inclination and enjoy a quiet, studious life, running the risk of being considered "degenerate" when others compared him to his forebears. Or he could go out and, as he put it, "fight one's way," to match his predecessors in denouncing "the tendencies of the age." If this was the choice, Charles conceded, then duty must force him out of seclusion. As a result, after 1840 his diary treats of the "anxiety" that he would "fail in maintaining the character and reputation of the family to which I belong."

Gradually, Charles Francis conquered his dislike for that enterprise he had long avoided—politics. He lacked entirely the hunger of John Quincy for combat, and sought office only out of duty to "the race," as he often called his own lineage. This feeling of duty had always been familiar in Adams family discourse. John Quincy saw Charles Francis' ambivalence; when, in 1840, his son entered politics in the lower house of the Massachusetts legislature, he wrote: "You have so reluctantly consented to engage in public life that I fear you will feel too much annoyed by its troubles and perplexities." He told Charles Francis "never to be discouraged or soured. Your father and grandfather have fought their way through the world against hosts of adversaries." To John Quincy the lesson was simple: "keep up your courage and go ahead!" (Compared to his father and grandfather, Charles Francis' time as a statesman was brief. It was sufficient, however, since he needed only to prove to himself that he could succeed in public life.)

The younger Adams was not much impressed by his father's exhortations. "If it is to be my portion to throw away my life in politics and squabbling," he said, "I am prepared to submit to it but not to rejoice in it." Actually, Charles Francis' entry into public service was similar to the initial appearances of John and John Quincy Adams. Like them, he began by offering his pen, contributing newspaper essays. In Charles Francis' case, the issue was subduing the Masonic organization. In doing this, Charles also attacked the powerful wing of the Whig party led by Daniel Webster, which was then opposing John Quincy's election to the United States Senate. These enemies of his family used tactics which encouraged Charles to believe that politics usually brought out the very worst in human nature. By 1837, he was sending essays to the *Boston Courier* which scolded both Whigs and Democrats.

This evenhanded policy soon became uncomfortable for Charles Francis. He recognized some virtue in the Whig opposition to Democratic policy in the economic depression occurring after 1837, and he could not be entirely unwilling to acknowledge the merits of partisan politics when the Whigs began urging him to accept their support for election to a seat in the Massachusetts House of Representatives. When he finally consented to stand for that office in 1840, he received the largest number of votes of any candidate for a Boston seat. Thereafter, for five years, three in the House and two in the Senate, Charles Francis toiled in the state legislature.

As a member of the increasingly fragmented Whig party, his feelings were mixed. "There can be no happiness in public life if the individual is guided by virtue and morals," he told himself, echoing both his grandfather and his father. Even in the state Senate, where he led in urging national opposition to slavery and its supporters, Charles was disgusted by the paradox he could not dismiss from his mind. As the world acclaimed him, he confessed, "my self-esteem has been going downward at the same ratio."

Charles Francis gained a few friends during this time with whom he could talk candidly, something he was rarely able to do outside his family. His closest associates were John G. Palfrey, the historian, and Charles Sumner, the rising statesman, both of whom shared with Charles Francis an intense dislike for the powerful slave bloc in American politics. Even with such friends and allies, he could not escape a serious attack of melancholia during 1845–46. He left the legislature and had difficulty even in writing newspaper essays. "[H]ow hard I find it to struggle with my utter disgust at the vanities around me," he cried. He was so depressed that he could not keep his mind on religious services. "My nerves have been so deeply shaken that I doubt I ever recover them in any tone. The illusions of life have vanished and I see it only in its dreariness prepatory to a better." The death of his son Arthur deepened Charles Francis' misery.

Even so, once legislative duty was exchanged for the quiet of his study, Charles Francis began to improve. He moved his political residence from Boston to Quincy, saying that he wanted "to continue the relation to the town which those have held who have gone before me." Then, in 1848, came an unsolicited tribute which helped him regain his composure. It carried no tainted comparisons or degrading drudgery. In August, he received the nomination for vice president of the United States on the Free Soil ticket. Through his newspaper writings, Charles Francis had become a leader among the Conscience Whigs, a faction opposed to those more conciliatory toward slavery, known as the Cotton Whigs. Joined in convention at Buffalo by dissident Democrats and others, the Conscience forces considered Charles Francis, who was a Massachusetts delegate, not only a forceful speaker in his own right, but also a representative of his late father, slavery's heroic foe.

With no effort on his part, then, Charles Francis found himself running on a ticket with Martin Van Buren, the old Jacksonian veteran of the wars against John Quincy Adams. Their platform opposed any further westward movement of slavery, demanding "free soil."

Charles Francis recognized, of course, that the nomination was mostly a tribute to his name and he was good-humored enough to be amused at the efforts of people clustered around him to make "a great man out of me." Nonetheless, the experience did wonders for his contentment. It was, he said, "honor enough for me." He even rejoiced momentarily in political battle, reminding himself that "it has been the fate of three generations of our race to stand as the guardians of Liberty in this Commonwealth against the corrupting principles of a moneyed combination." This brought him to pledge: "So long as I live, there shall be another Adams in the Commonwealth who will denounce every bargain that shall trade away the honor of his country." However, such outbursts were rare, and Charles Francis' skeptical nature was soon back in command. He spent election day in November 1848 tramping around the Penn's Hill area with surveyors, and was unperturbed when the Free Soil ticket ran a poor third even in Quincy.

Claiming that his "mission" was over and disgusted with the compromising politics of 1850, Charles Francis retired again to the family manuscripts from which he could exhume truth and virtue without the exasperations of daily political warfare. "To me public life with my principles should be a sealed book." He seemed relieved to have stepped back, gratified that he now had earned the "reputation" which permitted him "to stand upon something like a level with my family." For a time, Charles Francis limited himself to service on the Quincy School Committee, an assignment he relished. Only with great reluctance did he agree to stand for Congress in 1852. Although he was defeated, a fate he claimed to accept indifferently, the experience showed him something about himself. Charles Francis realized that he felt his ultimate vindication could only be in winning the congressional seat once occupied by his father.

Charles Francis' determination to appear again in public life arose less from his belief in the principles of the new Republican Party than from the goadings of family pride. He tried to pretend he cared nothing about going to Congress and that editing family papers was enough for him. However, his study and his books were not now as comforting as they had been, and by 1857 he faced the danger which had afflicted his father so often—boredom. "I can scarcely interest myself in any pursuit whatever," he confessed. "I hope this will not last long." It did not. A local Republican Party, filled with his old Free Soil allies, made it certain that he might be chosen for Congress.

Discovering that he was delighted by such a prospect, Charles Francis was half-amused. "I wonder at myself sometimes." Since he knew that he would be miserable in politics, he consoled himself that his appetite for victory was something special and really a matter of conforming to duty. "Nothing but the idea of what is due to my name" reconciled him, he said, to the "sacrifice" of his quiet life. Nevertheless, in the days preceding the election of 1858, Charles Francis behaved much as any other candidate, easily despairing at any fancied sign of losing. When matters were promising, he strove to be casual. It fooled no one. His son Charles said: "The Gov. is trying to look as if he didn't much care, but in reality he's tickled to death." Elected by an overwhelming majority, Charles Francis called the event a fulfillment "of all the wishes I ever had in political life."

Once the satisfaction was savored, however, Charles Francis' dread of political service reappeared. He had such nightmarish memories of Washington that returning there made him shudder—a place "so hollow, so tempestuous, so full of evil passions, so disappointing to the most laudable ambition." His family tried to encourage him. Henry assured his father that ambition and duty were, in his case, wholesomely united, while his son John begged him at least to try displaying some interest in proceedings within the House of Representatives. Charles Francis, however, persisted in his disgust, muttering in his diary "how much I should like to fly this pandemonium and give up public life forever." An evening of dining at the White House in 1860 reminded him of "how little substance there is at bottom to pay for the labor employed to get there."

His wife and his two youngest children, Mary and Brooks, went with him to Washington. There Abby's talents as a hostess made her a worthy successor to Louisa, a feature which helped Charles Francis overcome his discomfort. He was a successful member of the Thirty-sixth Congress; his most notable triumph was the single major speech he made, at the close of May 1860, in which he argued that through the Republican Party the nation could reclaim the principles of free government which men of his grandfather's generation had bequeathed the nation. It was an impressive speech, and it brought Charles Francis even greater intimacy with William H. Seward, the most powerful Republican leader at that time. He was also assured of reelection by his Quincy constituents.

To all of this, Charles Francis' response was predictable. He was gratified, he conceded, but only because the world regarded him as something more than "the son of my father." He recorded with pleasure that he had earned a reputation for character and capacity "not unworthy of my ancestry," while the large vote in his district stirred him deeply. "Neither of my ancestors in all his brilliant career ever received so brilliant and so feeling a testimony to his character and services, from his own townsmen, as I received at this time. . . . I had equalled the best of my race." Then he quickly fell back to his impatience with politics: in the political arena human nature took "painful shapes" and "all the worse passions" were aroused. "I look with dread at a continuation of this existence," he wrote on the day in March 1861 when Abraham Lincoln was inaugurated.

With the arrival of a Republican national administration, Charles Francis could not ignore gossip which spoke of his receiving a Cabinet post or a diplomatic assignment. He fretted that such developments would upset further the quiet life he preferred. When the suspense was finally over, he was appointed minister to Great Britain. The household at 57 Mount Vernon Street and especially Abby were much dismayed, although Charles Francis consoled himself with the realization that he was "third in lineal descent in my family on whom that honor has been conferred." A few weeks later, on 1 May 1861, he, Abby, and their three youngest children, Henry, Mary, and Brooks, sailed for London. The seven years which ensued before the Adamses returned to Massachusetts contained not only a personal triumph for Charles Francis Adams and one of the great interludes in American diplomatic history, but they brought something else almost as striking—a genuine sense of contentment. Even by his severe Adams standards, Charles Francis had fulfilled his hopes and his potential.

Thanks to his patience, outward calm, and knowledge of history and diplomacy, Charles Francis was successful in preventing Great Britain from abandoning neutrality in the American Civil War. Had she done so, the southern states would probably have received the measure of support from England that could have brought victory for the Confederates. His deportment helped greatly in making the Union cause more popular with the English government and with many citizens. The new Minister's first achievement foretold his eventual success. In late 1861, he managed to avert war between the United States and England. This crisis occurred when the U.S.S. *San Jacinto* stopped the British steamer *Trent* and removed John M. Mason and John Slidell, who were traveling to England as Confederate commissioners. Reports of this seizure produced an anger in Great Britain which required all of Charles Francis' tact and talent to subdue. Two years later, he persuaded the English government not to release ironclad vessels—the famous Laird rams being built in Great Britain—to the Confederate government. This signaled success for Adams and the United States. From that time on, he sought impatiently to be relieved of his post and allowed to return home.

There were three reasons why Charles Francis was so eager to be back in Boston and Quincy—his three eldest children, who had remained in America. He remembered what diplomatic missions had cost earlier generations. As late as 1866, when Sister Lou, John, and Charles had all reached age thirty, the Minister still chafed to be back, near them, recalling how "the same enemies of separation" had destroyed the usefulness of two out of three members in both the second and third generations. "I may be pardoned for my uneasiness," he said.

"Quincy is my paradise on earth," he wrote in his diary, with a fervent note his father and grandfather never quite managed. He recorded with quiet sincerity: "The ambition I had to make myself a position not unworthy of my name and race has been gratified. All the common conditions of man's life have been fulfilled. My children may rise to take my place."

The conflict and tension accompanying his delicate mission brought little of the gratification which similar excitement had often given his father and grandfather. "I am weary with this sort of profitless controversy even though I have the side of truth and justice," Charles Francis wrote at the close of 1862. He knew that "it is not my nature as it was my father's to exalt in the exercise of his power in strife; to dilate under the heating force of controversy." In mid-1865, he had another serious bout with depression, arising from "a sense that my absence from home is working unfavorably to most of my children and my property." Sometimes he distracted himself by thoughts of building a new residence on the Mount Wollaston farm, near the shore, but then his devotion to the Old House would intercede. When he was warned that the place would be too plain and cramped, he waved this aside— "the consideration that it is my own, as it has been my father's before me will overbear any other." He thought about home incessantly.

This was not lost upon the younger family members. While they were pleased with the acclaim given their father in England and Europe after 1865, they were chagrined to observe how sincerely Charles Francis yearned to leave London and the public life which he resented. Still, the children claimed that time in London had improved him, giving him, one son said, "a certain calmness of manner and control of mind which he did not formerly have."

Charles Francis' mood was certainly rare in the family annals. He had entered the world which had so abused other Adamses, met it on its own ground, and mastered it. He could retire to his cherished quiet life, having fulfilled duty and served the family name. It meant, Henry Adams observed, that his father was "less a creature of our own time than ever." He was "separate from the human race," perhaps the most gratifying tribute any Adams could have wished for.

Henry Adams was drawn close to his father during the London interval. He served as private secretary to the Minister, an association strengthened by their shared respect for history and literature. Together they labored to make the Union cause acceptable to England, and Charles Francis acknowledged that he could not have succeeded without Henry's companionship and help. The elder Adams was also grateful that, delicate though his public duties were, he and Abby had time to give to Henry and their two youngest children, Mary and Brooks. Charles Francis managed to take the family on excursions throughout the British Isles and Europe. It was a precious advantage for them, particularly as they were still grieving over little Arthur's death many years before. Such a loss may have made them more loving toward the children who survived.

This second career as parents, however, was almost as worrisome as their earlier one. Henry was still a special concern. Even before he had rejoined his parents late in 1860 after his two years of wandering in Europe, Sister Lou had reported to them how glum Henry had become. She likened him to a little old man. Going to London with his parents did not free Henry from depression. Gradually, he began again to write for American newspapers, but this did little to diminish the familiar Adams anxiety at having to choose an honorable path through life. "I've disappointed myself," Henry confessed to his brother Charles, "and experience the curious sensation of discovering myself to be a humbug." Failure stared at him everywhere. "My enemy is only myself." The specter refused to go away, and Henry suffered poor emotional and physical health until after 1865. "I grow stupider and stupider every year as my hair grows thinner," he said, in an effort at humor. "I wish I were fifty years old at once and then I should feel at home." Such a leap into age, of course, would allow him to evade the responsibilities of youth—military service, romance, and a proper career. But these milestones seemed to Henry either disagreeable or elusive. Growing up as an Adams had been made more difficult for Henry because he had a special degree of that sensibility which plagued most family members. From an early age, he was torn between his devotion to his parents and his yearning to be quietly away.

The stay in England soothed Henry somewhat because there were so many historic buildings and other reminders of a distant past. Their salutary effect on him was evident in 1863 when Henry and Charles Francis stood together entranced before the old English cathedral at York. Arm in arm, they returned to what Charles Francis called "a wholly different age and another class of thought." Father and son often talked of matters which, many years later, Henry made central to his mature outlook and to his literary works. They were especially struck by the "power" in Christianity which could produce magnificent cathedrals for worship. Sights and thoughts like these made Henry all the more estranged from his own time. If only he could have lived centuries before, he thought. "I should have become a monk," he said, and would have been "Abbot of one of those lovely monastaries. . . ."

Pleased with the archaic character of foreign customs and settings, Henry took readily to English clubs, parties, and country life, becoming what his father called "something of a favorite." This reassured the elder Adamses, who worried about the young man's excessive tendency to stand aside from the world. However, Henry's passion for "English living" became so great that his parents were at last alarmed. He no longer spoke morosely of tedium, a sign, said his father, that "we are all staying much too long." However, Charles Francis was not as close to Henry as he wished to be. Henry never really explained to his father his torment over his future in an uncongenial world. It was, Henry told his brother Charles, a simple prudence to keep his views to himself, which he did until he shared them with Charles, who spent his military leave in England during 1864. "Henry is much older, and, in his philosophy of life, not improved by life in England," Charles observed.

Misanthropic the young Henry Adams may have been, but he was also a gentle soul, always considerate of others. No Adams was less like the severe mold from which so many family members were formed. Even before the years in London, Henry lectured his brothers on kindness, saying, for instance, that he now realized "how excessively selfish and exacting we children always were toward mamma, and still more, how much she felt it." Grandmother Louisa, whom Henry came to admire, would have applauded as Henry urged his brothers to display affection, especially "in those little matters that a woman feels most." Sister Lou was starved for this kind of treatment, and she particularly responded to Henry. "He is just as sweet and gentle as ever and I should like to take him in hand and feed and look after him for six months," she wrote to Abby.

Henry's solicitous nature brought him an unwelcome reward. He was deputized in 1865 by his father to take Abby, Mary, and Brooks for a six-month holiday on the continent. The ordeal at times crumbled Henry's kindly exterior, at least in his howls of distress to his brothers. His mimicry of Abby's nagging style must have convulsed John and Charles. " 'Oh! Henry, how you do look with that beard! I really think it is wicked in you to go so, when you know how it pains me and disgusts me to have you seen so!' " Conceding that he should be contrite for so ridiculing "one of the most devoted of mothers," Henry said he could not be silent constantly. "I grow old and cynical," he told Charles, while he confessed to his father finally, "I abominate this family work." Was his life to be "the ancient and active part of sheep dog?" he wondered.

The ordeal made Henry so gloomy that Mary and Brooks began calling him "Mausoleum." Silence was Henry's way of containing his real feelings. Rarely did the young man share his troubled views upon life. He did acknowledge to a brother in 1866 that should his gloom overcome him, he would commit suicide. "I have even decided the process," he told Charles, but offered no specifics. Later, he informed his brother John that he wanted only to be left alone. Thanks to a bequest from grandfather Peter Chardon Brooks, Henry now had an income which promised him independence. He would need it, he was convinced. "I am little formed by nature to act the part of a combatant," Henry said. "If I were to express even in private all the opinions I hold, I should sacrifice my influence, the little I have, and perhaps my character, without stirring other people's opinion a hair's breadth."

With a mixture of affection and bewilderment, Charles Francis, as well as other family members, watched Henry try to be comfortable in life. Loving and considerate though he always appeared, this brother became a deepening mystery for even those nearest to him in the London years. Whether from cowardice, self-doubt, or disdain, Henry sought to keep life at a distance, although he never tired of observing it. It was the posture which his grandfather and great-grandfather had yearned to take. But their natures, so different from Henry's, would not allow it. In contrast to Henry's personality, the other two children accompanying Charles Francis and Abby to England were, at the time, remarkably uncomplicated. Rearing them had seemed a simpler task, since Mary and Brooks appeared as transparent as Henry was opaque.

Born on 19 February 1845, Mary Gardner Adams was the precise opposite of Charles Francis and Abby's eldest child, Sister·Lou. Mary became the most conventional figure of the fourth generation, although she seemed so stubborn and disobedient as a tot that her vexed mother vowed to give her up. Eventually, Mary submitted to family ways, becoming a solace and mainstay to her parents. But during the family's time in Europe, she was unwittingly the cause of an amusing squabble. Mary needed dental care and her parents decided to have the work done in Paris. On such excursions to France Charles Francis remained in London, letting Abby choose a Parisian dentist. Abby was strained by the task, describing herself as "wringing my hands and . . . half crazy." The problem was that she had ordered expensive treatments, and now Charles Francis objected. Abby defended herself: was she not using her own judgment, as "you always have wished me to do?" But it was almost too much for her. "God help me," she wrote, "I am pretty used up. Your note ought to be obeyed, but how can it when her teeth are half done."

The argument over dental care provides a revealing little glimpse of family life. While Abby begged forgiveness for arranging the treatment—"I fear you and your displeasure too much to be otherwise than miserable"—Mary, young as she was, struck a sterner note. After asking her father to stop reproving Abby, she stated: "As to me, I think that I am getting old enough to judge somewhat *for myself* in such a personal matter as my own teeth." Charles Francis retreated with as much grace as he could before the seventeen-year-old's logic. In another year, Mary had acquired the currently fashionable passion for horses, then was presented at court and entered society. She attended parties and attracted the attentions of a handsome young American officer, come to London after heroic military service—Oliver Wendell Holmes, Jr.

Charles Francis was gratified, on the whole, at what living abroad had done for Mary, with whom he took long walks when she was not riding horseback with Henry. "I know of no one of the family who has benefited so much from this adventure as she has," he said. "The sight of Europe has expanded her mind and stimulated her desire for knowledge." Henry was less confident, and possibly more discerning, about his sister than his father was. Mary, he said, "wants depth and wit and doesn't show so much quickness at catching and improving ideas as I would like to have her." This young lady was, of course, pale in comparison to Sister Lou or to the woman Henry eventually married. At best, Henry said of this little sister, "she is pretty, sympathetic, and good-mannered, not highly accomplished but tolerably well-informed." Mary Adams returned to Boston to become one of the most eligible young ladies in that society, but she waited a decade before marrying.

Brooks Adams was three years younger than Mary, which meant that he spent his early teenage years in London, acquiring from school and society such decidedly British ways that he was hustled back to Massachusetts in 1865, three years before his parents' return, in hopes that Harvard might make an American of him again.

Born on 24 June 1848 and named Peter Chardon Brooks Adams after his grandfather, the child was soon called Brooks rather than Pe-

Holograph letter from John Quincy Adams to Charles Francis, which proved to be a final testament. It closes with the words: "A stout heart and a clear conscience, and never despair." The old gentleman was soon dead. From the original in the Adams Papers. *Courtesy of the Massachusetts Historical Society.*

Louisa Catherine Adams after the strain of life in Washington had begun taking its toll. Portrait by Gilbert Stuart was completed in 1826. Louisa detested this painting. *Courtesy of the White House.*

Charles Francis and Abby Brooks Adams in middle life, perhaps 1850. These charming profiles long remained undiscovered in a desk at the Old House. *Courtesy of the National Park Service, Adams National Historic Site.*

Louisa Catherine Adams Kuhn, "Sister Lou," as she appeared shortly before her death in 1870. *Courtesy of the National Park Service, Adams National Historic Site.*

John Quincy Adams 2d, eldest son of Charles Francis and Abby, in a photograph taken around 1885 at the Glades, John's beloved ocean retreat. *Courtesy of the National Park Service, Adams National Historic Site.*

Henry Adams in a rare photograph made near the time he married Clover Hooper. He was then a Harvard faculty member and editor of the *North American Review. Courtesy of the Massachusetts Historical Society.*

Charles Francis Adams soon after his return from England in 1868. This newly found photograph is unusual because it shows him with nearly a smile. *Courtesy of the Massachusetts Historical Society and Stephen J. Kovacik, Photography.*

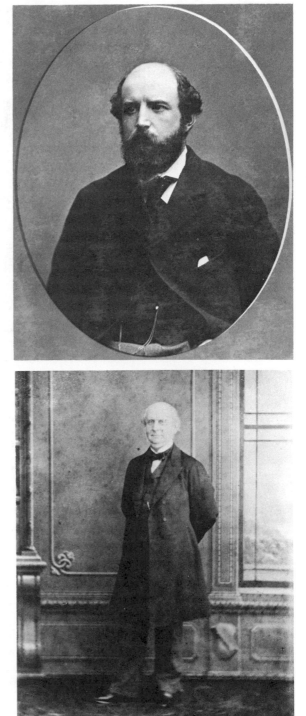

The garden of the Old House as it appears today and as Mrs. Charles Francis Adams designed it after her years in England. Building to the left is the library which Charles Francis constructed in 1870. *Courtesy of the National Park Service, Adams National Historic Site.*

Interior of the library which stands in the garden beside the Old House. Here Charles Francis Adams and his sons read and wrote during the summertime. Portrait on the left is John Adams; that on the right is John Quincy Adams. Desk at window to the right was used by John Quincy in the House of Representatives when he was fatally stricken. *Courtesy of the National Park Service, Adams National Historic Site.*

ter. While Brooks behaved well at his November christening, the style soon left him, for he was a very difficult lad. He was hyperactive, wild, inattentive, and made a pest of himself by incessantly asking questions. When he was two years old, he began showing the first signs of the trouble which lay ahead. One day, refusing to wear his apron, he "made it into a *ball,* and like a flash threw it into the fire." Abby's report of this alarmed Charles Francis, who feared the two-year-old's violent ways. Surely Brooks had done this deed accidentally, he asked? No, said Abby, "He fully intended doing just what he did. He said before hand: 'Brooks will throw his apron in the fire and burn it up.' " Thereafter, he refused to say that he was sorry, though Abby managed to extract his promise not to repeat the deed.

Charles Francis gave much attention to this youngest child, hoping to subdue and shape the lad. Before Brooks was ten, he went with his father to exhibitions, concerts, the theatre, while, of course, they read together, hiked, fished, and swam. This persisted even in London, where their special project each Sunday was to find a different Anglican church in which to worship, preferably another of Christopher Wren's buildings. Brooks always retained joyful memories of these jaunts. However, his most enduring recollections were of his nurse, Rebecca Blanchard, who had tended a succession of the children before him. She must have been a most attentive mother to Brooks, for when she grew old, Brooks defended her against charges that she was no longer useful, and he visited her regularly when she was dying in 1868. Not only had Rebecca cared for Brooks, but she had had the agony of nursing Arthur when he died and Henry when he came close to doing so. The family remembered her with a pension, but Brooks kept her most in mind. It was her name which he muttered in his final illness, in 1927.

His two older brothers treated young Brooks rather harshly, despite Abby's intercessions on his behalf. "I know he has no tact and is a tiresome boy," she said in 1858, when Brooks was ten, "but I wish they did not dislike him quite so much." Kind-hearted Henry did what he could in his little brother's behalf, but Charles called him "a disagreeable little sculpin." Henry rejoined that the lad was "really a first rate little fellow, apart from his questions, and we ought not to snub him so much." But the greatest problems occurred in London; the family watched Brooks apprehensively, fearful that he would prove uneducable. There Henry absorbed much of the questioning and argumentation which had become the lad's hallmark. Consequently, it was Henry who first detected Brooks' erratic brilliance.

Thanks to Henry's patience and his father's forebearance and attention, Brooks did reasonably well in an English school at Twickenham,
twelve miles from London. He succeeded in what interested him—history, literature, cricket, and boating. However, he acquired little of that
self-discipline advocated for so long by the family. During 1864, debate
began over when the young man should return to America and study
at Harvard. Writing from Quincy, John opposed such a step, perhaps
because he would be responsible for his difficult brother. Claiming
Harvard had done him no good, John was certain the school would be
even worse for his brother. Better that Brooks pursue a business career, he said. Charles Francis overruled this. Brooks would try Harvard, and John must find a boarding place "where manners are
preserved." Brooks had acquired in England a wholesome disposition
and his father feared what Harvard life might do to it.

So, Brooks went back to Massachusetts and Harvard in 1865. Charles
Francis immediately began sending him letters filled with much the same
admonition which John Quincy had sent to Charles Francis when he
was a student. Brooks was, after all, the only member of the fourth
generation who attended college when the parents were away from
Boston. The father reminded the son of what the Adams name had
meant to Harvard, of the need for hard work and righteous living, and
of the importance of exercise. Meanwhile, Brooks' letters were sternly
criticized for carelessness. This shortcoming made Charles Francis yearn
to be nearer his son. He urged Brooks to "think of me, as if ever at your
elbow advising you to all that is good, and detering you from all that
leads to evil." These were almost the words John Quincy used with his
son George. Was Brooks avoiding "tippling shops" and was he attending church regularly? When Brooks complained about the unreasona-
ble expectations the college seemed to put upon him because he was
an Adams, Charles Francis replied at once: indeed, the world had come
to expect from a youthful Adams accomplishments "attained by few at
the end of their lives." However, he, Charles Francis, had learned to
be more reasonable. He wished for no "great share of political distinction" for his sons, but only an honorable and upright station in life.

While still in London, Charles Francis glanced back across the years
to assure himself that he had tried to raise his large family without
being excessively severe. Never far from his mind were the expectations he knew had been heaped upon him and his brothers by John
Quincy, with such tragic results in two cases. The thought made Charles
Francis more conscious of the fact that English families remained to-

gether. Consequently, he was all the more eager to hasten back to Massachusetts where he yearned to draw his children around him. The Old House was a "centre" where he could gather "all my children who have been wandering about for so many years." Quincy was the place where the rising Adams generation must remain, he felt, to carry forward the town's association "with the fate and fortune of the family ever since the first settlement of the place."

There was a practical side, too, for Charles Francis had been examining the stiff elegance of many of the English country mansions he and Abby had visited. For him, the Old House now took on a special charm because one could feel it was never "too good to use." He was also homesick for the simpler life-style and its financial restraint. He complained loudly that the glamor required in diplomatic life had by 1864 cost him $15,000 above his salary. It encouraged him all the more to return to the plain residence in Mount Vernon Street and to the Quincy farm house into which John Adams had moved when he completed his duties as Minister to England.

While Charles Francis was in London, his family's history was always in his mind. He thought especially of John Adams, whose special qualities never failed to interest him. He also found strength by reviewing his father's career. "He was right when he judged that the true climax of the political health of our republic was reached during his administration," Charles Francis wrote in his diary. "What is it now and what is it coming to?" The Minister made pilgrimages to scenes associated with his parents' life. He visited the house where John Quincy had settled his family in 1815 when he was minister to England, and where they had briefly been so happy. "I remember nothing since that came back quite so sunny to my heart," Charles Francis observed, with the sober afterthought, "Life has rolled away since I was here. The future which then seemed illimitable is now little or nothing."

However, the most stirring moment of Charles Francis' London stay came when he entered the church where his parents had been married in the parish of All Hallows Barking. There he wept over the "trials and vicissitudes" which haunted his father and mother's union of half a century. He found himself as overcome by thoughts of his mother as John Quincy had been by memories of Abigail Adams. Louisa's tenderness and affection moved her only surviving son to tears, and he emerged from the ancient church "feeling as if I had bathed my face in the light of her blessed memory."

Yet by the end of his London stay, Charles Francis knew that his achievement there was great enough that he no longer need consider himself unproven or inadequate. Once more he shrank at the thought of reentering politics when he returned home. Instead, he sincerely hoped that the nation would forget him. "The country has had enough of me, and I have had enough of its service. . . . We are quits," he said, "and both of us content." A few days after writing this, Charles discovered to his dismay that the 1868 Democratic Convention would be meeting in New York City at the very time his boat was scheduled to arrive there. It looked, he felt, "as if I meant by it to invite them to consider me a candidate." He promptly selected a later date for sailing.

Charles Francis was in the midst of preparing to return to Massachusetts when he celebrated his sixtieth birthday. The occasion put him in a pious mood and he looked back in review on his past life. "Have I not had infinite reason to be grateful for the supervision that guided my steps back to the straight path when I was most led astray. And shall it not be so again?" he asked. "Let me go on strong in the faith," he prayed as he hoped for "cheerful confidence in the continuance of mercies far beyond my deserts." On 4 April 1868, when the great men of England gathered at a banquet to pay him homage and say farewell, Charles Francis thought of how "my wildest dreams of youth" had been far surpassed. "After this," he said, "the idea of returning to take a part in the forum of contention in America seems like madness." Instead, he departed from London determined to be secluded with his family at home. "How grateful I am to Divine Providence for thus showering its blessings upon my unworthy head." Looking back over his good fortune as a man whose active life was over, Charles Francis asked himself: "Was there ever any thing like it?"

Charles Francis was not the only person happy to leave London behind. Pleasing though life had been there, Abby joined her husband in pining for home. She had, of course, made good use of her naturally sociable nature, prodding her husband into making the American minister's house a festive background for politics. On 17 March 1864, a typical gathering included the Archbishop of Canterbury and his daughter, the Duke and Duchess of Argyll, Lord and Lady Lyndon, the Lord Chief Justice and Lady Turner, and the poet Robert Browning. Nonetheless, by 1866 Abby was so homesick that she became physically ill and seriously depressed. Charles Francis was not overly alarmed, however. "I fear she is getting rather too fat for her own comfort," he observed, while Abby sought various cures in Europe, complaining that she wanted her husband to send her bourbon whiskey. All she could find at hand was scotch and brandy, neither of which suited her taste.

When the long-sought recall home arrived from Secretary of State Seward, Abby's health improved at once. She began walking three miles daily, and, to her husband's delight, "she takes things easy, a strong proof of her increase of physical strength." The tributes to her from many of Britain's leading ladies also delighted Charles Francis. He praised her "excellent nature" and "her discretion through all the trials of which her situation has been full." Abby beside him had been of inestimable value, said he, concluding with the familiar refrain, "God be praised for all his mercies," a chant which bespoke the near-euphoric mood in which Charles Francis and Abby sailed for home. They needed all this spirit and more to withstand the disappointments and renewed family burdens which greeted them almost the very moment they reached New York on 7 July 1868.

It was raining in torrents when Charles Francis and his family arrived, and there was no one to meet them. "The whole affair then burst upon us with such a sense of the ridiculous that we burst out laughing in each other's faces," he recalled. However, "it was full of annoyance and vexation. What a reception for us after our long term of service abroad!" On the next day, they boarded a steamer for Boston, and passed, enroute, a deputation heading for New York, too late to greet the triumphant diplomat. At the Quincy station, a dismal rain continued to fall, and only a few friends were there to welcome them. Charles Francis consoled himself with the reminder, "I never fancied demonstrations."

The retired Minister began at once to renew the familiar domestic duties, while trying to cultivate the old-fashioned family spirit. It was spurred on to this end by the malignant talk which political foes circulated in hostile newspapers about the family's success and purported wealth. This was, for Charles Francis, simply another reason to preserve humility. Every Adams must be "prudent and cautious," particularly in Boston where critics of the family never stopped saying that John Quincy Adams had betrayed the Federalist Party. "Many have doubtless been watching for some evidence of exaltation which would prove that our heads were turned by prosperity," Charles Francis commented in amusement. "Thus far I flatter myself they have been disappointed." He predicted that the family's retirement "will deprive the subject of any further interest."

Charles Francis proceeded to concentrate his energies upon home and books. The Old House appeared unchanged on the outside, except that the trees had grown; but the interior was dilapidated, a startling contrast to their London surroundings. Abby was particularly distressed, finding home scenes "much worse than I expected." Even Mount Vernon Street was in disrepair. "Oh desolation of desolation," Abby moaned. "I broke down and was discouraged and for the first time wished I had never come home and would gladly go back." When her husband discovered the cost of improvements and rehabilitation demanded by Abby, he was almost as chagrined as she.

Since Abby would not consider staying in Quincy for the winter, Charles Francis had to forget his plan to build a year-round home on the Mount Wollaston land, and to put his money instead into reconditioning 57 Mount Vernon Street. At this, Abby's spirits revived. Now enjoying what was to be her last period of robust health, she energetically supervised the remodeling, so that by November 1868 the family was comfortably installed. Charles Francis prepared to settle down for good in his study, the room he loved and the largest in the house.

His retirement was premature, however, and he once more had to leave his books and the plans to edit John Quincy's diary. More summonses came to reenter public life. Most of them failed, especially those from liberal Republicans who were unhappy with the drift in American affairs and claimed that an Adams was the solution. Not all family members helped Charles Francis resist. In 1872, when dissident Republicans talked of nominating him for the presidency, his sons worked manfully to push their father into the campaign. He was quite unwilling and grumbled, "I begin to fear that something of a trial of my fortitude is impending." When his sons begged to be allowed to go to the Cincinnati Convention of Liberal Republicans in the spring of 1872 in order to advance his cause, Charles Francis was immovable. "I will not consent to be a party to such an auction." To his genuine delight, he escaped the call. His sons were disgusted.

All four of Charles Francis' sons tried again in 1876 amid more talk that the nation needed Charles Francis Adams as president. When he stood fast against this, there was then family pressure for the aging diplomat to consent to be governor of Massachusetts. Charles Francis tried to make it clear that he was not interested. His only duty must be "to purify the rotten morals" in the Republic's politics. And there was no chance of that, he was convinced. He was equally certain no Massachusetts party would ever nominate him for governor.

To his great discomfort, the Democrats did so. While his sons labored in his behalf, Charles Francis waited calmly for defeat. On election night, he was alone in his study on Mount Vernon Street reading and looked up only momentarily when news arrived that he had lost. When his youngest son, Brooks, later tried to involve Charles Francis in the 1876 controversy ensuing from the Hays-Tilden presidential election, his father said he had no heart for it. He did try to be good-natured about the pressures from his four sons, although he predicted, "Another year like this would kill me."

The summons he could not escape, however, was to return to diplomacy. In 1871 and 1872, Charles Francis went to Europe twice in connection with settling the maritime claims the United States made against Great Britain. His role in these meetings added even more to his luster as the nation's leading diplomat. Still, Charles Francis had wavered; he yearned for family haunts. "No, there is no path for me but the straight line of duty," he finally decided. After much agonizing in 1871, it was agreed that Abby must be spared an autumn ocean crossing, so she was left behind—a fearful mistake, as Charles Francis soon learned. Abby had pushed him gruffly away when they parted, although she quickly wrote to say that this was only to hide her tears and her yearning for fifty kisses. She wished to have him "by my side so longingly."

This was November. By mid-February 1872, Charles Francis was back in Massachusetts, summoned by reports from his children about Abby's increasingly distraught state. She had grown frantic with worry and loneliness, while friends were murmuring about the brutality of keeping the poor woman from her husband. When news came that Charles Francis was coming home, Abby's recovery was instantaneous. "It paid me amply for all my trouble to find her so relieved," Charles Francis said. In the spring when he returned to Geneva for renewed negotiations between the United States and Great Britain, he took Abby with him. Still, she complained: "I am wretched at being left and wretched at going." Leaving the family "is a trial almost beyond me," she confessed, using sentiment which sounded much like Charles Francis himself: "At my age, home and quiet is all I want."

From May to November 1872, the elder Adamses were in Europe, with affairs going well enough. Charles Francis used his shrewdness and firm tactfulness to extract major concessions from England. The United States was claiming that there should be compensation to Northern shipping merchants for damages by British-built Confederate raiders. Charles Francis secured over $15 million from the tribunal. As a result, he felt that the Geneva experience was "the pleasantest of my whole existence so far as external influences are considered." However, he went back to his family obligations determined never again to let the outside world intrude, "I have a right to rest," he sighed.

Beyond his career in international affairs, there had also been some other assurances for Charles Francis that he had not failed the Adams lineage. In 1869, he was offered the presidency of Harvard College, a position proferred with unanimous enthusiasm. He declined with his own special joy, noting that this recognition was new, for Harvard had never been eager to show "good will" to the Adams family. Honorary degrees from their alma mater had been awarded comparatively late and grudgingly to John and John Quincy. "Yet I doubt whether it has a more brilliant line of four generations in its list," said Charles Francis. His decline of the honor offered him gave him a rare moment of gleeful pride: "The account is now squared," be noted.

A renewal of family recognition which he did accept was the presidency of the American Academy of Arts and Sciences in 1873, a position his father and his grandfather had held in their time. "I like to follow in their footsteps in all honorable enterprise," Charles Francis observed, with considerable understatement. After his triumphs in London and elsewhere, he was aware that he had performed his duties well.

❧ 12 ❧

Struggles

At no point during the successes of London and Geneva did Charles Francis forget that his children were, in turn, facing their own duties. He and Abby were acutely mindful that their eldest children's personalities and careers were maturing back in America while their parents were abroad. Each child—John, Charles, and Sister Lou—presented a special set of concerns.

John and his bride Fanny had been placed in charge of family affairs during Charles Francis' absence, a reasonable step since John's behavior after his marriage had begun to suggest that he might eventually be worthy of his name. Yet with both their eldest son and the family estate in precarious circumstances, the parents in London sent back no end of admonitions, embroidered in affection. The effect upon John's stability and assurance was damaging, for he was dependent upon his parents; he had been responsible for managing all the family's economic interests but only under his father's watchful eye. Equally troubling for the young man was his name. To serve as another John Quincy Adams was a challenge no one could easily meet. It was, John said, "a grievous heavy name to bear." He saw himself a mere errand boy for his father. "I should be grateful once (but I know it is useless) if I might in any one thing be considered as an individual and not as a Son or Grandson."

In an attempt to establish a reputation of his own, John accepted a largely honorific position on the staff of Massachusetts Governor John A. Andrew which allowed him to run errands for the Governor and to appear with him on public occasions. Unfortunately, not even this role was satisfactory, as John quickly discovered when he was called upon to speak while he and Governor Andrew were in Amherst. To his dismay, John heard himself introduced to the crowd as a descendant of three men who had either been president or ought to have been and as a prospective president himself. "What can a man say when he is thus absolutely beaten over the head with ancestry?"

Clearly, John had mixed feelings. He was, he acknowledged, proud enough of his lineage to take pains that portraits of all his great forebears were kept in a safe place, and he perused his grandfather's diary with interest. But he also announced his determination to discourage "the vile family habit of preserving letters." He burned his own correspondence as well as his diary, claiming "the less weight you carry in life the better." It was equally important, John told his father in 1864, for families to recognize that the talent of one generation did not necessarily pass to the next. To perpetuate a name was a natural hunger, John said, but he urged Charles Francis to agree that a family, when its vigor was gone, should not pretend otherwise. For such diminished families, said John, "let them go just as soon as they can for only so long will it be well to have them remain."

The letters which Charles Francis and John exchanged weekly during the former's London sojourn were dominated by the son's self-doubt and the encouragement, often mixed with impatience, which the father offered. The relationship was never easy, for Charles Francis was fearful that he had left the family property in the hands of a child who would turn out to be careless and indifferent. However, considering the enormous gap in spirit and talent between these two, they got on remarkably well. John was paid $750 a quarter to collect rents, record income from investments, pay stipends to relatives, and watch over the land. Charles Francis had no alternative but to leave these duties, which he himself had performed for his father, with John. His second son, Charles, who might have helped in this task, was determined to be a soldier and, according to his father, "he seems to have an idea that management of another person's property is derogatory to the dignity of an independent citizen."

Occasionally, John wrote that he yearned to be off to battle and that he resented family duty, but most of his limited energy as a correspondent was spent on financial matters or politics. His father, meanwhile, looked eagerly for signs that John could become the family leader, one time going so far as to assure him that he would continue to manage the family when the elder Adamses returned. Such unaccustomed words of confidence made John blunder. He suddenly spent considerable money on the Quincy farms and planned more expenditure in converting Charles Francis' former residence, up the hill from the Old House, into a comfortable year-round domicile. These projects were mistakes, for when Charles Francis learned of them, his confidence sank. John was ordered to halt his plans, and the father decided to resume family management after leaving England. Charles Francis continued to be kind, and spoke of his satisfaction with John's stewardship, but John knew quite well what had happened. He hastened to unload his responsibilities onto Charles when the war ended. "I have turned all the work over to Charles . . . and I can't say I bother myself just now at all," John wrote his father.

Then, either in desperation or for consolation and respite, John turned to politics. Even here, however, he rarely seemed serious. It was as if he were on a lark. The public endeavors of the second John Quincy Adams bore little resemblance to the political style of the first. Occasionally, he embraced an unpopular cause, as when he enlisted in the Democratic Party to oppose the harsh punishment dealt to the defeated southern states by the Republicans. The policy of reprisal was generally applauded in Massachusetts, so that once again an Adams took the minority view.

In 1866, John was a Republican representative in the lower house of the Massachusetts legislature; in 1869 and 1870 he sat with the Democrats, a bedraggled party which offered him as candidate for governor four times between 1867 and 1870, an experience John enjoyed because he knew he could not win. A splinter national faction of the Democrats nominated him for vice president in 1872 in another doomed cause. The ticket drew 30,000 votes. This kind of gallant excitement seemed to appeal to John as much as he shrank from the humdrum, unpleasant grind of political labor which would come with victory. John wanted to lose.

He might have turned out differently had the family let him alone, but misgivings about John's enterprises and his judgment made Charles Francis and others heap counsel and criticism upon him. Thus, trading family business for politics brought him no escape, for an Adams in public life was most assuredly on a family mission, and John found again that his work was not satisfactory. Charles Francis rebuked him for making an "unwise preliminary challenge of all the popular fancies of the day." He predicted, too, that John would soon lose interest in politics. John's "inertia and self-indulgence" would overcome his "sense of duty," Charles Francis said. John himself confessed to his father that he considered his legislative seat "a mere temporary frolic," and that "politics except just at election time had not much attraction for so lazy a devil as I am."

Soon afterwards, John assumed a rather cynical pose, speaking of politics as ego gratification, of genuine duty as being to one's wife and children, of the fickle nature of public appreciation, and of his own retiring style—"I don't want to make acquaintance nor do I care to be known." He stopped talking of working for causes, shrugging off the attention he had received as merely a "compliment to the name I inherit." John's most consistent note was a wish "to be let alone." The refrain rang with an immature and timid sound, and it struck Charles Francis that this was John's way of covering a fear of failure. "You really have more ambition than you imagine," the father wrote the son. "Like a good many other people you do not choose to probe that part of your character too deeply, for fear that it might create hopes destined to disappointment." To offset the younger man's mood, the elder Adams cited passages from family scripture about duty, diligence, and determination.

Eventually, John grew almost angry when his relatives implied that if he could only taste a little success, he would then rush happily ahead into the world of affairs. To underscore his search for a separate path, John wrote to Charles Francis, "I will not be a public man nor lose my life in a vain struggle for nothing. I am determined to live a life to suit myself. . . . Secluded, quiet . . . somewhat bucolic . . . with a little law and much ease and luxury. There you have it. . . . I want to be left alone." The marvelous irony in this confrontation between father and son apparently dawned upon neither, for John was confessing the very desires which for many years Charles Francis had privately hankered after. However, now Charles Francis was able to assure John that ambition, duty, and public attainment could be worthily combined. Had not the father left behind his own desires for privacy and found public success gratifying?

John remained unimpressed. He continued to let his name be used politically for a time, not having the courage to withdraw completely. He still enjoyed notoriety without responsibility. However, public excitement began disagreeing with his digestion, making him complain of biliousness. His brother Charles observed, "I don't understand John— he seems to grow old and change so fast . . . he is as nervous and anxious and tied by the leg as an old woman." When efforts were made to bring novelty into John's life by having him visit his parents before they returned from London, he resisted. His wife, children, farm, sailing, and role as man-about-Quincy were all he wished. To Charles Francis' continued prodding, John replied: "I am afraid you like most parents overestimate your children. I am no consequence here under Heaven except to my home."

Only rarely did John display the more complicated struggle within himself. Although he kept repeating, "I am nobody and I know it in my heart and I am sure to be found out," he occasionally confessed how it pained him when someone would tell him of having known his grandfather and of anticipating John's similar triumphs. The moments when he recalled old John Quincy Adams were, young John conceded, "My abiding torment." At such times he took comfort by sailing or by planting trees on the Mount Wollaston farm. There he began building a palatial home, with help from his wealthy wife, Fanny.

John and Fanny's new house, completed in 1872, put them far enough away from the family mansion that John's parents saw less and less of them. The move, said Charles Francis, "is one of my griefs," while John kept stressing that his "natural sluggishness" was as great as ever and that he wished only to be "part farmer; part loafer; part real estate agent; and part country lawyer." John's few surviving letters reveal a man of forty who doubted himself profoundly, insisting that what he undertook beyond his home was of trivial importance. John kept maintaining that "any man who really sets his heart upon anything like political success here must be a knave or a fool." He even quit tinkering with politics as a game.

Then came an event which blighted John's most notable quality, his lively good humor. His brothers and sisters always readily conceded that he had a charming nature, which Adamses traditionally lacked. "He is the only one of the family," said Henry, "who can make one laugh when one's ship is sinking." In April 1876, however, diphtheria carried off two children, a son and a daughter, with such cruel abruptness that the entire family was left in shock. Charles Francis and Abby were doubly grieved by fresh memories of their loss of Arthur. John, who admittedly "lived only" for his children, suffered in a way which, his father acknowledged, "moved me to tears the more that I knew he had set all his happiness" on the children—"the rest of life will be to him a blank."

It was a terrible ordeal, momentarily bringing all the family together, as John's brothers carried the children to their graves, the coffins covered with rosebuds. At the cemetery Charles Francis was so overcome "that I prayed to be taken away. . . . Why was it not I to be taken before the clouds came down over me?" The old gentleman rallied, however, determined to try to rescue the grieving John. He and Fanny sat for hours beside the fresh mounds of earth near the Mount Wollaston farm, and Abby reported that "John looks like an old man, the lines so deep and the expression so unhappy." Fanny was also changed, "but differently, so sad and quiet. Oh dear, how hard life is."

The elder Adamses coaxed the grieving parents into accompanying them to Philadelphia for the 1876 Exposition in hope this might restore their interest in life. They had little success, for John was going through the same prolonged anguish his father had experienced when Arthur died. In a rare surviving letter, one at Christmas 1876, John confessed his distaste for life: "I suppose the time will come if I live long enough when some tolerable patience may replace the intolerable misery." He did survive another eighteen years, but he never quite recovered himself. He remained withdrawn and more willing to depend upon others, particularly his dynamic brother, Charles. There was one cheering development. The dead children were, in a way, replaced by another son and daughter, Arthur born in 1877, and Abigail, in 1879.

Charles Francis fretted that John's Mount Wollaston home was too solitary. He coaxed the recluse to come out and accept election to the Harvard Corporation, a role which eventually grasped John's imagination sufficiently for him to serve as treasurer. It had required some familiar pep talk from the father. He urged the forty-four-year-old John to bestir himself in hope that "he could develop the talent which I knew him to possess." The step led nowhere, really, for John still spent most of his time at home, sailing, a quiet figure who carefully destroyed all of his papers. On Sundays Charles Francis would often pay a visit to Mount Wollaston, and usually found his son and grandchildren out on the water, a situation he called "a bad omen for the name. I can only grieve and say nothing."

When he thought more about it, however, Charles Francis concluded that there was a shred of comfort in the habits of his children. At least, he assured himself, "they have thus far avoided the vices of those who fell from grace." In their day, John and John Quincy Adams had not been able to say this, a fact in which Charles Francis took considerable comfort, sorely needed as he watched his second son and namesake, Charles, develop in ways as disturbing as John's.

Soon after the elder Adamses had settled down in London, Charles began writing of his determination to enter the Civil War as a soldier. Neither Charles Francis nor Abby offered encouragement. "I fear that he has sealed his own fate," said the father. "If he escapes the dangers of the war itself, he will scarcely recover from the dislocation of mental or moral habits consequent upon such a change of life at his years." The young officer never forgot nor forgave these cold parental objections to military service, not realizing how much disdain Charles Francis had for the folly and wastefulness of war. While Charles waited vainly for a wholehearted blessing, his father said merely that he must do as he chose. Charles Francis privately shook his head—"none of his predecessors have been soldiers, why should he?"

Charles argued that an Adams above all others should take up arms against slavery, the foe John Quincy had fought so tenaciously. However, he also had a more personal reason for entering military service. He wished to escape from Boston where he saw himself a "failure as a lawyer," pitifully sinking "into a real estate agent." Equally irksome to Charles, who had not outgrown his youthful need to match or excel his brothers, was that as long as he remained in the Boston area while his parents were in London, he must defer to John's leadership. It galled him bitterly to acknowledge in his diary that "my brother has become the head of my father's house and I may remain in it while I submit myself to his will and that of his wife."

So goaded, Charles went off to war at the close of 1861 as an officer in the First Regiment of Massachusetts Cavalry Volunteers. While his service, notably in South Carolina and Northern Virginia, was mostly without excitement, he performed courageously whenever he saw combat. He was at Gettysburg in 1863. Later, Charles entered General George Meade's headquarters staff, where he could observe General Ulysses S. Grant's style during the long Federal campaign to take Richmond. Midway in 1864, Charles became a lieutenant colonel and was transferred to a regiment of black troops, the Fifth Massachusetts Cavalry. Six months later, he was a colonel and commanded these troops, for whom he developed no respect whatsoever. At their head he rode into the fallen Confederate capital in April 1865, the first of the Yankee soldiers to arrive on the heels of President Jefferson Davis' departure.

Throughout these four years, Charles' primary qualities of impatience, bad temper, and abrasiveness proved almost as serious an enemy to his military career as the Confederate army. His digestive ills continually disabled him and left as much permanent harm as many

battle wounds might have done. For both advancement in rank and for
health he often depended upon John's help—"My damned bowels
played Hell with me and will again if I don't get your pills," he wrote
to his elder brother in 1862. When John paid him a visit, he discovered
what military life was doing to an Adams. "I find Charles looking so
rough, rugged and sun-burned that I did not recognize him at all,"
John reported to Charles Francis. Nevertheless, the headiness of command and of following his own track made Charles exultant. He told
his brother that even when he was tired, rain-soaked, shelterless, and
sleeping at his horse's heels, he had only to think of the family office
in Boston to make his army life seem "paradise." "[N]ature meant me
for a Bohemian," he wrote to Henry.

By 1863, Charles began planning a professional military career, from
which he intended to emerge as a person in his own right. After the
army, he would pursue a combination of literature and politics, hoping
that his military attainments might help him command attention. There
would be many advantages from service as a soldier, he believed. His
family, however, saw the military influence as less beneficial. When
Charles visited his parents in England during a leave in 1864, he
shocked them with his rough ways, and he, in turn, drew back from
society. "Ye Gods, how tired and restless I am here in London. How
much happier I was in the Army," he said in his diary. Charles Francis
spoke sorrowfully of the deterioration he perceived in his namesake.
The army had increased "the defects of temper and manners which
always stood in the way of his advancement at home."

Had not Colonel Adams suffered a collapse of his digestive system,
he might have kept his resolve to remain a soldier. He certainly dreaded
the approach of civilian life which for him held no prospect save working as a family lackey. Hoping to escape this, Charles was preparing to
take his regiment to duty in Texas when illness forced him back to
Massachusetts, virtually helpless. However, in returning to civilian life,
Charles now had more than the faithful John to aid him. He had recently fallen in love and was engaged to marry. This prospect was a
new spur and he determined to escape the shadow of dependence upon
his parents.

Mary Ogden of Newport—Minnie—was the daughter of Edward Ogden and Caroline Hone Callender. She had been born in New York
City in 1843, making her eight years younger than Charles. Her Massachusetts connections were few, which pleased Charles. In fact, Minnie
took Charles' fancy when he first met her while visiting in Newport

during a military leave. "She runs in my head infernally! she persuades [me] that my sensibilities were not played out." Minnie was a woman of great personal beauty and quiet charm whose delicacy appealed to her suitor's latent fastidiousness. Soon after Charles fell in love with Minnie in 1864, John took him to a gentlemen's dinner where the soldier reported himself disgusted with the "profusion of obscenity. I don't like it!" he said. "Dirty stories disgust me and I am ashamed that waiters should overhear them." He proved to be an awkward lover, scolding himself for saying all the wrong things, but the Ogdens liked him—he replaced their only son who had been killed in the war.

Momentarily Charles built his hope for independence upon plans for taking Minnie to New Orleans where he expected to manage the considerable Ogden investments there. But after the war these were found to be worthless, and Charles' betrothed brought nothing to speed his emancipation. The marriage took place nevertheless, to the delight of Charles' parents in England, who hoped the step might calm and refine their second son, as it had their first. There was "much sterling merit" in Charles, said his father, who, in a moment of unusual frankness, told his son that when he heard of his decision to marry, he had wept— "but let that pass. Nobody saw the Minister and he will tell nobody out of the family." The groom was promised $3000 a year support as soon as he was settled; and, Charles Francis added, "I wish I could do more." He was very careful to inform Charles that it was no less than the amount John received.

The young Charles Adams was by no means so bent upon independence that he refused his allowance. In fact, he pressed to know just when payments would begin, and Charles Francis indulged him by arranging to start the support even before the wedding. After an autumn ceremony in 1865, the couple departed for a six-month tour of Europe which Charles announced he deserved. This trip marked the beginnings of a lifetime of self-indulgence. In England, the bridal pair stopped to see the elder Adamses who were as pleased with Minnie as they were chagrined by their son's continuing rude and opinionated behavior. Charles Francis conceded to John that he could easily see "how much you and your wife had to put up with during his worse period." Fortunately, said the elder Adams, Minnie "has character enough to understand him."

The second Charles Francis Adams remained a contrast to his brother John. In every quality—personality, aspiration, manner of life—they were wholly different, as epitomized in Charles' eagerness to leave a

huge written record about himself and John's determination to obliterate all traces. Charles left letters, diaries, and autobiographical material,
although he carefully sifted through this legacy to see that only what
he wished survived. Not even the greatest care, however, could disguise
that as he grew into middle life, Charles was largely unchanged from
his younger days—a man determined to succeed in areas beyond the
traditional Adams role of statesmanship.

Upon returning to Massachusetts after their honeymoon, Charles and
Minnie looked after the family mansions in Mount Vernon Street and
Quincy until the family in London was back. John had moved out to
seek his own place. At first Charles Francis was cheered, if a bit startled, by the younger Charles' flattering remarks about the Old House—
"the most charming old place of residence I ever yet lived in and spoils
me for more modern and convenient houses. I should like nothing
better than to own the old place." But Charles was Charles, and soon
proposed that his father spend $25,000 to improve the building and
grounds, and advised his parents that their Boston residence was now
hopelessly beneath them and should be sold at once.

This clamor was an early symptom of Charles' determination to ride
the materialist whirlwind he gleefully foresaw in the nation. Said Charles
Francis about his son's ideas: they "simply frighten me." To rush into
new debts, warned the cautious elder, "is simply madness." Instead, he
arranged for his son to be offered a splendid position in the law firm
of R. H. Dana, Jr., an opportunity Charles declined. Were he to be
reduced to a lawyer's life, he preferred it on his own and then only
because of "stern necessity."

However, it was not the law but a determination to be modern which
occupied Charles after his military service. He chided his father for
taking a pessimistic view; Charles preferred to preach of man's capacity
for unlimited improvement through the railroad, the telegraph, and
the press. The American people, he assured an astonished Charles
Francis, "are chock full of moral vitality," and only "wider education"
was needed to usher in "a grand new age." Charles Francis was not
impressed when his namesake told him how much he regretted "that
you should have allowed yourself to drop out of sympathy with the
passing age." However, he knew when to turn a child loose, so he sent
Charles on his way, warning him that he would make many enemies in
the business world because of his curt manner.

Charles paid little heed, for he looked to a different member of the family for help and support. Abandoning his father and John as sadly caught in the quiet ways of the past, Charles turned to his brother Henry with whom he had carried on an intimate correspondence during the previous ten years. Charles proposed to lead America into the new industrial age by writing articles on social and economic issues. In his determination to be different, Charles failed to note that this path into public life had been followed by his father, grandfather, and great-grandfather.

Still eager to help this child, Charles Francis warned him that such an enterprise could be frivolous. Essay writing, reviews, and speeches were "as evanescent as the perfume of the rose," said the elder Adams. His son shook off the advice: "My pen is the one weapon I can use effectively." This was braver than it sounded, for Charles in his heart doubted himself and was impatient for Henry's return, since the younger brother already was an experienced essayist whose reports from London during the war had appeared in Boston and New York newspapers.

"Henry is what I want." Charles tried to shout across the Atlantic. "He ought to be here and lend me his head and hand—if he were I should have no fear and very little doubt of the result. As it is, I fear a *fiasco.*" There was, briefly, an intellectual partnership between the brothers, Henry sharing for a time Charles' interest in reforming the railroads, something the elder brother decided was the greatest issue of the future and thus should be his province. The pair wrote several essays after 1867 in which they denounced corruption and called for public action. Charles stressed that the states must create agencies for overseeing railway development and management.

Soon after Henry was back in America, he and Charles drifted apart, mostly because of differing personalities and tastes. In 1869, Charles felt confident enough to stand on his own without Henry's intellectual stimulation. By then, he had been named a member of a new body he had helped create, the Massachusetts Railroad Commission. He worked vigorously for ten years on this assignment, striving especially to organize the facts about railway operation. For another five years, rail transportation was his primary activity. He served as an arbitrator for trunk-line railroad companies that were seeking to govern themselves by association. Charles also gave a large share of his enormous energy to investments and to historical writing. Meanwhile, his domineering ways overpowered his brother John who was nominally in charge of family

business once more when Charles Francis Adams became incapacitated
by a premature senility which destroyed his mind after 1880. Charles
could not be a collaborator to this task. He had to be in command.

The difference between Charles and John was evident in the way
they established themselves as residents of Quincy. John chose to build
apart from the family, selecting the shoreline remoteness of the venerable Mount Wollaston farm Abigail had inherited three-quarters of a
century before. Also on family land, but at the top of the hill across
from the Old House, Charles erected a palatial home. From this site he
looked down on everyone, including his father, who remained content
in the ancient mansion below. His perch on a hilltop symbolized his
approach to life, both within the family and beyond it. John's retreat
to Mount Wollaston was equally representative.

Charles, of course, gave up tranquillity in order to make use of his
aggressive nature. There were times as his aspirations faltered that he,
too, claimed to have preferred all along the setting his forebears had
also professed to crave—the quiet of the study and the proximity of
books, quill, and foolscap. But the temptation of public opportunity
was just as irresistible for him as for many earlier Adamses. Charles
even went to Vienna in 1873 to attend a trade conference for Massachusetts, an extended excursion which confirmed him in his love of
Europe. Soon afterwards, in the bitter cold of February 1875, he delivered the Lowell Lectures in Boston, a series of addresses on the railway system. His father observed that Charles seemed to have some capacity to hold an audience "even on dry topics," for the family naturally
attended his lectures.

Like his father, the younger Charles managed to endure the depression, the nervous tension, and the digestive problems which came with
trying to be known and affluent. Charles' parents saw the strain, worrying as much at this boy's overwork as they did at John's passiveness.
The grim, driven air about Charles could not be put aside even when
he visited his parents, which was frequently. "I am afraid his mode of
life is injurious to him," Charles Francis said. "He needs repose and
moderation." Meanwhile, as late as 1877 Charles claimed that his father
still treated him "as if I was about 10 years old." The elder certainly
had not disguised his apprehension about his namesake's reckless
spending. The younger man occasionally acknowledged his own misgivings, as when he wrote in his diary in 1874, "I am always getting
myself into new scrapes. Here I stop!—Cash or nothing." It was a pledge
he invariably broke.

Charles and Minnie had their first child in 1867, but to the new father's dismay this baby was a girl. The infant, named after her mother, was born in the Old House. Charles was so graceless upon learning that his child was a girl that he announced he and Minnie really had not wished to have a baby at this time. Thereafter, two more daughters arrived, Charles greeting each with louder distress, although with the third girl's appearance the new father ruefully commended himself. "I didn't swear and shout, but my disappointment was bitter." Soon Minnie was pregnant again. Abby was alarmed, while Charles was convinced that once more the male "heir" would fail to appear. To everyone's astonishment, in July 1875 Minnie produced not one, but two sons, one weighing nearly seven pounds, the other almost eight. The birth occurred a month early. "My sense of relief is inexpressible," the father reported, after which he memorialized the appearance of the twins by asserting that while there were bodies for two, there were probably brains enough only for one.

Charles' remark made a vivid impression upon Brooks, who had learned early how sharp this brother's tongue could be. However, in striking off on his own route in life, Brooks developed a personality even more bruising than Charles'. Brooks' troubled career had begun uneasily. It had seemed for a time in 1865 and 1866 that he would not meet Harvard's requirements for admission, despite careful tutoring. Then, in October 1866, the family had word that Brooks was permitted to matriculate. "A rejection would have mortified him for life," mused Charles Francis, who was still in London.

Thereafter, reports to Brooks' parents spoke of modest scholarly success along with remarkable social triumph. "He is tremendous on society and young women and is learning to dance with mighty efforts," said the younger Charles of his twenty-year-old brother. Brooks told his father that he felt quite at home in Boston and was invited to dine in the finest residences. He talked of staying forever around a college so that he might think about medieval times where he seemed at home. "A gothic cathedral, a ruined castle," these were his spiritual haunts, he said.

After graduating from Harvard in 1870, Brooks followed his forebears' path by claiming the law as a profession and by commenting on the world's ills as an essayist. He was, at the time, a genial person, but the flair for argument which later became his trademark was already evident. Briefly Brooks replaced Henry as cohort with brother Charles— a truce these two strong temperaments could not keep for long. Charles

Francis wrote of Brooks in 1877: "He is thoughtful and kind hearted though singularly brusque in his manner. What explanation can be made of individual character in families?" He contrasted Brooks with Henry, whom he designated "gentle and easy." Who could say, marveled Charles Francis, "which of the two was the most true to his family or his name."

Although he moved out of Mount Vernon Street and the Old House, the grown-up Brooks looked in upon his parents frequently and took many meals with them. In November 1877, however, he came home for more urgent reasons. He spent a night in the Old House in order to be comforted by his father over a rejection he, at twenty-nine, found insufferable. He had just lost his first race for office—a seat in the state legislature—by only two votes. The loss could be blamed upon his uncles, Chardon and Shepherd Brooks, who had voted against him. "He was a good deal excited," Charles Francis observed with massive understatement. "I gave him my views . . . but I doubt his disposition to pursue them."

For a time Brooks remained active in politics, seeming now to be reform-minded. By 1880, this had bred a profound pessimism and a propensity to be harshly critical. In one of his rare surviving letters from that era, Brooks told his intimate friend, Henry Cabot Lodge: "we may talk and we may talk but the world will go round and if we are bound for the deuce, to the deuce we shall certainly go." To this nascent determinism, he added a misanthropic note, for which he later became notorious. "Rich or poor, men are just the same," he advised Lodge. "They will rob and oppress their neighbors when they can. It all boils down in the end to a question of who has the power." In the early 1880s, Brooks' viewpoint was hardly softened by a painful, if mysterious leg injury which left him glumly hobbling around Europe and America for several years. In 1881, he wrote to Lodge from Paris, "I rather wonder at myself when I think how very keenly I enjoyed the life ten years ago. After all ten years makes a difference in the way a man looks at the world."

However, the largest influence on the development of Brooks Adams must have been a tragic romance. Only the most general information survives about this liaison, but it appears that Brooks fell deeply in love with a woman of Canadian background named Heloise, a teacher of French in the Boston school system. Little more about her is known; she was acceptable to Brooks' parents and she and Brooks were very close, spending summers in Europe together after delicately crossing on different boats. She was a joy to the youngest Adams, helping him to a better command of French by first reading drama in the language aloud with him, and then taking him to theaters in Paris to hear these plays. Brooks wanted desperately to marry her, but Heloise refused. Perhaps she took the view that the social gulf between her family and the Adamses was too great. Marriage would embarrass all parties, she believed, and force Brooks into being a social outcast. No matter how Brooks tried, he could not alter her view. Some members of his family believed Heloise was very foolish to hold to this purpose. And still the pair remained devoted, so that Brooks forever afterwards claimed that a private pledge of understanding was as moral a bond between two people as the legal tie of marriage. Heloise died of leukemia, probably around 1885. The legend that she is buried in the cemetery in Milton, a village near Quincy, cannot be confirmed. Brooks claimed she was interred there, but no record of the grave exists. In his old age, Brooks insisted that he wished to be placed next to her, to the discomfiture of his nephews who feared for the family's reputation.

With the years, Brooks Adams grew deeply troubled and his writing developed the theme that forces superior to mankind propel the course of history. Brooks' personal style became abrasive and he argued anything endlessly, cluttering his language with profanity even in the most polite settings. Few persons, and least of all his brother Charles, wanted to hear Brooks say that men and society were pawns pushed by forces as relentless as scientific laws. Charles, in fact, came to dread seeing his brother. An entry in his diary in 1886 illustrates this—"Brooks here, parading more Brooks than would seem possible."

In 1887, Brooks published his first book, *The Emancipation of Massachusetts,* perhaps the most telling of his many efforts to demonstrate the puniness of man and society. The way he used the Puritan era for his subject was sure to offend many readers, especially when he depicted Massachusetts under John Winthrop and Cotton Mather as a place where the clergy were tyrants who hindered reform by playing upon injustice and fear. He himself described his purpose with what may have been a sarcastic gleam: "I am, for my sins, trying to write something about this state." The point of this book, however, though it was lost in the sensational nature of the subject, was Brooks' determinist view of mankind. The work was intended, he insisted, to show humanity as controlled by habit and instinct in a mechanistic universe. Not even his friends, however, or brother Henry saw fully what he was trying to do. Brooks reacted badly as reviewers disapproved what was called his distorted treatment of society and thought in provincial Massachusetts. He raged that no one recognized his purpose—an inquiry into the action of the human mind in civilization. Massachusetts was only a case study. Brooks' career as social critic was thus spawned of complex emotional and intellectual impulses.

No matter what its intent, *The Emancipation of Massachusetts* outraged the community the fear of whose disdain had kept Heloise from marrying Brooks, and no one more so than brother Charles who sneered, "poor stuff!" Once more, an Adams was estranged from the prevailing sentiment of Massachusetts. Brooks told a friend, "I'm a crank; very few humans can endure to have me near them, but I like to be with you, and I suppose I like to be with those who are sympathetic the more since they are so few."

Beyond any family member since John Adams, Brooks became estranged from the larger world. His private disappointments and his brilliant, if erratic, capacities of analysis set him apart. In consequence, he turned for comfort to the traditions and legacy of his family—there, at least, Heloise had been welcome. After her death, he spent most of his time with his mother and with his sister Mary, whose life was a calm contrast to the stormy, frustrated, and grief-stricken existence of her sister and brothers. She was the one thoroughly conventional member of this fourth generation and seems to exemplify what every admirable daughter in well-to-do circles of Victorian America was supposed to be and do. She entertained nicely, accepted polite visits from several gentlemen, and by 1875, at thirty, seemed destined for a comfortable, if colorless, maidenhood. In 1871, Abby had grumbled worriedly to her husband, "I wish Mary would get engaged, that would settle all questions, but there is little hope of that at present." Soon Abby was reduced to describing her daughter as "a wonderful woman" who "takes life easy."

Then in June 1876, a startled Charles Francis recorded in his diary that "young Dr. Quincy" had called upon Mary. Six months later, on 2 January 1877, there was another bulletin—"Mary this evening communicated to me the fact that she had accepted an offer of marriage made by Dr. Henry Quincy, the son of my friend Edmund." Charles seemed pleased, especially that the pair shared an ancestor eight generations back, an ancient Edmund Quincy. Mary and Henry's brothers and friends, however, were surprised that these two seemingly settled single persons should have turned to matrimony. One person who extended astonished congratulations to Dr. Quincy covered his lapse by assuring the prospective groom that he had "drawn the first prize in the Centennial Matrimonial Lottery." Mary's brother-in-law, Charles Kuhn, told her how startled he was by news of her engagement, but that he was reassured by recalling that she was a "young woman of calmness and perspicacity, who had waited some time before making a choice." John Hay, a close friend of Henry Adams, told Henry Quincy that "you have simply *everything* that is lovely and of good report in a wife."

Henry Parker Quincy was not obliged for financial security to practice medicine, thanks to family substance. He seemed the perfect male copy of Mary, tender, pleasant, and rather colorless compared to the exciting figures in the Quincy and Adams circles. His father, Edmund Quincy, had taken extreme political positions before the Civil War, and was now still notorious for what Charles Francis uneasily called extraordinary religious opinions. It was thus somewhat in character when Edmund, never one to make others feel easy, startled everyone by dying shortly before the wedding. Family debate concluded that the event had best proceed as scheduled. So, on 20 June 1877, Mary Adams married Henry Quincy in the Long Room of the Old House before sixty guests, in an environment of rubber plants, orange trees, and passion flowers which Abby had insisted on for the occasion. An aging Charles Francis had very mixed emotions. "My daughter is no longer what she has been for so many years," he said, remembering her role as "the joy of the household." Still, her marriage was a blessing, he thought. "For to have left her alone when her parents must go would have been only a source of anxiety for us."

Saved from the perils of nineteenth-century spinsterhood, Mary Adams Quincy had a long and uneventful life. She and her husband spent summers in Dedham, a village easily reached by carriage from Quincy, while for several years the pair lived in the winter with her parents at 57 Mount Vernon Street. Eventually they moved nearby to 79 Mount Vernon and, in 1886, to Clarendon Street, and finally to Beacon Street near Massachusetts Avenue. Mary remained close to her parents in their last years, usually accompanying her mother on the daily rides that were so important to Abby. Abby was elated when, in 1885 Mary gave birth to a daughter, Dorothy, and, in 1888, to another girl, named Elinor. The diary fragment Mary left for these years shows a lady and matron who accepted the social mores of the 1880s without question and who moved easily through the polite visits and carriage excursions of each day. Her troubles were mostly difficulties with servants and her pleasures were in horses and dogs. No one in the fourth generation of Adamses became a defender of convention except Mary, showing this particularly in her impatience with the extraordinary views and behavior of her brothers. As an unexceptional person, she was an unusual Adams.

Mary's marriage was only one in many developments that showed Charles Francis and Abby how their offspring were on diverging paths. By 1870 the family ties, seemingly so secure for previous generations, had begun to loosen. Not that there was lack of attentiveness from the children to Abby and Charles Francis. They were dutiful, on the whole. It was more that a spirit, a shared outlook, was missing from this generation as it reached middle life. The contrast in temperament among the youngsters who had once filled 57 Mount Vernon Street had become the rule rather than exception. Charles Francis observed in 1871, "The whole thing convinces me how vain it is to anticipate any durability in families in this country. Mine has lasted three generations and perhaps may go on to the fourth, but that will I suspect be the end."

This observation was made after a terrible loss afflicted the Adamses, a catastrophe which Charles Francis contended was inevitable when a child avoided the duties and proper behavior which he believed were part of an admirable life. Not surprisingly, it was Sister Lou whose route in life led to this disaster. She had come back to America in 1860, determined to show that she was unspoiled by European indulgences. With her husband Charles Kuhn, she lived in Newport, where they found no contentment; Louisa's unhappy moods and mysterious illnesses kept Kuhn away most of the time. Sister Lou did try army hospital nursing briefly, and she sought diversion in passionate worries over her brother Charles' military danger, but these exposures to the male world out of her reach seemed to upset her.

Once, earlier, Sister Lou had unburdened herself to her father, who was then far away in London. She had found, she wrote, no place where she felt her efforts in life could have meaningful outlet. She acknowledged that she was wholly undisciplined. Hers was a hard fate, Sister Lou concluded, to "have been sent into this world with certain characteristics in which my will has no chance and that every circumstance of my life appears calculated to punish me for being born with any heart, impulses or feeling." This was 1862, by which time she realized how much emotional satisfaction she missed in her marriage. "My poor little dog who jumps up on me and kisses my hands when I speak to him gives me the only real affection I ever get," she wrote. Had she possessed "maturity or character" when she was married, she thought she might have shown her husband what she needed. Now she recognized that she was permanently imprisoned in a desperate state—"a Steamboat" she called herself—a nature which she predicted, accurately, would not be altered "short of my coffin."

Four years in America left Louisa an emotional and physical wreck, so that in 1864 the Kuhns emigrated once more to Europe. There, Sister Lou and her husband hoped to find those "gay, brilliant, and amusing" people whom her brother Henry said his sister needed to have around her. After a moderately successful visit with her parents in England, she and Charles Kuhn were soon back in Italy in search of pleasure. Her family was not hopeful, sharing Henry's view that "Loo will never be happy anywhere anymore." How, Henry wondered, could "any frame" endure "the tortures of so restless a spirit." For Louisa's parents, there was, as Charles Francis put it, only "great grief" over their daughter's outlook. "She feels herself as capable of being more and better than she is or will be," said her father.

Before they departed for America in 1868, Sister Lou's family visited her in Florence. Her life there made Henry only more pessimistic: "Lou is near insane, and will kill herself ultimately, I think." Her father began a new effort to understand the causes of her plight. He, of course, did not imagine that a nineteenth-century woman could be satisfactorily independent. Yet to be so had always been Louisa's desire. Her father, however, accepted the era's conventional explanation for female misery—she had made a poor match. She needed a husband who could curb her as well as guide her. In Charles Kuhn, the family believed, she found none of this. "Such is the lot of us all," Charles Francis mused. "Marriage makes some and mars as many."

He saw additional proof of this as Louisa grew obsessed with having her own money at her disposal. Her last surviving message to Charles Francis in 1870 insisted that he send "every red cent" that could be wrung from funds the family had invested for her. She was determined to own a pearl necklace and earrings at any price, despite her husband's plea that she not be rash. "I am tired of waiting," she said, as if she knew how little time remained to her.

In 1870, living in Italy, Louisa began driving her own carriage again, something she had always enjoyed. On one such outing in late June, her horse got out of control, and she was hurled from the vehicle, which then passed over her leg, breaking it and crushing her foot. For a time she seemed to be recovering readily, so amputation was rejected as unnecessary. But then tetanus began its inexorable course, and Sister Lou died of lockjaw on 13 July 1870, at the age of thirty-eight.

The ordeal of his sister's final illness always remained with Henry Adams, who interrupted a holiday in England to go to her deathbed. He came upon a scene which roused all his old admiration for Sister Lou, for she rallied her enormous energy and resisted death. "[S]uch a struggle for life," he wrote, "is almost worth seeing." She held entranced the dozen friends who kept at her side during the twenty-day ordeal. She joked through rigid teeth, and commanded everyone's admiration and astonishment when she appeared to return from the dead on one occasion and calmly asked for food. Henry acknowledged that the courage his sister showed amid her convulsions helped him grow out of his own cowardice. But finally even Sister Lou's will had to give way and she died, leaving Henry wordless. He only wanted to be "away from society and condolence."

During this ordeal, Henry kept the family at Quincy informed of Sister Lou's condition. The shock and horror of Louisa's end erased many distasteful memories. In dying, she earned for herself the admiration and redoubled affection which she had believed life denied her. Even her sardonic brother Charles stood in awe at reports of her bravery and, at her death, confessed in his own diary that "some of the light went out of my world." Charles Francis and Abby were shattered when a cryptic telegram told them that their daughter was dead. Only later did they have details from Henry's letters which moved so slowly across the Atlantic.

Charles Francis had the news first. He had come home from the notable July 1870 meeting of the Harvard Overseers which elected Charles Eliot college president and was seated alone in the Old House when the telegram arrived. Overcome, he walked out into the pasture where he tried to recover the control needed to share the news with Abby. Finding he could not face his wife, he summoned John to tell her what had happened. Abby, stricken, cried out, "Oh, my God be merciful to us," speaking of Sister Lou as "our darling, our pride, our fascinating, brilliant child." On the other hand, Charles Francis' way was to see a lesson in the terrible bulletins from Florence. "The strong religious feeling which I inherited from my father, and which has ever been my stay in the great trials of life, will in the end subdue my spirit to the proper level of humility and resignation," he wrote in complete sincerity. He told himself that man's story was the trivial one of a worm.

While Charles Francis did his best to call up pleasant recollections of

Sister Lou, especially an encouraging talk he shared with her in Paris during their last reunion, he could not dismiss the distressing fact that an Adams had fled from the burdens of life. In taking the path to Europe, he said, his daughter had found the "attractions for the consumption of time which saved her from preying upon herself." Yes, she had been his most difficult child, but still, Charles Francis insisted, "I think she had at her outset the opportunity to become the most accomplished as well as brilliant woman of her generation." These troubling meditations, he wrote, turned him "to stone." He continued to pray for the tears that might relieve his memory of a child's life wasted in "self-indulgence." Over and over, he recalled her childhood, trying to determine what he might have done to save her from running "to waste." His burden was heaviest when he remembered his failure to instill a religious outlook in this eldest child, now dead in a foreign land. He wondered if a merciful God might "measure the extent of her merits by the severity of her trials to govern herself."

Eventually, Charles Francis found some solace as he studied the "craving void" which had been Sister Lou's life. It occurred to him that his daughter's nature had been very much like that which had caused the first Louisa Catherine's distress. With this recollection of his daughter's resemblance to his mother, the tears finally came to comfort him. There was, however, one last disappointment from his daughter. Word arrived of Sister Lou's instruction that her body remain in Italy. She did not wish to return to the Old House and the ancestral vault. Instead, Charles Kuhn supervised the building of her tomb in Florence, and there she was placed on 30 January 1871.

Although this foreign interment may have symbolized how far apart the fourth generation of Adamses was moving, it did not grieve Charles Francis and Abby as much as did the restlessness of their son Henry, who, like Sister Lou, eventually made his escape to Europe. Henry had felt a special fondness and admiration for both the Louisa Catherines, his sister and his grandmother. And both these women had spent their few really happy moments in Europe. Henry's nature, however, was even more complex than his sister's or grandmother's. For him, being an Adams proved more of a struggle than it did for any other member of his generation.

❧ 13 ❧

Tragedy

The death of Sister Lou had an enduring effect upon the Adams family. Her brothers and sister thereafter built their domestic recollections by using Louisa's tragedy in Italy as the pivot. For Charles Francis and Abby, however, their grief was somewhat assuaged when Henry agreed to try to live in Boston. As he watched his sister die, Henry had been appalled by her wasted life. He also knew his parents feared that he shared Sister Lou's impulse to escape and roam. Indeed, they had good reason to think so. Restlessness had beset Henry after he graduated from Harvard and during the time he was secretary to his father, when Charles Francis was the American minister in London. After coming back to Boston in 1868, he distressed his parents within a few months by departing for Washington and a career as journalist and political critic. This "break-up of our old home relations," his father had said, was "a sad trial to us all." It was a breach Henry made deliberately, telling Charles, "My path is a different one, and never chosen in order to suit other people's tastes, but my own." Henry liked to stress his need to be different. "I mean to be unpopular and do it because I must do it."

So he went to Washington in 1868, saying that affairs in Boston and Quincy were on "too small a scale." Henry predicted that presently he would have to find inspiration in Europe or the Pacific. "I am a finished debauche, and must have excitement," he said. The political life in Washington, where he wrote essays for such journals as the *Nation, North American Review, Edinburgh Review,* and the New York *Post,* soon proved insufficient for Henry's needs. Nor were his friendships with attractive politicians and journalists such as William Evarts, William H. Seward, Samuel Bowles, and Henry Watterson enough. Even his analysis of public issues such as the tariff, the currency, the reform of civil service, and the general plight of American statesmanship failed to hold his interest. Washington itself grew less comfortable for Henry with the rise of U. S. Grant and his associates.

At a loss over what course he ought to take, Henry decided in 1870 upon the distraction of a trip to Europe. When he urged his father to travel with him, Charles Francis bluntly dismissed the plan as "the height of folly." Henry went anyway, and there, watching Sister Lou die, he found he could no longer brush away family memories and a sense of duty. There seemed no way out for him but to give life in Boston and Quincy another chance. He now accepted appointment as assistant professor of history at Harvard, a position President Charles W. Eliot had unsuccessfully offered him earlier. When Abby heard the news that her son had changed his mind, she assured him that his life would be "more useful and settled," a statement which may not have given him much comfort. Meantime, Henry was admonished by Charles Francis, who sensed that his son still wished to be influential, to remember that a teacher could have much to say in the shaping of modern America.

In the autumn of 1870, Henry began a bachelor's life in Cambridge sharing quarters with his brother Brooks, who was still in school. They often visited their parents on Mount Vernon Street, especially near dinner time. At Harvard Henry was a complete success, at least in everyone's eyes except his own. He proved to be a fine teacher in the area assigned to him, medieval history and institutions. Later he taught recent European and United States history. The need to be creative encouraged Henry to introduce such novel practices as seminars where students became colleagues in research. Also, Henry appeared once again to be putting himself at the center of family affairs, working closely with Charles and Brooks on essays which peppered their favorite target, misconduct in business and government. During the winter of 1871–72 the Adamses were even more together, trying to calm Abby who was distraught by loneliness because Charles Francis was in Europe preparing for the Geneva arbitration conferences. Amid this fellowship Henry kept a secret. It was not until his father had returned early in 1872 that he disclosed his plans. He was going to marry. Astonishing news indeed, for everyone seemed convinced that at thirty-four, Henry was not the marrying sort.

In February 1872, Charles Francis came home late one evening from dining with Ralph Waldo Emerson and James Russell Lowell and was met by an excited Abby. She told him that Henry was engaged to a twenty-eight-year-old Boston woman, Marian Hooper, whom everyone called Clover. Most of the family did not know her. Charles Francis, who thought he understood Henry best of all among his sons, confessed the announcement "filled me with surprise." In fact, it was two days before the cautious parent reported in his diary exactly who it was his son was marrying. Stating that this "Miss Hooper" seemed to have a good reputation among her friends, Charles Francis demurred: "I trust the issue may be propitious to both." Abby and Henry's brother Charles had particular reason to be astounded, for while he was dining with them only a few weeks earlier, Henry had categorically denied that he had any plans to marry Clover Hooper. As it turned out, the two were already secretly engaged.

Once his intent was known in Boston, Henry sent word to Brooks, who was in Europe. Be prepared for a shock, Henry warned his brother: "I am engaged to be married." Who was the lady? Not the likely prospects, teased Henry, tossing out the names Brooks might have guessed— Clara Gardner and Nancy Wharton, particularly. No, Henry said, it was a person whom Brooks himself had mentioned a year before when they were walking in Cambridge. "Yes! It is Clover Hooper." Then Henry showed how controversial he knew his choice to be. He appealed to Brooks to "stand by me and bear me out," for the marriage "will make a row, and people will discuss it to please themselves." As long as the gossipers left him alone, Henry claimed, he was undisturbed by the tumult. "I know better than anyone the risks I run. But I have weighed them carefully and accept them." For Henry's fiancée had a history from her mother's family of emotional and mental instability, and it was thought that she might have witnessed an aunt's suicide.

To Brooks and to other close friends, Henry confessed himself hopelessly in love—Clover "has completely got the upper hand of me." She was irresistible—"a clever woman who chooses to be loved." She was one with whom he was confident he could go a different way, a woman of financial means, whose father had reared her to be independent in outlook. As for the risk which Henry incurred, he was confident that he was marrying Clover Hooper before the shadow touched her. It was this aspect of Henry's choice which Boston could whisper about. In fact, when Charles first thought that Clover might be Henry's bride, he had exclaimed, "Heavens!—no! —they're all crazy as coots. She'll kill herself just like her Aunt." Thirteen years later, after Clover had indeed committed suicide, Charles acknowledged, "I certainly never forgot my brutal prophecy."

In terms of social acceptability, Clover Hooper had genealogical credentials to satisfy the expectations of Henry's family. Her ancestry was as notable as that of any other person brought into the Adams circle by marriage, except, of course, the peerless pedigree of Abigail Smith. Clover's father, Robert William Hooper, was a Boston physican whose inherited wealth allowed him to remain casual in his practice of ophthalmology. Robert was born in 1810, one of nine children. His father, John Hooper, a prosperous banker in Marblehead, came from a line which had landed in Massachusetts in 1635. John had married a distant cousin, Eunice Hooper. Robert Hooper married Ellen Sturgis, the sister of his brother Samuel's wife, Anne.

These Sturgis women were themselves part of a large and wealthy family whose success in mercantile affairs was well known. Also widely discussed, however, were the neurotic tendencies in some female members of the Sturgis clan. These same women were often exceptionally brilliant, so that Marian "Clover" Hooper's wit, talent, restlessness, and melancholia are qualities readily traced through her mother's relatives. Her mother, Ellen Hooper, however, had died of consumption at thirty-six, before the time when emotional trauma usually struck the women in her family. Little is really known of Clover's mother, except that she was a poet, schooled in the tradition of Margaret Fuller and the Transcendentalists.

At Ellen's death Dr. Robert Hooper was left with three children, five-year-old Clover, being the youngest, born in 1843. The oldest was Ellen, often called Nella, born in 1838, and then came Edward, known as Ned, born in 1839. None of these children escaped the legacy of emotional weakness. Indeed, the "brutal prophecy" of Henry's brother Charles was borne out to an extent so ghastly that all who watched were dumbfounded. Yet in their early lives the three motherless Hooper children were apparently given every loving encouragement by their father, who left professional life to attend the duties of parenthood. Clover, her sister, and her brother were surrounded by affection, a high regard for literature, an encouragement to discover themselves, and a comfortable existence. Life in Boston was eased by summers on the North Shore, as the ocean front between Boston and New Hampshire was called, or by travel to Europe.

The eldest, Ellen, married a much-beloved historian and administrator at Harvard, Ephraim Whitman Gurney. They never had children—nor did Henry and Clover, who probably fell in love at the Gurney house in Cambridge where both often visited. Gurney died from leukemia in 1886, a year after Clover. Ellen Gurney survived her husband by only a year before depression drove her to suicide in 1887. Ned Hooper lived to 1901 before perishing in a mental asylum after attempting suicide. In 1864 he married Fanny Hudson Chapin, who bore him five daughters before she died in 1881. These five sisters eventually played an important part in Henry Adams' later life, for he became close to them and helped them and their father after Clover's death. Ned became one of Henry's few intimates. He was admired for his quiet manner and his abstracted way of contemplating the heavens. He received early and ample warning that he must struggle if he were to escape the legacy of depression. Henry Adams did all he could to help Ned, but at last he admitted that he knew his brother-in-law was doomed.

This was twenty years after Clover's collapse, however, and at the outset of Henry's marriage, he was confident and in love. He believed he had reached Clover in time to protect her from her ancestral handicaps. As it turned out, she never really was able to separate emotionally—or with any success, physically—from her father; and when Dr. Hooper died in 1885, Clover collapsed, even with all Henry's love and good will. Appropriately, most of what we know about her comes from the letters she wrote regularly on Sunday to her father. Little else of her is available, and even these glimpses survive because Henry unaccountably spared them when he erased almost all other vestiges of Clover's presence. The letters to Clover from Dr. Hooper evidently are gone, but whether Clover or Henry destroyed them is unknown. All her life, Clover seems to have been filled with self-loathing, and she was so critical even about her person that she made certain no real photographic likeness of herself existed. As a skilled artist with a camera, she had taken at least one excellent picture of herself—which she promptly destroyed. The primary source of our knowledge of Clover Adams is the letters to her father, which were published in 1936 by the husband of one of her nieces. Even this, however, is only partially revealing because most passages treating of depression, insanity, or suicide were silently omitted by the editor. Although she died when they were tiny, her brother's children cherished the memory of their Aunt Clover, and as mature women, they set out to obliterate any recollection of her except that of a brilliant and vivacious creature happily married to an adoring husband who was, in his turn, known for his wise, gentle ways.

Henry and Clover were married in her father's summer home at Beverly Farms on 27 June 1872, in a ceremony as startling to those few Adamses present as the fact that the bride and groom had spent the previous month under Dr. Hooper's roof. Henry expressed himself much relieved that his parents were in Geneva for the arbitration conference and would not be at hand for the wedding. Charles was there, still bothered by the story that Clover as a child had seen an aunt die by consuming arsenic. John was also present, and both Charles' and John's wives, Minnie and Fanny.

Charles called it "no wedding at all . . . [T]he ceremony was like the engagement—peculiar." The occasion certainly signified how intent the couple was on doing things differently. Thirteen persons, including a clergyman, who was allowed only the most essential words, gathered in Dr. Hooper's parlor. This was the first time John had ever met Clover. In fact, he was startled to see her, having had in mind some other woman entirely. Ever gallant, however, he reported to his mother that he found Clover fascinating.

Not so Charles, who described the bride as highly agitated and the occasion itself as very dull and "commonplace." Warm champagne was served in very limited quantities. John, with his mother's love of convention in mind, tried to make light of the contrast between Henry's wedding and those of the other children by praising the weather and the scenery—the luncheon, he said, was "cosy," and the occasion generally "sensible." "I think Henry is a very fortunate and he certainly seems to be a very happy young man," he said, "and I trust he will thoroughly enjoy the wife which he has earned by hard work." Henry and Clover soon left the scene for a relative's cottage and departed shortly on a lengthy honeymoon tour of Europe and Egypt. John and Charles smoked cigars with Dr. Hooper and Ned, after which the day closed.

Later that year, the newlyweds briefly joined Charles Francis and Abby at the family's temporary headquarters in Geneva. Clover did not improve her image on that visit, persisting in doing what she chose rather than following the proper behavior which Abby, a superb diplomat's wife, considered essential. Henry professed to adore his wife's style, but Abby was not amused when Clover refused to dress and attend a ball. Clover acknowledged that her mother-in-law "is quite disgusted with me," but her husband's adoration seemed to grow. When Charles later visited the pair in Paris, he was indignant at the completeness with which Henry was in love.

Some explanation of Henry's choice may lie in the fact that until Clover entered his life, the two women he admired most intensely were Sister Lou and his grandmother, whose private papers he had studied. Each of these Louisa Catherines had possessed a brilliant mind and rebelled against the lot assigned her in life, each had been a sharp critic of society, and each had possessed a fascinating but unstable temperament. Henry had found in Clover a woman curiously like these two. Neither of these women was ever entirely comfortable as a member of the Adams family, and Clover proved no different in that respect.

At first, the family tried to welcome Henry's wife. Charles was the only exception; perhaps because his younger brother's romance had suddenly disrupted the close intellectual partnership he had depended upon, he seemed determined to dislike the bride. She had few of the qualities that Charles admired in women—submission, reticence, and beauty. When Charles caught up with the bridal couple in France, he found Henry and Clover depressed and ill, huddled in tiny quarters and generally making Charles "want to yell and scream and tear my hair," especially when they gave him a room he called a kennel. "Damn their Turkish coffee—and I wish it and they were in Turkey now and I were alone in Paris." Henry was "too much married and *they* tire me savagely," he said, and left Paris in disgust.

To his wife Minnie, Charles particularly revealed his desolation at losing Henry. "My brother has grown to be a damned solemn, pompous little ass, and his wife an infernal bore; —they are the most married couple I have yet seen." Clover was always talking, always holding Henry's hand, he said, no manners for a proper woman in his view. Why, on the day he joyfully escaped Paris, he had risen early, and found he had "knocked the mature Henry out of the arms of his Clover, for he's always in clover now (Joke! ha! ha!)!" Two months later, when Charles was in London while Henry and Clover were also in town, he carefully avoided them. They "*are* antipatica—not a doubt of it!"

Like Clover, Henry was half a rebel, eager to criticize, yet reluctant to be involved or committed. Both were self-centered, with one important exception—Henry was gentle and solicitous. He might wish to withdraw, but he was also prone to help and to teach. Just as he had hoped to aid Sister Lou, whose wit and irreverence he had relished and whose talent he had admired, so he wanted to bolster Clover against the penalty which her nature and the inheritance of temperament might extract. As bittersweet as his marriage proved, Henry could still say, four years after Clover died, that "the only Paradise possible in this world is concentrated in the three little words which the ewig Man says to the ewige Woman." Yet Henry saw on his honeymoon that love might not be sufficient to sustain Clover's delicate equilibrium.

From Europe, Henry wrote to Dr. Hooper, telling him discreetly that Clover was uneasy and unhappy—could Dr. Hooper join them for the rest of the honeymoon? If not, Henry said, "it may be wise for us to come home much earlier than we had expected." Clover's letters at this time, November 1872, show her struggling with depression and self-rejection. What confronted her was an alarming turn in the nuptial trip, one taking the pair away from the busy distractions of European life which turned Clover's mind away from herself. She was afraid of the loneliness of a small boat they had rented to go up the Nile, a trip Henry was eager to make. The situation left Clover nearly mute. "I seem to have softening of the brain whenever I touch a pen," she told her father. Confessing that she was utterly demoralized, she added: "I cannot write except to you who are used to my stupidity and shortcomings."

Traveling up the Nile, confined with her enemy, boredom, Clover went into a serious nervous collapse which Henry ever after recalled with despair. She concentrated on her perceived inferiorities, seeing others they met on the trip as comparatively bright and eager. "I get so little," Clover complained to Dr. Hooper, "while the others about me are so intelligent and cultivated that everything appeals to them." Not until the ordeal of the Nile trip was over did the melancholy and apathy lift for a time. Henry and Clover went to the more familiar comforts of England and then they returned to Boston, where they had bought a house at 91 Marlborough Street, just around the corner from Dr. Hooper's residence at 114 Beacon Street.

Back in Massachusetts, in the summer of 1873, Henry resumed his teaching and editorial roles at Harvard, and Clover tried living with the demands of being a devoted daughter, a wife, and a daughter-in-law. Trouble was not far off. Most of her time was spent near her relatives; and Henry broke the time-honored Adams custom of summering at Quincy. He and Clover preferred her family's property at Beverly Farms where they soon built a summer residence. Curiously, upon returning from his honeymoon, Henry learned that his early love, Carrie Bigelow and her new husband, George W. Amory, had leased Charles Francis' Quincy house where Henry had lived as a child. How much the memories of Carrie may have discouraged Henry from summering in Quincy is unclear. However, Charles Francis reported that the Amorys turned out to be unfriendly tenants. Henry and Clover, meanwhile, rarely appeared at the Old House, and then usually for perfunctory afternoon visits.

For two years, 1873–75, Henry and his wife took a small place in the Adams circle. Thereafter, a growing breach between Clover and her in-laws developed until by 1877, when Henry took Clover to Washington, there was open hostility. Abby was the first to lose patience with Clover, but Charles Francis was also soon obviously troubled by his daughter-in-law. The elder Adamses deplored Clover's strange ways, her caustic talk, and especially her unwillingness to share Henry with them. Abby had very decided notions about a woman's role as wife, which included cleaving to her husband's family, a proper devotion to the church, a discreet tongue, and the ability to bear children. Clover failed by all of these measures.

References in family papers to visits of Henry and Clover after 1873 become scant. In October of that year, Henry removed his financial affairs from Charles' care and handed them over to Ned Hooper. Nothing, however, deterred Charles Francis and Abby from visiting Henry and Clover in their new home on Marlborough Street. They dined there with Louis Agassiz, the most famous American scientist of the day, and his wife. Abby left one of the few descriptions of the home on Marlborough Street, marveling at the possessions Henry and Clover had collected while honeymooning abroad. "Their house is a little picture, filled with no end of pretty things," she wrote. She found their house "a small museum." "Henry glories in a fascinating study filled besides books with all sorts of fancy and rare things." The water-colors, persian carpeting, china—all caught Abby's knowledgeable eye. Furthermore, Clover had been "clever, sweet-tempered and *refined*," and since Henry was such a "lovely" person, "who would not be happy with him?"

Henry's preoccupation with Clover, and theirs with their affairs and particularly with Clover's relatives, persisted, although Charles Francis tried to keep the path open by calling on the couple after church every Sunday. These were brief visits, as was the time Abby and Charles Francis stayed with Henry and Clover in Beverly during July 1874. On that occasion, the older couple soon sensed that their presence created an awkwardness, and within three days they departed. "We left our children with regret and yet with a feeling of relief," Charles Francis recorded in his diary, "as fearing we might have been a burden."

Two years later, in 1876, the distance was even greater between Henry and his parents. In New York City for medical attention, Abby wrote frankly to Charles Francis of her indignation at what marriage and life away from the Adamses were doing to Henry. He had become showy and distant, she said, fretting that her son never left home "or his Clover." In addition, Abby was outraged when she learned that during one of Charles Francis' calls upon Henry, Clover, withdrawn upstairs, had insisted that Henry come to her while Charles Francis was shown the door. "She is always on the bed or sofa with *that* pillow," the energetic Abby sniffed. "To let you go so cold as it was and at your age was too bad." The best she now would say was that "Clover is a queer woman." When the young couple appeared in New York while Abby was still there, the two generations stayed at different hotels. "Their coming," said Abby to her husband, "don't matter to us, they are so far off and their occupations so different. I believe I have nothing more to say." Meanwhile, Charles Francis' sabbath visits at Henry's residence ended, for Henry found it wise, evidently, to visit his father instead, and he came alone.

In the late summer and fall of 1877, Clover joined Henry for a brief visit or two at the Old House, but a major change in family affairs was by then in sight. With inexpressible chagrin, Charles Francis learned that Henry had decided "to abandon Massachusetts for the fluctuating excitement of the Capital, where none of our name have ever prospered." Noting that Clover had made herself very pleasant on the eve of their move, her father-in-law gloomily saw the reason. Henry was "now taking a direction which will separate us from him gradually forever." Often Charles Francis expressed the wish that he and Henry could be close, "but we cannot have every thing as we desire it in this life." After 1877, a few visits took place in the summers and once, Charles Francis and Abby went to see their son's new seaside house at Beverly Farms. But generally the distance between Henry and the community of Adams became nearly as great as Sister Lou's had been. Charles Francis began speaking of Henry as he had of his dead daughter. "He has but one fault: the passion for roving." Henry and Clover were "afloat." Charles Francis reported gruffly in his diary in 1880: "He gets no aid from me."

Henry professed himself unable to explain why he was happy under these conditions. "My heart bounds as I tear myself away from family, friends, the familiar streets and boyhood associations of the best of all cities," he confessed to John Hay, his good friend, and said of Washington: "I wish to clasp in my arms this corrupt and corrupting tomb-

stone of our liberties." To which he added: "Don't tell my secret." While
Henry claimed to be bored with his existence at Harvard and in Boston
and eager to seek his fame as a master historian in Washington, he had
another motive—the delicate nature of his marriage. Clover required a
place where her uneasiness with self and world could be channeled
safely into her brand of gossipy sarcasm, the sort which had made Boston and Cambridge unsafe for her. So they departed in 1877, risking
the separation from Dr. Hooper, to whom Clover was as deeply attached as ever.

Once the pair removed to Washington, there was a revealing development. Word reached Mount Vernon Street that Clover had not delayed in openly ridiculing Massachusetts and its society. Charles Francis
felt he had no choice but to rebuke his son. In January 1878, he noted
in his diary: "Finished a letter to Henry which he will not relish." The
epistle was blunt. Henry and his wife certainly need not live in Massachusetts if they did not like the area and its population, he said, but if
they chose to leave, it was still "not worth your while to be sarcastic
about them before others who do not comprehend your drift." When
Henry replied that they might have been misquoted, Charles Francis
let it stop there, only remarking that all his children appeared to be
unduly severe with others. "We are all of us imperfect at the best," he
reminded Henry. "Do not let the habit grow upon you." Meantime, as
Henry began hinting at plans for long sojourns in Europe, Charles
Francis was not surprised. "He has been a rover from the first."

In the aftermath of this bad feeling which Clover had established,
Henry's uncle Sidney Brooks died, leaving handsome legacies to Henry's brothers but cutting him off without a cent. Although this drastic
step surprised Charles Francis, he confessed that he found the will judicious and well-meant. A year later, in May 1879, as Henry and his
wife prepared for a stay in Europe, the embarrassment and vexation
caused by Clover were still acute. The climax occurred when Charles
Francis and Henry faced each other in the Old House over what the
parent called "the unpleasant business caused by the indiscreet conversations of his wife." Her talk had "nettled all the family." Although
Henry evidently bore the scene well, Charles Francis was relieved that
Abby was absent. "I am afraid that this commotion is driving Henry to
exile in Europe," said he; the change was "a great source of pain." Still,
he was certain "that separation is far more safe for all the family" than
to bear more of the "thoughtless malice" which he attributed to Clover's jealousy. While he had "no feelings but those of affection and

love" for Henry, Charles Francis carefully gave the Adams family's opinion of Clover: "I pity rather than dislike his wife. . . . I must regard her only as a marplot and a subject of commiseration."

But Henry had not yet escaped. When he returned to Quincy two days later for more farewells, Abby kept at his side during the entire visit, and would talk of nothing but what Clover was doing to him. When Henry finally fled, she collapsed, prompting Charles Francis to note that Abby's "affectionate temper joined in her constitutional anxiety create needless distress." To calm his wife, he agreed to go into Boston and talk with Clover. Whether this was to try to reason with his daughter-in-law, to confront her directly, or to ask that she not oblige Henry to go to Europe Charles Francis did not record. At any rate, he was spared the encounter, for when he went to Dr. Hooper's house, where, of course, the couple was staying, Clover was not in. "I left a card and felt relieved," Charles Francis confided to his diary.

From the autumn of 1877 until Clover's death in 1885, she and Henry spent their winters in Washington, living in rented houses near Lafayette Square—first at 1501 H Street, and then at 1607 H Street, very near the site of the house they were building when Clover died. From spring 1879 until autumn 1880 the couple lived in Europe, where Henry consulted documents he needed to prepare his history of the United States. Otherwise, Henry and his wife regularly summered in their house at Beverly Farms, a rambling building on a wooded hilltop with a marvelous view of the sea.

Later, looking back on those years of marriage, Henry was to be ambivalent. He recalled great happiness and love, but in rare moments he also conceded that there had been terrible times, especially after 1882, as he watched Clover worn down by her emotional burdens. He had sought to provide ample affection and concern for Clover while he determined to rival Edward Gibbon and Thomas Macaulay as a great historian. It was a difficult balance to sustain. He methodically gave a part of each day to horseback rides with Clover and participation in the famous social aggregation which they drew about themselves, a group filled with beguiling and sympathetic people who enjoyed the gossip and topics which Clover's sharp wit relished. For Clover, Washington apparently provided the release which Florence had afforded Sister Lou.

Meanwhile, whether in Europe or Washington, Clover kept her life-supporting tie to her father. Dr. Hooper regularly visited Clover and Henry, while Charles Francis and Abby never appeared in Washington.

She tried to be happy sharing Henry with his desk, manuscripts and books, pen and paper, allowing him to write with the same duty-bound regularity which marked his ancestors. The results were impressive. Henry's first task in Washington was to edit a collection of the writings of Albert Gallatin, one of America's great statesmen and diplomats, a person whom John Quincy Adams had much admired. Henry also wrote a biography of Gallatin. In 1879, with the materials made available to him by Gallatin's son, Henry completed *The Life of Albert Gallatin* and *The Writings of Albert Gallatin,* the latter in three volumes.

While working on his next project, a life of Aaron Burr, Henry paused in 1880 to publish anonymously a novel, *Democracy,* whose keen and caustic observations on American society and politics fascinated and dismayed the circles in which Henry and Clover moved. Henry never said how much Clover might have helped in writing the novel, and the secret even of his authorship was kept until after his death. The Burr venture was less successful, for reasons not clear. Apparently the publisher shared Henry's uneasiness with the manuscript, so it was put aside. This study may have been one of Henry's most imaginative accomplishments, but it has vanished, either burned by Henry along with so much else from the period or simply lost.

Next, Henry prepared a biography of a figure unpopular in Adams family tradition—John Randolph, the brilliant but unstable Virginian who had helped to trouble John Quincy Adams' presidency. This book appeared in 1882 as part of the American Statesmen Series. Now, however, Henry pushed beyond these comparatively small enterprises, turning to the majesty of his ultimate goal, a lengthy history of America during the administrations of Thomas Jefferson and James Madison. To this end, he and Clover had wandered among the archives in Europe, after which he and his scribes sat at length in an office in the Department of State, examining records of the United States. In 1884, a private edition of the volume dealing with Jefferson's first term was circulated among a half-dozen friends, and the next year a second volume was similarly distributed. However, most of this great project was not completed until after Clover's death.

There was one other work from Henry's pen during the years with his wife, his second anonymous novel, *Esther.* This is the most puzzling thing he wrote. Published in 1884, it is a novel of ideas, specifically an examination of the question of religious belief. The models Henry used for his characters in *Esther* were probably people he knew, including Clover, but the resemblances are lost in the story. There seems no valid basis to argue, as Otto Friedrich has, that Henry used the character of Esther to explore the disturbances in Clover's psyche, and particularly her dependence on her father. Friedrich's reasoning led further to the claim that Henry's writing of *Esther* forced Clover to suicide. It has never been certain why Henry wrote the novel. *Esther* was as inconclusive as *Democracy* was straightforward, but Henry was usually much clearer about politics than about religion. *Esther* proved a great popular failure, though Henry himself preferred parts of it to anything else he published. All his writings showed a range of interests and depth of appreciation which mark Henry as one of the best Adams minds.

Clover evidently tried to encourage Henry's work, but complicated ideas and information had little appeal for her. Henry needed and sought solitude and contemplation to draw forth energies and ideas welling up within him, and Clover did not find her own company amusing, as she liked to say. In these things she and Henry were unalike, a contrast which did little to improve her view of herself. Life for her was an effort to avoid boredom, since that mood forced her to look in upon herself, a view from which she shrank. In other aspects, however, he and she were curiously alike. They were nearly the same size, she five feet two inches and he an inch taller. They had kindred tastes in art, furnishings, and food. They enjoyed nature in much the same way—horses, wildflowers, the ocean.

Clover seems to have come closest to happiness safe in a group of friends who could share irreverences, satire, and even destructive talk about other people and events. Clover had a wicked tongue which she used in ways sometimes delightful, sometimes cruel. This manner continued her alienation from the Adams family and from the Boston community. Washington was filled with transients who thrived on this kind of social life, so that a salon for Clover's amusement and nourishment was always available. She could include or exclude guests to a degree impossible in Boston, which was filled with vulnerable relatives and in-laws. In Washington, therefore, the Henry Adams residence became famous. Some persons were bidden, others rejected, the measure being whether they were lively, interesting, or attractive. It was the same sort of circle which Sister Lou needed to have about her.

The tiny group most intimate with the Adamses, called the "Five of Hearts," was composed of Clarence King, John Hay and his wife Clara, and Clover and Henry. There were other close friends, of course, including Henry James, Nicholas Anderson, Mr. and Mrs. George Bancroft, Senator and Mrs. J. Donald Cameron, John La Farge, H. H. Rich-

ardson, and Carl Schurz. But the people in the "Five of Hearts" shared
something special, for each was handicapped by an unhappiness or
restlessness which seemed to require a special irreverence, outrageous-
ness, wittiness, or social satire as antidote. The Adamses cherished this
same vulnerability in others, such as Henry's brother Brooks, John
La Farge, and Ned Hooper.

Of the "Five of Hearts," Clarence King was considered the most brilliant. A well-known geologist and explorer, roamer and spendthrift, he
appealed to and touched the Adamses, who nevertheless knew nothing
of his deeper private life which included a black common-law wife, with
whom he had children. King died eventually in despair and alone. Clara
Hay bore the burden of being the rich wife of an attractive and quix-
otic man. She had little charm herself, and her father had committed
suicide. Silently for the most part, she watched the hijinks of the others
in the group, a prim and orthodox person who must have known that
she was tolerated because her wealth made John Hay's successes possible. Later, after her husband's death, she was a severe censor when
Henry agreed to her request to prepare an edition of Hay's life and
letters.

John Hay was Henry's closest friend, and he adored Clover, who,
with her masculine swagger, her brusque talk, and her irreverence, was
such a contrast to his own wife. He also was drawn to Elizabeth Cameron, wife of the senator from Pennsylvania. Her youth, beauty, and
unhappy marriage gave her a special appeal for him. Hay wanted to
be a dilettante. After serving as President Lincoln's secretary and helping write a biography of his great chief, he wished only to enjoy people,
politics, letters, and travel as his wife's money allowed. Easily discouraged and depressed, he needed Henry's encouragement and ideas.
While the talents of King and Henry Adams awed Hay, he himself
possessed a charm, openness, and wit which endeared him to them.

Much of what Clover sent in her letters to Dr. Hooper consisted of
versions of what had been said and done by the "Five of Hearts" and
others in the community she and Henry drew about them. Thus, the
letters often disclose the caustic side of her that so distressed the Adams
family. She also reported to her father her spells of depression, self-
disgust, boredom, and illness. There were even dreams; she wrote Dr.
Hooper of one nightmare in which a black man had entered her room
and was clutching her by the throat. Her loneliness for her father and
the scenes around him also came through in these letters. They were
written with absolute regularity on Sunday when the rest of Washing-

ton was at church. While Dr. Hooper himself remains mostly unknown, since his replies to Clover have vanished, he did send her a warning in 1881, for Clover in her reply, mentioned the danger he had spoken of in her preoccupation with such "horrors" as insanities and death. She only sought information about these troubles for others, she protested.

Anyone who reads the letters of Clover Adams can see how restless, unhappy, and depressed she often was, and how these moods seemed to force her to destroy people around her. Only the pathetic sufferings of others brought out her sympathy. These were circumstances with which she could identify. More evident, though, are the reasons why Clover's tongue had created trouble for her in Boston and with the Adams family. While the published letters often display a cruel, even vicious quality, those portions which the editors expunged from the edition often offer even more vivid evidence of this trait. On 3 April 1881, Clover told her father that Senators Dawes and Hoar "show themselves greater asses every day—its a shame to represent the Old Bay State through such poor white trash." She did not hesitate to call people drunkards and blackguards, and when she learned that an acquaintance in Boston was marrying, her retort (omitted from a published letter dated 6 May 1883) was to be glad the girl had chosen a husband outside the Beacon Hill crowd for "a little brains will improve her stock."

Another theme, often missing in the public version of the letters, was Clover's dread of emotional breakdown, mental asylums, and death. Beginning in the spring of 1882, she anguished over the plight of a Bigelow cousin, a woman who suffered nervous disorders and who had to be sent to the mental institution at Somerville, Massachusetts where, Clover had heard, the circumstances were most dreadful. She begged her father to do what he could to save Adie Bigelow from this confinement, but he could not. Clover did not shrink from acknowledging the presence of mental disorder; she willingly conceded that Adie suffered, but she blamed two of her own favorite scapegoats, mothers and religion. Clover said that the pioneer scholar in mental disorder, Dr. S. Weir Mitchell, whom she knew, ought to write on "The Divorce of Mother and Daughter." Clover proposed that Adie's mother be committed to the asylum and the daughter released.

Clover was often drawn to women caught in a genuinely pathetic situation, as was Elizabeth Cameron. Clover also took a liking to Anne Palmer of New York, a young woman fourteen years her junior. From this close association came the other revealing group of Clover's letters that have survived. In them she often thanks Anne for having tried to cheer her. "My path to the grave seems to be strewn with lovely roses by you." Beverly Farms was described to Anne as "quieter than any average grave," a comment made during the time Henry was becoming alarmed over his wife's moods. Meanwhile, Clover wrote Miss Palmer of her admiration for those who bore up under affliction and "didn't commit suicide."

Henry sought some diverting hobby for Clover, trying photography in the early 1880's, an art which had interested them both briefly during their honeymoon trip on the Nile. For a time the idea appealed to Clover, and she made many pictures, carefully recording the technical details, the dangerous chemicals used in treating the plates, and her evaluation of the results. These were in many instances better than Clover acknowledged, as the file of her photographs preserved in the Massachusetts Historical Society shows. Her greatest activity seems to have been in 1883, when she took many fine likenesses, including the one of Henry that follows page 310. Other successful pictures take as their subjects Henry's parents (both by then pathetic, Charles Francis lost in senility and Abby crippled by arthritis); Charles (in a very unflattering pose); Brooks (who successfully hid behind his horse); the Old House; and old friends such as John Hay, George Bancroft, and H. H. Richardson. There are numerous pictures of Henry and of Dr. Hooper.

The distractions offered by the camera were only temporary, however. Thinking that a change of scene and some days with Miss Palmer might cheer Clover, Henry urged her to go to New York for a visit which helped briefly. The letters to Anne nevertheless continued to talk of the nearness of death. She said of her life with Henry that "we take the little that is perhaps left of life daily in our hands. Its not more than a handful and enjoy it much." When, late in 1883, Henry acquired land on H Street for their new Washington home, Clover wrote that she anticipated building "a modest mausoleum." Once it was completed, she announced, "we *can* have nothing to look forward to beyond the grave." And when Henry seemed especially preoccupied with writing the *History* in late 1884, Clover conceded to Anne that it left her "not very well so that I must write to you or blow up." Her Washington circle, she now complained, seemed made up of "a few very old, infirm and near insane women."

Clover greatly admired in others what she felt she herself lacked— courage in the face of life's rigors. A family crisis was particularly difficult for her, as when, in 1881, her brother Ned Hooper watched his wife die, leaving him with several young daughters. Clover had bravely promised to come up to Boston from Washington to help, but when the event came to pass, she lost courage and remained at a safe distance. "We should be a bore," she told her father, and it would mean tearing Henry from his work.

She kept looking for fresh distractions. In one of her last letters to her father, she wrote that "this is a good time for health of mind as one never has time to think of one's self." It was December 1884 and the social season was at its height, but Clover continued to implore Anne Palmer to come for a visit and cheer her. A few months later came a personal challenge she could not avoid. In March 1885, heart disease brought Dr. Hooper into his last illness, thus beginning the tragedy which moved relentlessly to Clover's suicide in December of that year. When Henry took his wife to her father's side, he knew a price would be exacted. This is evident in the letters he wrote constantly, encouraging her, epistles he allowed to survive while he destroyed other papers from that time in his life. It was as if Henry wanted to let some scraps endure to show his devotion to Clover.

The messages he sent from Washington to Boston strove to be cheerful. The first one set the mood. "As it is now thirteen years since my last letter to you, possibly you may have forgotten my name. If so, please try to recall it. For a time we were somewhat intimate." At the close, he scrawled on the side after the signature—"just a little crumb of love for you, but you must not eat it all at once." Usually his daily epistles began with the salutation, "Dear Mistress." However, a week with her dying father was enough momentarily to break Clover, and she sent Henry a telegram crying for help. The wire does not survive, nor do any of the letters Clover dispatched to Washington, but Henry said it "frightened me out of my wits." His architect and friend H. H. Richardson said to Henry "that he would give anything in the world if his wife should send such a telegram to him." So Henry allowed himself to be calmed enough to keep a dental appointment before he went up to Boston to comfort Clover. By the close of March, Henry was back in his Washington study, writing to "Dear Angel" as she waited for Dr. Hooper's death, perhaps thinking to distract her by chatter about construction on their house and especially about "our sleeping den."

Henry, however, recognized he was not succeeding, so he went once more to Boston in early April. "I must take a hand at the nursing," he told Clover, "for I see what I feared has happened, and that the refusal of a trained nurse has made the nursing a very serious matter." Dr.

Hooper and his children had resisted having professional help. The stricken man was in the hands of his son and two daughters, themselves emotionally frail, as Henry very well knew. Again, however, Henry returned to Washington, evidently at Clover's insistence. He begged her to summon him once more. "Uneasy as I am about you, and unable to do anything here, I go on from hour to hour and make no engagements at all." His effort to cheer and reassure her produced on 12 April a glimpse of the way Henry adored his wife. "How did I ever hit the only woman in the world who fits my cravings and never sounds hollow anywhere?" he asked her. "Social chemistry—the mutual attraction of equivalent human molecules—is a science yet to be created." For him, said Henry, their instance was "my daily study and only satisfaction in life." The next day, 13 April, Dr. Hooper died, and Henry went again to be with Clover. By 20 April, she was back in Washington, "in better condition than I feared," Henry reported to John Hay.

Soon Clover was explaining to Anne Palmer that she and Henry would not go near Boston or Beverly Farms in the approaching summer, planning instead a long camping trip in the Rocky Mountains, concluding with a stay in Yellowstone. She said she was "tired out in mind and body." Telling of her grief at losing her father, Clover said, "no one fills any part of his place to me but Henry, so that my connection with New England is fairly severed as far as interest goes." Meanwhile, Henry made a hurried visit to his parents, and then he and Clover began what was to be a vacation in search of her renewal, away from what Henry called that "gloomy spot," Boston and Beverly. Within a few weeks, however, something happened to change their minds—no one knows what it was—and by mid-July, Henry and Clover were after all in the very place she had wished to avoid, their summer house in Beverly Farms, so near her father's cottage.

Henry spoke only of "various domestic necessities" as the cause for abandoning the wonderful flight to Yellowstone. Probably Clover had fallen into a near-paralyzing melancholia, in which she wished to be nowhere but near her father's haunts. What it cost Henry to acquiesce we do not know, but he confessed to his aide and secretary, Theodore Dwight, that he was getting no work accomplished that summer. "We are rather draggy." To another intimate, he reported on 30 August, "My wife has been out of sorts for some time past and, until she gets quite well again, can do nothing." He did travel to Quincy once a week to visit his parents, and in September, he insisted that Clover join him in a trip to Saratoga. Two decades later, Henry said, "After that May in 1885, I never for long years lifted my head, or cared to lift it."

It may be, ironically, that Clover's own outlook during this time was captured in a letter to her from her cousin, Sturgis Bigelow, who had also inherited a generous share of the family's emotional weakness. He wrote to her from Tokyo in July 1885: "Leaf by leaf the roses fall . . . I don't see that there is anything left for me . . . but the long mechanic pacing to and fro—the set grey life and apathetic end." Nor was Clover comforted, probably, when, in early October, Henry had to pack her up for a return to Washington—"always the most depressing moment of the year," Henry confessed as he did the last chores. On the train, they met Charles, who was moved to try to speak to his sister-in-law. "It was painful to the last degree," he reported. "She sat there, pale and care-worn, never smiling, hardly making an effort to answer me, the very picture of physical weakness and mental depression." According to Charles, "her mind dwelt on nothing but self-destruction." She had lost the one mainstay in her life, her father.

Before leaving Beverly Farms, Clover got off a letter to another friend, Rebecca Dodge, whose fortitude in a relative's illness brought an envious remark from the stricken Clover—"you are so sweet and brave and strong that you never seem to lose courage." There were few letters now to Anne Palmer, who had announced her wedding plans just at the time when Dr. Hooper died. When Anne evidently asked for reassurances, Clover managed to reply, "No, I'm not sorry. I am very glad." Returning from Beverly Farms, Clover brought with her a fragment of something written in French in her father's hand and found in his desk: it captures the strong spirit which he had displayed before his death and which Clover had mentioned so often to Henry—"The thing is that they knew how to live and die in that time," Dr. Hooper said of some bygone era. "If they had gout, they kept on walking all the same and without making any grimace, concealing their pain by the means of good breeding."

Her father was evidently Clover's main anchor to life. She could not cling through religious faith, or through social or political interests except as they were absurd to her critical eye. She had no children, and she lacked the capacity to shift entirely to Henry for support—she wrote Anne Palmer that Henry filled only some part of the void left by Dr. Hooper. Now she was alone in life with Henry, watching the house they were building, a step which may have seemed to Clover a commitment to life and convention more affirmative than was bearable. During autumn and the early winter in Washington she remained in a seriously depressed state. At the close of November, Henry barely lifted the curtain when he wrote to one of his closest friends, "I hardly want you to come here just now, for my own plans and prospects are a little unsettled and I could not enjoy your visit as I would like." Clover, he said, had been "a good deal off her feed this summer and shows no fancy for mending as I would wish." The pair did venture to the nearby home of Don and Elizabeth Cameron, where Henry talked privately with these friends of his worry over Clover's "nervous prostration." Clover brought Mrs. Cameron flowers on Friday, 4 December, and there seemed some hope that she was improving. It was false.

About this time, Clover seems to have begun a last letter, in effect her final testament, to her sister, Ellen Gurney. In it, as Ellen quoted the letter, Clover wrote, " 'If I had one single point of character or goodness I would stand on that and grow back to life. Henry is more patient and loving than words can express—God might envy him—he fears and hopes and despairs hour after hour—Henry is beyond all words tenderer and better than all of you even.' " Upon reading this note so filled with self-reproach, Mrs. Gurney recalled Clover's pleas for help during the agonizing summer just over in Beverly, and especially her anguished talk of having nothing within herself which she could respect. To Ellen, Clover had said, " 'I am not real—Oh make me real—you are all of you real.' "

On 6 December, a dental problem took Henry, briefly, away from home. He felt Clover was better, for she had roused herself sufficiently to remind him to write to Henry Cabot Lodge. Then, upon returning to the house, he discovered Clover lying on the floor of her room. She was dead, having swallowed some of the chemicals she used in developing her photographs. Henry summoned his family and Clover's. Charles received a telegram and took the late night train to Washington, stopping briefly in New York, where he read the first newspaper accounts of his sister-in-law's end. Reaching Washington, he found Clover's brother and sister already there.

Characteristically, Charles took charge of the situation, sending telegrams and making arrangements. The next day, he took Henry away from the house for a long walk, where, according to Charles, "Henry told me the whole story—and sad enough it was." Henry said that Clover had been improving, "and that, could he have saved her then, she would have come through and again been well." Charles himself doubted this, believing that by this time Clover had become so pathetic that she was "as responsible for her own act as an infant." The tragedy made Charles more charitable in his recollections. He spoke of how in Clover's lucid intervals, she and Henry had lived "very happily." She adored Henry, said Charles, "and well she might, for his patience and gentleness seemed inexhaustible." In return she joined in his pursuits "and made his house just what he most liked." "Me," Charles acknowledged, "she never liked; nor can I blame her much for that—I trod all over her, offending her in every way."

On Wednesday, 9 December, it rained heavily. Henry's brother John arrived with his son George, completing the Adams delegation for Clover's funeral. Brooks remained away, in all likelihood absorbed in his own fresh grief over Heloise. Sister Mary had just been delivered of her first child. Although Clover had passionately rejected all elements of religious faith, Edward Hull, a Unitarian minister conducted a service at Henry's house, after which all that remained of Clover was buried in Washington's Rock Creek cemetery. Whatever guilt Henry may have felt over the manner of Clover's dying he kept to himself. Brooks later spoke of being bewildered that Henry had made suicide so easy for Clover, since he recalled how Henry had emphasized he was well aware of Clover's taint by family insanity.

Whatever Henry's feelings may have been, he forecast the nature of his career thereafter when he told his neighbor and friend, George Bancroft, shortly before Clover's funeral: "I can endure, but I cannot talk, unless I must." On the same day, he said bitterly to John Hay, "Nothing you can do will affect the fact that I am left alone in the world at a time of life when too young to die and too old to take up existence afresh." Henry was now forty-seven. To Henry Cabot Lodge, he announced that he was entirely "too much occupied in thinking of her" to do anything else. He must try to find sustenance from the "twelve years of perfect happiness" with Clover. On the other hand, friends like Hay and Lizzie Cameron heard him say that his only hope now was to push on "straight ahead without looking behind," that his nerves were nearly destroying him. It was miraculous he could sleep, he said, and he brooded over "my poor wife" and "the wreck of her life."

There was further anguish after Clover's tragedy. First, Clover's brother-in-law and Henry's friend, Ephraim Whitman Gurney, died in 1886, after which Charles Francis died, also in 1886; then, in 1887, Ellen Gurney committed suicide by throwing herself under a train. Thereafter, in 1888, Ned Hooper barely survived a desperate battle with his own depression, and finally, Abby died in 1889. Caught in this cavalcade of loss, Henry said that nothing in life thereafter could surprise him—not even if the moon should forsake the earth for another planet.

Inexplicably, the only fragment of Henry's diary apparently to survive is from this period, including entries for some of 1888 and 1889, times in which he helped nurse his dying mother. It displays a laconic and restrained style which may be the clearest view we have of the real Henry Adams. The pages show how he tried to face his own severe depression. "I am weary of myself and my own morbid imagination, but still more weary of the world's clack and bustle and the dreary recurrence of small talk." He said that in his heart he never really wanted "to see Washington or home again," and he began resolutely to destroy his early journals, saying, "I mean to leave no record." As he prepared to burn the pages of his past, he read them. "My brain reels with the vividness of emotions more than thirty years old"—all the more reason, he said, to keep pegging away at "my demise." Henry did, however, take time for a trip or two to Beverly Farms to see Ned Hooper and his daughters, whom Henry was now determined to help.

Henry survived Clover by over thirty years, an era in which his thoughts about her were often associated with the famous monument by Augustus Saint-Gaudens which Henry commissioned to be placed at the grave site in Rock Creek cemetery. He was thrilled by the artist's success in capturing the sense of peace he wanted to surround Clover, and he was inclined to talk eagerly about the monument and his antic- ipation of joining Clover there. If this work succeeded in showing that, "at the end of philosophy, silence is the only true God," Henry an- nounced in 1891, "I can be quite contented to lie down under it and sleep quietly with her." Thereafter, he spoke of "my permanent resi- dence under the protection of St. Gaudens' figure at Rock Creek."

The monument stirred Charles to say that after Clover escaped a load "greater than she could bear," she became the person in Saint- Gaudens' figure. It was a metamorphosis about which Charles became poetic, writing of Clover that "at the very instant the burden dropped from her, as, the journey done, she sat down at the grave. The pain is all over, —the weariness gone, —the disappointment is felt no more, —every wrinkle is relaxed, and the tired eye-lids fall on tired eyes." Since Clover saw to it that no clear photograph of her survived, the presence in the cemetery came to be the world's portrait of her, al- though Henry gave no sign that he was aware of this.

He did, however, strive to beautify the site, struggling to get shrub- bery to grow, while he became annoyed at the "high-pitched, sharp- nosed women" who intruded on him as he worked. In 1910, he re- ported being disappointed in trying to find "a minute's content and repose" at the monument. There, he said, "the ocean of sordidness and restless suburbanity has risen over the very steps of the grave, and for the first time I suddenly asked myself whether I could endure lying there listening to that dreary vulgarity forever." Could he, he wondered, "forgive myself for condemning my poor wife to it"? Henry took some consolation in remembering that his great-grandmother Johnson had been resting in Rock Creek cemetery for over a century, and eventually he seemed at peace in his visits to Clover's tomb. Only in going "to see my St. Gaudens" was it possible, Henry said, to "fancy myself back in 1878," when he and Clover had just begun their new life in Washington.

Other memories were less consoling. In 1898 he made the mistake of returning to Egypt and going past many of the sites he and Clover had visited on their honeymoon. The effect was nearly overpowering. "[T]he tears rolled down in the old way and I had to get off by myself for a few minutes," he wrote to Mrs. Cameron. He also dreaded the coming of each December which brought memories of Clover's final desperation. "This season of the year always demands some human sacrifice from me," he observed, noting how many of those dear to him had collapsed in December. There was no doubt about it, Henry said, the "cerebral troubles" of others "wrecked my life, and cost me much more than life is worth."

14

Memory

Both Charles Francis and Abby suffered much indignity from old age. Toward the end of his life, Charles Francis had to race to complete his duties before his powers failed. He lived another year after Clover died, but his mental functioning had already changed. Loss of memory, lethargy, and depression had occasionally troubled him even before he served as minister to Great Britain between 1861 and 1868. Such difficulties increased upon his return to Massachusetts, adding to his distress as he watched his children suffer. For Charles Francis, the only recourse was the one his elderly father had taken, striving to finish the labor at hand before it was too late. As he hurried to do so, he said it made the years in London "like a wild dream."

Not only did Charles Francis feel changes in himself; Quincy, too, was different. The beloved scenes around his home were by 1873 no longer soothing. Change was everywhere, and much of it was due to Charles Francis' own efforts as a Quincy banker and businessman. Now the clamor of the quarry business and other trades had overcome the rural serenity Charles Francis had cherished, as had his father and grandfather. Acknowledging dourly that all such development enhanced "the mercantable value" of his land, Charles Francis was still homesick for "the wild, careless, simple features of the landscape to which I look back with pleasure in the days of my youth."

Some tasks, however, he found still fulfilling. For a time he particularly enjoyed enlarging the Old House with more kitchen and servant space to the north. He took down the unsightly and dangerous building behind the mansion which had served as office, library, and overflow living quarters since the days of Abigail and John. The books and family papers were at last housed in a splendid stone library he built at the northeast corner of the garden, an addition that fulfilled a pledge dating from John Quincy Adams' time.

These were the last pleasing developments, however, for Charles Francis soon took no pleasure from his property, observing, "I sometimes wish the whole burden off my shoulders. The revulsion from the great interest I took in it twenty years ago is one of the curious features of my life." His final project, a huge stone stable and carriage house near the Old House, distressed and wearied him. He quarreled with the architect and fretted that all these improvements actually harmed the family's surroundings. Also disquieting was the fact that by 1874 Abby's health required long absences from Quincy in the summer. How serious matters had become was clear when the family moved into Boston for winter. No Adams had ever relished this annual step, but in 1874 Charles Francis announced, "I am for the first time glad to go to town and leave the place to take care of itself."

More troubling for Charles Francis, however, was knowing that there was a chance he would become too confused to complete his duties to the family. As soon as he could break away from carpenters and stonemasons, he resumed screening the family letters, a chore he had interrupted to go to England. Once more, he tried to destroy items he deemed obsolete, useless, and occasionally too personal. "Nothing gives me more scittish feeling than such work," he admitted, although he found the task easier whenever he threw away remnants from his own youth and the era of his brother George's decline. "On the whole it is well to expunge all traces of the evil of that day," he commented. "The time has come when I ought to apply the knife unsparingly to all this."

Absorbed by these family manuscripts, Charles Francis often thought of the days when he sat on the Old House piazza, with so many kinsmen in John Adams' time and in the days of his father. "And so here am I alone in their places!" he lamented, although he reminded himself that his life had been much happier and less tormented, than those of his forebears. All the more reason, he told himself, that he must set about "rectifying history," so that the Adams name might be properly appreciated.

For Charles Francis, this meant ensuring that John and John Quincy Adams took their rightful places among the nation's truly remarkable men. "[T]he native excellence of their characters" placed his ancestors "among the Catos and Scipios" of the American republic. Thrilled though he was by these glories, the Adams in Charles Francis himself also made him marvel at how drawn his great forebears had been to the world beyond their family and books. "Public life is a very fascinating occupation," Charles Francis said, and his family's experience showed that "it is like drinking brandy, the more you indulge it, the more uncomfortable it leaves you when you stop." He saw added evidence of this as he prepared his father's sizeable diary for publication.

Nothing in life gave Charles Francis greater satisfaction than editing family papers. In the case of his father, however, he felt he was too old to write a biography; he sensed how little time he had left to work. He preferred the comforting notion that it would be best if his father spoke for himself—certainly no one else ever had spoken for him, and Charles Francis shrank from doing so even a quarter century after John Quincy's death. Instead, he began to pluck from his father's enormous diary those sections which the public ought to read. What Charles Francis wanted was for posterity to "see in my father the only picture of a fully grown statesman that the history of the United States has yet produced." Jefferson, Hamilton, Webster, Marshall, all these may have been talented in some singular way, he conceded, but each had deficiencies made up in the well-rounded John Quincy Adams.

Reading his father's diary so fascinated Charles Francis during the 1870s that he often forgot his social obligations, including whist parties, of which all the Adamses were very fond. As he followed his father's minute penmanship, Charles Francis' admiration grew, until it reached what he acknowledged was "reverence" for his parent's character and power of "self-examination." However, some of John Quincy's candor, his son believed, must never be published. The world, he said, "needs no admission to the interior of a wounded spirit in its private home."

Charles Francis was particularly reluctant to publish materials from 1829, when the family was in a period of despair. Here he closed the curtain, saying that these things must be "strictly private," although he did acknowledge that John Quincy's character and career would never be fully understood without including these personal torments. Reading how his father agonized over problems created by his children, Charles Francis discovered that since John Quincy tended "to dwell on the dark side of things," he had had to find distraction. "I do not wonder therefore that he accepted politics." Keeping the private travail of the family out of the published diary meant that Charles Francis omitted perhaps half of his father's pages. Even so, the manuscript grew to alarming proportions, bringing complaints from Lippincott, the publisher. Charles Francis held fast, determined to use his own money, if necessary, "for this last and greatest of my undertakings."

Unmoved by Lippincott's appeal for condensation, Charles Francis struggled through the seemingly endless pages his father had left, making pencil notations to guide his copiers so that they would take only what he wanted the world to know. Meanwhile, he was impatient, sensing that his failing memory might betray him before the cherished work was finished. When Lippincott seemed lethargic in meeting printing schedules, Charles Francis grumbled, "I am sorry I ever engaged with him but it is too late to mend."

Finally, the early volumes appeared and sold well. Charles Francis was elated, hoping that now history would do justice to his forebears. "My life has been largely spent in the duty of placing two men on their proper pedestals," he wrote in 1876. Including twelve volumes of his father's *Memoirs,* he counted twenty-two books of family material he had bequeathed to the American people.

Hastening to finish editing his father's diary, Charles Francis talked more and more of his "presentiments" of mental incapacitation. However, he won his race against time. There were days when his confidence waxed and he spoke of the joy which "literary labor" brought him. A momentary surge of his old vigor briefly encouraged him to plan an autobiography for posthumous publication. As late as 1877 he looked over some of his own diary, preparing to recount his career. Such optimism was unusual, however, and Charles Francis realized he was lucky to have completed John Quincy's diary. The goal was reached in the summer of 1877. He was deeply moved. "It has pleased the Master to let me acquit myself of this great work!"

"The rest of my life is of no consequence," he said to an old friend; now he understood how a man felt who was released from prison after a long term. The responsibility for family history which had hovered over him for decades was lifted. On 10 August 1877, Charles Francis received the last volume of his father's memoirs from the printer. He wept as he cut the pages. Overcome by a sense of all duty completed, an outlook no other Adams had attained, he felt, he said, that he was now ready to die. "My mission is ended and I may rest. . . . the public is satiated with Adamses." The only task left was to place all the family manuscripts "on deposit for the future."

There were but two years remaining before his mental powers failed in 1880. Consoled though he was in this tribulation by his religious

faith, Charles Francis sat saddened and guilty as he watched his children remain indifferent to spiritual matters. Worried about their
grandchildren, he and Abby persuaded their son Charles to bring four
of his youngsters to the Old House for baptism. The grandfather
thought perhaps here was a sign "that my sons are not absolutely insen-
sible to all religious impressions," but the hope was groundless. As
Charles Francis put it over and over, his children's "total secession from
all religious worship" implied "a degree of negligence in me of their
early education which I take to heart."

Despite their advancing infirmities, he and Abby continued to appear
at Sunday services in Quincy's old church or, when in Boston, at King's
Chapel, the last of a series of churches Charles Francis had attended
when he was in the city. The venerable King's Chapel had become a
place for Unitarian worship and was called only "Chapel" by the
Adamses, who had not lost the old distaste for royalism. However, while
Charles Francis may have appeared Unitarian, he was most stirred by
his mother's Anglicanism. As he aged rapidly, he found that the Episcopal prayers never failed to move him. He grew less inclined to attack
the great theological issues, preferring a simple faith. Such matters as
foreknowledge and free will he called a "fearful mystery," to which, he
said, humanity ought to be "content to await the solution in another
sphere."

One of his last good deeds for the Quincy community was to struggle
out from Boston in a heavy snow to try to make peace among warring
factions in the family church. The problem was created, Charles Francis grumbled, by the minister, "a busybody who meddles indiscrimi-
nately with what is and is not his business." But an Adams spoke and
the minister was defeated. "So much for parish tumults," said he. "At
my time of life to fall into such shallows is provoking enough."

More taxing for Charles Francis were the family fiscal burdens. En-
croaching senility led him to make big problems of small matters, and
he dismissed the advice of his sons, who tried to calm him. He worried
because the family fortune was committed to real estate in Boston, where
he now owned the Boylston Hotel as well as a newly constructed office
building on Court Street. These investments entailed debts, which
Charles Francis had once so courageously admonished his father to
avoid. "My wonder," he marveled now, looking over his property, "is
how it became so large, for I never had any spirit of enterprise." At
the close of 1879, he confided to his diary, "I am at my wits' end."

To his family it was clear that Charles Francis was failing rapidly,
though they tried to reassure him as he grew querulous and fearful.

John especially tried to help the old gentleman realize that there was a new financial age at hand in which debt was taken for granted. However, it was difficult to cheer Charles Francis, for his thoughts rarely strayed from his diminishing mental capacity. He could no longer remember the text of a sermon halfway through the address, and he found he hesitated to trust himself to keep the business accounts he once had maintained so meticulously. In June 1878, while presiding at a meeting of the Harvard Overseers, Charles Francis had to be prompted by the secretary in recalling members' names. "This ought to be a warning enough to conclude while there is time," he observed. Finding thereafter that it was even difficult to converse intelligently, he began withdrawing into silence. "I almost shudder at the change which is passing over me." His "duty" now was "to prepare myself for the decline of my days."

"Must I shut myself completely up?" he wondered. It would gravely vex Abby, he knew, for she still made him the center of her life. "Let me pray heaven for mercy and patience," he said. He did not have long to wait for yet another, ominous sign. In the spring of 1880 he struggled unhappily to fulfill a commitment. In the spirit of family tradition, he had agreed to make a centennial address to the American Academy of Arts and Sciences. Trying to write his speech he went, after church on Sunday, 23 May, into the stone library next to the Old House to work. Soon Abby heard sounds of distress and hastened to him, to find Charles Francis in despair, for he realized his manuscript had become hopelessly muddled. Abby summoned their sons to help her persuade Charles Francis to abandon the oration, and in the midst of this family council, Charles Francis agreed to retire. John was dispatched to Boston to explain to the Academy, while the old man was left to ponder his remaining duty, "the preparation to stand before my God." Settling into "a line of life as private as it can be made," Charles Francis continued that summer to make brief entries in his diary, largely from a habit stretching back over half a century. He grew stoical in facing what he called "my metamorphosis," although he made little comment upon one tragic aspect—that his father and grandfather had kept their vigor of mind virtually to the moment they died. Here he was, barely past seventy and the most reasoned and deliberate of the family. Yet his mental powers had failed. By October 1880, he no longer tried to visit the bank and in November, abruptly, he closed his diary forever. He knew that some recent entries had been lucid while others were confused. It was, Charles Francis said, "a proper place to terminate."

Abby and the children watched this wreck of a once-powerful mind.

Charles wrote Henry that their father "has completely lost all power of sustained effort." While the old gentleman might live peacefully for a time, "the end is none the less here. We must all make the best of it." Charles had a consoling hypothesis about his father's plight. Reminding Henry that Charles Francis Adams was now the only surviving statesman from the great controversies of the Civil War, he pointed out that all of their father's associates had broken in old age under the residue of the strain they had endured so long. "He is the last of the race, and the extreme tension to which he was subjected at 50 is telling on him fatally now at 73." Despite this theory, however, in later life each of Charles Francis' sons entered his final years terrified that he would become as helpless as his father.

By the spring of 1882 Charles Francis could no longer be left unattended. This realization came about through an embarrassing incident that was well covered in the press. Charles Francis had been in the habit of going alone on his cherished walks around Boston Common. On 28 March he was persuaded by two crafty strangers to accompany them to a card game where they coaxed the befuddled old man to sign a check for nearly $20,000. Releasing him, they rushed to the bank to demand payment, calculating that so distinguished a family would never challenge the transaction, preferring to avoid scandal. The pair mistook their quarry, however, for although Charles Francis had gone home oblivious to the true nature of the event, the bank notified his sons, who immediately ordered the swindlers prosecuted. They were apprehended and brought back to Boston for a trial in late May. The jury convicted the man who planned the crime and sent him to prison.

Charles Francis, who only once recalled the episode under Abby's prodding, would have been pleased when witnesses at the trial testified that, at the moment he was surrounded by card sharks, the old man had momentarily realized what was happening and called out that he had never gambled in his life. John spoke for the Adamses at the trial, testifying to his father's regularity of habit, his physical fitness, and the fact that all Boston knew him; this, he said, had deluded the family into believing he could go unescorted, even though the old gentleman lived mostly in the distant past.

The naturally nervous and fearful Abby was especially stricken by this moving proof of her husband's helplessness. Charles Francis had most dreaded this about his fate—its impact upon Abby. Her own physical condition had begun deteriorating in 1871. A sprained ankle suffered in a fall in the Mount Vernon Street house brought on severe sciatica. Her emotional state, always delicate, left her fearful and distraught over her debilities, especially when they required lengthy stays in New York City for electrical nerve treatments, which were then a medical fad. In all this, Charles Francis had been patient and helpful despite his own worries, for the elderly couple were very close. "We plod on to the end with that increasing feeling of tenderness and mutual dependence which softens the regrets of decay," he had written in 1873.

Even in her tribulation, however, Abby kept up her lifelong struggle to shake her husband out of what she considered his Adams quality of measuring everything and everyone too much by his own demanding standards. In 1876, when she was in New York and her husband was struggling to handle the household in Boston, he had trouble with a servant. "So you have just found out that Joseph is a fool, have you?" she wrote him. "I have always told you so and you have always said it was my impatience. Bless you, if you kept house you would be in hot water all the time. You are ten times more impatient in such matters than I am." When Charles Francis failed fully to appreciate Abby's loneliness in New York, she told him, "You judge me by yourself. You ought not. We feel things so utterly unalike. Things which kill me you never think of twice. However, we are too old to change either your nature or mine."

There was clearly much affection between these wholly dissimilar persons. Abby loved to tease "my old darling," and probably delighted him as he neared his seventieth birthday by telling him, "Oh how I hate to be away from home and you. I always want you so and feel a sort of rest when you are in the house." One moment exasperated—"You and I are so unlike we cannot understand each other"—Abby would in the next try to lure Charles Francis to New York with promises of a nice bedroom. "I had rather be lame with you than well alone," she said, teasing that he did not "long for me as I do you."

As for Charles' mental decline, for many years Abby simply refused to acknowledge that her husband was troubled by more than absentmindedness. "I laugh at the idea of your mind," she told him in 1877; "you never was better in your life." She urged her "dear husband" to recognize "how well and strong in body and mind you are." Vigorous in body he surely was after seventy; he could easily have walked into Boston from Quincy. But the problem was that when he tried to do so, he now lost his way.

In 1878, the couple celebrated their forty-ninth wedding anniversary alone in the Old House, talking together of their blessings and trying to recall details of that day long before in Medford when they were married. They counted their many friends who were dead—"our turn must come soon." On that anniversary night in September 1878, Charles Francis recorded the most moving glimpse he had ever allowed into his private world. He and Abby, he said, "slept in each others arms." The next year, when the Old House rang with the celebration of a Golden Wedding, Abby's husband was entering a living death before her eyes. It seemed to leave her pathetically alone, the only child of Peter Chardon Brooks to survive, for her brothers Sydney, Edward, and Chardon all had died recently. Upon arising during the anniversary dinner to say a few words before his family, all assembled but Henry and Clover, a tearful Charles Francis could remember nothing of what he had in mind and so sat down again in silence. What he might well have thought to mention, perhaps, was the astonishing fact that he, his father, and his grandfather had each been so blessed, to sit in the Old House rejoicing in fifty years of marriage.

Charles Francis' final years, between 1880 and 1886, were unspeakably sad. At first a male attendant managed him by following him about at a distance. Eventually, however, he had to be carried everywhere. His only flicker of recognition, according to family myth, was when Abby entered the room, at which he attempted to rise. Yet, as Abby sat at his feet, weeping and begging him to speak to her, Charles Francis would only stare vacantly over her head. Scenes such as this made the sons implore her to spare herself and them such agony. Henry gave perhaps the gentlest description of his father's condition: "Nothing is left of him but the form, and the perfect quiet and dignity of his manner."

Mercifully, despite his rugged physique, Charles Francis did not long linger. In November 1886 he was obviously failing rapidly. On the twentieth the family began assembling at Mount Vernon Street as his daughter Mary reported in her diary: "My father dying. Quite unconscious, only breathing." At 2:00 AM on Sunday morning, the twenty-first, Charles Francis Adams died alone, leaving the news to be sent out to Quincy after dawn. Later that day, all the family gathered, save Henry who arrived on Monday.

With a kind of gentle relief, the Adamses attended Charles Francis' funeral, "the end of the seventh generation on this soil," as his son Charles put it. It was a gloomy Tuesday, with rain dampening the small group of mourners who went from the Quincy church to the Mount

Wollaston cemetery a mile away—for no longer were the Adamses being interred in the old vault where so many forebears rested. "On the whole," said Henry, "everything was as little painful as could be expected."

There was some talk in Massachusetts that a tribute should be gotten up to mark the passing of another great Adams. A eulogist was selected, but there was little vigor in the effort since Charles Francis had, in spirit, been gone for so long. Speaking for the family, John sent word to those planning a ceremony that it would be best to let the matter alone. His father, said John, "had been forgotten in his lifetime. . . . We feel in a word that the time has passed when such an occasion could have been made successful or interesting."

Certainly Charles Francis would have been relieved at this. No Adams could welcome a half-hearted tribute from a community which had never understood the family, even as far back as when President John Adams had insisted upon keeping peace with France. Not that Charles Francis had never received local tribute. In 1869, even he himself had been stirred by James Russell Lowell's poetry honoring him at a commencement dinner at Harvard. On that occasion, as Charles Francis stood beside portraits of John and John Quincy Adams, Lowell had spoken the lines:

> Third of a stalwart race, to him is due
> No smaller debt than to the other two;
> Behold, they brighten from the canvas
> To feel their praise renewed in praise of him!

After this, there was not much to add that would have been convincing to him. However, Charles sought to enlarge the case by writing an inscription for his father's tomb:

This stone marks the grave of Charles Francis Adams, son of John Quincy and Louisa Catherine Johnson Adams. Born 18 August 1807. Trained from his youth in politics and letters. His manhood stretched by convictions which had inspired his fathers, he was among the first to serve and among the most steadfast to support the new revolution which restored the principles of liberty to public law and secured to his country the freedom of its soil. During seven troubled and anxious years Minister of the United States in England. Afterwards Arbitrator of the Tribunal of Geneva. He failed in no task which his government imposed, yet won the respect and confidences of two great nations.

Dying 21 November 1886 he left the example of high powers nobly used and the remembrance of a spotless name.

Charles Francis also left a lonely widow who survived him for nearly three distressing years. Abby had lost her lively sparkle. Those charms which had once so intrigued and inspired her introverted husband were replaced by physical and emotional torment. She suffered from sciatica and arthritis, and was vexed by a weakened bladder. She seemed terrified by the absence of her husband, and she worried constantly that her money was vanishing. In the middle of a quiet night at Quincy, she would command a servant to rouse her son John at his house and bring him to her. The drowsy John would try to be patient, telling Abby repeatedly, "Ma, you've got barrels of money. You couldn't possibly spend it all." Her only calming distraction was her daily carriage ride, taken even in poor weather and in spite of her physical handicaps. Ever the mother and grandmother, she insisted on having Mary's little girls with her during the winter outings.

Inconsolable, increasingly unmanageable, Abby presented a very real problem to her children. As Henry described her, "she is in a state which requires constant attention and infinite patience." To help in this, the family had the good fortune to remember Sister Lou's close friend of long before, Lucy Baxter, whom Abby had liked. Lucy was now a spinster living in genteel poverty. John approached her and, with elaborate graciousness, asked her to be his mother's companion, telling her that by this service she would greatly help the family. Abby was, they warned her, "sensitive and peculiar." She would accept almost no one near her, but she thought of Lucy much like a daughter. To everyone's delight, the plan succeeded—as well as any could. Lucy, said Henry, "looks after my mother to the relief of the family."

Still, the children could not turn away entirely from this scene. Charles' diary records his brief, dutiful, impatient visits to his parent, usually around 5:00 PM. Mary also looked in, as did John. But these three had families of their own. Through no fault on their part Brooks and Henry had none, so they were elected to superintend the home scene. Brooks did so in the winter season and Henry in the summer of those last three years. Neither labored with much grace, their mother's disorders being of an exceedingly vexing and embarrassing sort to both of them. Henry described his mother as "not a good subject for masculine care." He tried to get away from some of the painful scene by working in the stone library next to the Old House. Here, for five hours a day, he sought to forget Clover and Abby by writing steadily on his great history of the Jefferson and Madison administrations. "Of all things I dislike most to be useful, but evidently I am needed here," he observed in the summer of 1887.

Brooks was somewhat less patient, worrying especially over his mother's feminine disorders. "Oh God, my dear fellow," he wrote to Henry after a long winter of attentiveness, "why were women made as they are? . . . I pass my life trying to manage them and its harder work and wears me more than . . ." He left the sentence unfinished. Perhaps he was trying to justify his customary flight to Europe for the summer. Henry so dreaded being confined to the Old House in the summer of 1888 that he was tempted to return, instead, to his cherished Beverly Farms house where he had spent his last summer with Clover. "I would rather roast myself over a red hot gridiron" than do that, he confessed to Lucy Baxter, but it might be better than being at the Old House around the clock.

Abby became steadily more difficult. Early in the spring of 1889, Charles reported: "I found the old lady nervous to the verge of insanity." Charles kept his visits perfunctory, rushing home in relief to the comfort of his own dinner, leaving Lucy to try to calm his mother. In May, he sent word to Henry that the "final crisis" was beginning. Abby could not long sustain the "nervous excitement" afflicting her, he said; her physicians were obliged to use very heavy doses of drugs. Henry must come at once and aid Lucy when Abby was taken from Mount Vernon Street to the Old House—for the last time, Charles predicted hopefully. "Her existence now is a horror to herself and to everyone with whom she is brought in contact. Anything more wretched it is hard to conceive."

As an additional distress, Abby was gripped by the fear of death which had troubled her as a young woman. Even John was moved to reinforce Charles' call for help from Henry. Their mother, he reported, was an insane woman, a statement which many years afterwards the family disputed. Brooks later claimed that heavy medication had unwisely been given to Abby and had actually caused what all her children agreed was "a state of querulous, constantly complaining senility." Henry took his time about responding to his brothers' summons, and as a result he was too late; when he arrived he found his mother unconscious.

Abby fell into a coma just before Henry appeared, and she died in the evening of 6 June 1889, at the age of nearly eighty-one. Brooks had already left for Europe, and John, Charles, and Mary were at their homes, leaving Henry the satisfaction of being able to say: "I was present." The funeral on the late afternnon of 8 June, a Saturday, proceeded through a rainy gloom. The occasion was quite private, and there was a downpour when Abby was laid by the side of her beloved Charles Francis. Together, they rest on a pleasant knoll in the Mount Wollaston cemetery. The next day, Charles made the remarkable concession of claiming the place in the Adams pew at church—"it was not cheerful," he reported, and he did not return.

In writing his mother's epitaph for the stone placed over her and her husband, Charles inscribed:

> His companion and support in private life
> and public station, loved and honored,
> trusted and true.

After his mother's funeral, Henry was left alone to roam the garden and the fields around the Old House, scenes which had so inspired his forebears. "I recall distinctly the acuteness of odors when I was a child," he wrote his friend Lizzie Cameron, "and I remember how greatly they added to the impression made by scenes and places. Now I catch only a suggestion of the child's smells and lose all the pleasure." Less pensive, Charles began arranging distribution of what personal effects from their ancestors that the heirs wished to have. The choices were made in a September gathering, all present but John who sent a note that he was too busy sailing and adding: "do not postpone on my account as I prefer to be absent." He wanted to pass through the episode with as little involvement as possible. "I do not care what comes to me as long as the matter is settled satisfactorily." Thus were the precious pictures, books, furniture, and other personal items passed on, all the relics of a century of family life. On 1 October 1889, the second Charles Francis Adams shut up the Old House for what he thought was forever. Three weeks later he took grim satisfaction in going to 57 Mount Vernon Street "to remove the last picture and crossed the threshold for the last time. It was 46 years ago that we moved into that house, and I have not one pleasant association connected with it. I went out of it today with a sense of relief."

Loosened though the family ties seemed in 1889, neither Charles nor his brothers and sister ever found it easy to escape their ancestors. Separated and often antagonistic to one another, the fourth generation carried the burden of being Adamses to the end of their days, when, at last, the long and troubled story of family ambition quietly concluded. An important reason why the fourth generation had to think about family matters was their legacy from Charles Francis and Abby. In part, few family duties had so troubled Charles Francis as the transference of the estate after his death. Never before had an Adams had great wealth to bequeath and now, also for the first time, there were several heirs of sound mind and body.

The will, which was finally completed in 1871, never satisfied Charles Francis, despite the many hours he had devoted to it. Certainly the document tried to treat the fourth generation even-handedly, for with Abby "so amply provided for under the trust created for her benefit by the will of her father," Charles Francis left instructions that his wife encourage their sons to keep the considerable Adams real estate undivided. The chief concern, however, was the Old House and its marvelous heirloom furniture, paintings, and china. After Abby was dead, her husband's will directed that the eldest surviving son should move into the mansion as earlier generations had done, yielding it in turn to his oldest male child. Charles Francis' dream was that this be continued indefinitely.

Along with the cherished Adams land, which was to pass intact, the fourth generation received the treasury of family papers. These Charles Francis wanted to be owned by all the children, and to remain in the library he had built next to the Old House. He spoke to his sons, all now executors, from his final testament, praying "that they will not dishonor themselves or me by indulging in unworthy jealousies or contentions on the subject or by failing in that first essential to all prosperity in families, harmony and mutual good will." He left only one stipulation. Each of the four male heirs was to provide from his share the amount of $40,000 for his sister Mary. The total of $160,000 would comprise a trust for her, the proceeds paid to her "free from all marital claim or right."

When the will was proved on 5 January 1887, it was apparent how fruitful the long stewardship of Charles Francis had been. The real estate was valued at $916,959 and the personal items at $109,619. The largest element was a package of three Boston properties in the heart of the city—the Adams structure on Court Street, the Melodeon building, and the Hotel Boylston—this alone appraised at $314,000. Various bequests and expenses reduced the total received by the legatees to just under a million dollars. Immediately, and obediently, Charles Francis' sons put property valued at $800,000 in a business combine, and Mary chose not to take her bequest as a monetary fund, but to keep it in what she and her brothers called the Adams Real Estate Trust. When their mother died in 1889, another half million dollars was added to what became a family business in land and buildings. Thus far, the final hope of the third generation was realized. Where the fourth generation failed Charles Francis was in his request that his heirs live in harmony.

The dissolution of a tightly knit family was hardly surprising; Charles Francis himself had sensed it, and dreaded it. With his departure, a new phase of the Adams story began, mainly because the fourth generation contained not one strong individual, as had been the case for three generations before, but several varied and vigorous personalities. Each of these characters—John, Charles, Henry, Mary, and Brooks—differed sharply on what it meant to be an Adams. As a result, in its final days of greatness, John Adams' family story became one of conflict and contrast.

⚡ 15 ⚡

Charles

Nothing could have encouraged the contrasts among members of the fourth Adams generation more than their efforts to carry out the legacy of their parents. At first the two oldest brothers, John and Charles, were involved, for Henry, Brooks, and Mary were distracted by other matters in the 1880s. Brooks and Mary had turned much of their properties inherited from Peter Chardon Brooks over to John and Charles to manage, in addition to the family trust left by their father and mother. Henry, however, partly because of practical considerations arising from his marriage and partly because of his uneasiness over the abilities of John and Charles, had entrusted most of his private means to his brother-in-law Ned Hooper, who also looked after Clover's money.

This arrangement need not have led to trouble except that Charles was still determined to prove that only one great Adams was possible in each generation. He was convinced that such a role had now fallen to him. It seemed easy at the outset. Since Civil War days, Charles had dominated his older brother John, who, even when he was titular head of the family during Charles Francis' senility, looked to Charles for guidance. John had remained wholly unlike the first John Quincy Adams, who had never been happy in domestic quiet, had never stopped fighting real and imaginary foes, and had rarely betrayed his inmost feelings.

Appropriately, it was Charles who claimed a tie with old John Quincy. "I loathe vegetation," he acknowledged. "My brother John, on the other hand, did not know enough to keep himself well by movement. He got into the home habit." As Charles later described him, John never lived, only "existed." A trace of loyalty, perhaps, made him blame some of John's apparent indifference upon his wife, Fanny, who recognized that her husband was not constituted to be an authentic Adams. She encouraged his tendency to relax, to enjoy the children, and generally to avoid the bustle of life which so intrigued Charles.

As husband and wife John and Fanny took a proper if modest role in Adams family affairs before 1889. However, Fanny wanted her husband for herself and away from pressures at the Old House, and John made no objection. He confessed to a friend in 1890 that while he had not escaped the Adams inheritance of "a strong taste for politics," when his turn came to use it he "failed ignominiously." "The fault," John said, "was my own and was incurable." Thus, "I swore off." John was aware that when he put politics behind him, he began living more happily than had perhaps any other member of the four generations. There was no sign that he felt guilty about the endless hours spent puttering as a farmer or sailing.

As the changes in Quincy life made the once sleepy hamlet less agreeable to John, he drifted away from family scenes. In 1888, two years after his father's death, he gave up his twenty-year career as moderator of the town meeting. He bought a house on Commonwealth Avenue in Boston's Back Bay for winters, and he even left the Mount Wollaston farm during much of the summer to live by the ocean at the Glades, a communal resort property in which he and Charles each bought a share in 1880. There they joined the Sharp, Saltonstall, Sturgis, Lovering, Ames, and Codman families. South of Quincy along the shore, near Minot, the Glades was the spot where John and his brothers once had fished and frolicked with their father. Here John was happiest, for he could be entranced by, as Charles indulgently observed, the profoundly trivial subjects which were increasingly his only interest.

Some observers believed that John's withdrawal from a vigorous life could be traced to the devastating loss of the two children in 1876. Charles claimed that for John the death of the youngsters was "a blow from which he never fully recovered." The elder brother did fill two roles thereafter, but neither obliged him to act independently. He was treasurer of the Harvard Corporation, after his father had persuaded him to accept election to the board. And, of course, John attended to family business, but never without close supervision, first from his father and then from Charles. John's life was so unambitious that at times

Charles grew annoyed and even outraged: "I might just as well try to make an empty bag stand on end."

This contrast between the two eldest sons made it easy for Charles to take command in the family, and thus to refashion its character. Charles welcomed the rising materialism in America and he saw himself in this new age as the ideal central figure for the house of Adams. He certainly seemed to fit well into an America where great wealth was coveted. During the last years of his parents' life, he looked down from the handsome residence in Quincy which he had built on President's Hill. He lived there in the summers, although after 1880 he and his family spent much of August and September at the Glades. In 1887 he faced winter in a palatial Back Bay residence on the corner of Gloucester Street and Commonwealth Avenue which had taken him twenty-two months to complete.

From these locations Charles went forth to conquer the world, hoping to do so by railroad management, social criticism, historical investigation, and finally—disastrously—speculation in western lands. His need for a glamorous, triumphant life had troubled his parents and, in his own heart, distressed Charles himself. When he felt this way, he thought of John Quincy Adams. "My experience as to bauble chasing," he claimed, was not the first in the family's story. "More melancholy reading than my Grandfather's 'Memoirs' I know not where to find," Charles said. "He chased the phantom reputation, driven to it by the demon of duty, until he fell down four score."

Charles spent the years between 1877 and 1889 seeking wealth—and debt, influence—and disappointment, gratification—and unhappiness, but like old John Quincy, Charles too usually had the grace not to deceive himself. He knew himself to be imperious, abrasive, and cruel. "Nothing tells like being contemptuous," he once boldly wrote Henry in an effort to explain his style. Yet at heart, Charles always remained the miserable boy who had begged his parents to treat him more like the brothers he envied. He suffered often from depression. At the end of 1886, Charles took stock of himself and did not much like what he saw. "I am merely interested in a large game and to it devoting my life. Yet whether I win or lose it will not much matter. To it I am, as I well know, sacrificing things worth incomparably more. But I am in the game, and must play it out."

Even in the cruelest winter weather, Charles often walked on Sundays from Boston to his empty Quincy home, triumphantly noting how he needed less than the three hours most mortals required for this trek. After building a fire, he settled down for hours of reflection and writing—usually about himself. At times he gamely tried to be proud of his radical departure from the Adams philosophy. Never, however, was he able to escape the nagging suspicion that, alluring though his life might appear—travel in a private railroad car, frequent audiences with presidents of the United States—it was all wrong.

"What does it come to?" he forced himself to ask, and answered: nothing more than "mere living life out." Watching how he was caught in an amoral society, Charles made one of his most perceptive confessions to himself in 1889. He lacked at heart "the thing for which the world gives me most credit," he said, which was courage. He even had trouble facing a weight problem, as he gained eight pounds in less than six months. For Charles, 1889 was made depressing by much more than his mother's death. It was then that he realized he was failing in a bid for national eminence in the world of affairs.

In the 1870s and 80s, Charles had invested heavily in land and business, borrowing extravagantly to do so. By 1883, his interest as well as his investment in the Union Pacific Railroad earned him election to the board of directors. A year later he became president and undertook at once to improve management, expand the road's power in the west, and reduce the federal government's role in company affairs. The result was a complicated situation in which Charles correctly gauged neither men nor conditions. In particular he overestimated his own power and ability while underestimating that of Jay Gould, the leader of the Union Pacific who had bowed out to watch Charles try his hand.

By 1889, the Union Pacific was in great peril; its circumstances were, not surprisingly, very much like Charles' personal affairs. The firm had heavy debts and its resources to meet these commitments shrank relentlessly. Charles had no choice but to surrender to a much more practiced and shrewd manipulator of people and finance, Jay Gould, who set out to save the company from the effects of Adams' presidency. On 26 November 1890, Charles resigned his post and once more had to face a familiar quesion—what should he do with himself? He found his views badly mixed.

Should he leave the business world he now professed to loathe? Charles said of Wall Street, "I never walk down that thoroughfare of knaves and knavery without a sense of chill and a wish that I might never find myself there again." Should he enter politics? "The experience of my family for a hundred years ought to be a warning to me there. They had all that political preferment could give. Were they made happy or contented by it?" Yet something in his life must change, Charles knew: "I would not relive any of it." With a yearning for contentment now foremost, Charles sought asylum in the calm of the writing chamber. Here for him, as for all his forebears, there might be consolation and refreshment. "I love literary work for its own sake," Charles said. "I like to sit among my books in the sunlight of my library and investigate and write of the past. Here only have I got pleasure heretofore; here only shall I find it hereafter."

Once more, however, Charles was divided within himself. Part of him perceived that to enter his study was mere evasion: "I go out of the present world which I can't manage, into the past where I am master." There he moved from topic to topic, craving the attention of the reading public. "Damn'em though! I wish they had a little more correct idea of what is really good." Charles remained an unhappy wanderer, like his grandfather after the election of 1828, but without John Quincy's discipline, courage and moral vision. He did spend most of the rest of his life in his library, but it was not by choice. His failure as president of the Union Pacific Railroad left him no less eager to dominate others than before.

Charles' nature was reflected in his style as a husband and father. He was by turns exasperatingly jovial and cold, kind and harsh, concerned and indifferent. Devoted to his wife and children, Charles still never attained the passionate regard for their well-being which had characterized the fathers and husbands of previous Adams generations. He seemed aware of this in 1888 when he took his twin sons to Groton to begin their education. "I tried to say a few words to them—telling them to be honest and manly, and if in trouble to always come to me," he wrote in his diary, "but it didn't go very well and I had to give it up. So we rambled along talking of all sorts of things until I got my feet very wet and we had to go back." As usual, Charles could glimpse what he wanted to be or do, but he could not get there.

Soon after his failure as Union Pacific president, perhaps to reassure himself, he switched his zeal to advancing the family's wealth left by the recent deaths of Charles Francis and Abby. Out of his frustration, Charles blended his ambitious designs with the future of the Adamses, unveiling this strategy before his gullible brother John. Family wealth must be risked "on the growth of the country," said Charles. This meant investment in western lands where Charles assured John they would make huge fortunes for themselves and their relatives. More impor-

tant, said he, from such fresh vigor the Adams family would become
for the nation what the Astors had been to New York City, the inspiration and leader for growth. Proudly, Charles claimed that he was the
only Adams with the experience and insight essential if the family was
to be a dynamic force in the new America. Only wealth now counted
in the United States. "Those who went before gave to my family all the
eminence it needs," said he. "I can give to what they did such family
permanence as things in this country admit of."

With John as an ally, Charles drew upon the family fortune, which
together they were managing, as the collateral for his plan to make a
splash in the nation's Gilded Age. In one way, Charles was correct. No
Adams before him had preferred enterprise to statesmanship and letters. As if to start afresh, he took to reading the diary he had begun
as a lad. Afterwards, he destroyed it—"worthless, mortifying trash." He
did preserve and continue his smaller pocket diary, which was left at
his death. With the larger diary burned, Charles commemorated his
new ambitions by beginning a weekly practice—usually on Sunday—of
commenting upon himself and current affairs. The result over many
years was a series of essays he called "Memorabilia." Here, and in his
pocket diary, he talked frankly of his determination to lead the Adams
family into a different national role.

There were alarming signs, however, in Charles' own finances. While
he claimed to be a millionaire, he conceded that in 1889 he also was
shackled with over a million dollars of debt. As for his 1889 income of
more than $100,000, mostly derived as president of the Union Pacific,
Charles had spent it entirely in pursuing the good life. In that year he
purchased $14,000 in art works. The next year he paid $9200 for his
daughter Mary's wedding to Grafton Abbott. He confessed to Henry,
"I freely acknowledge to having reached that point in life at which the
best is none too good." When his wife chided him on being so restless
about money, Charles told her that she did not understand him, "and
I don't think you ever will. . . . I am an energetic man, but not a
restless one."

At any rate, in a spirit of aggrandizement, and using his brother
John as a kind of genial clerk, Charles tried to make himself and the
Adams family eminent as the nation developed the Far West. The others of his generation were unaware of his plan. Unfortunately, Charles
lacked two vital attributes needed if his grand design was to succeed—
wisdom and luck. Although he announced after resigning the Union
Pacific presidency, "The simple fact is I have been tried in the balance

and found wanting," and that he was "responsible for my own failure," Charles did not think twice before investing the money his family jointly held. "I shut the sneering world out," he insisted, as he approached family money with the same extravagance that had characterized his private spending. While conceding that all signals in 1891 indicated caution, Charles still pursued his plan, pushing the family debt over two million dollars. He borrowed vast sums to purchase lands in the West, particularly in Idaho, using as collateral the real estate owned by his brothers and sister under the family trust.

The Adams financial empire was therefore in serious trouble even before the national depression in 1893. Then came the terrible crash of that year, severer than any previously in American history. "For the first time in my life I could not sleep," Charles recalled. By late June he could see that catastrophe was inevitable. John was aghast at what his casual complicity had done to the family fortune. He collapsed emotionally and physically, "a mere bag of meal," Charles said. Admitting that he himself was "a financial wreck," Charles surrendered. The only thing to be done was to summon the rest of the Adamses, for "any dirty linen must be washed in the bosom of the family." So humbled was he for the moment that he acknowledged to Henry how "awfully clumsy and reckless" he had been. In early August 1893, the Adamses gathered in the family's old office in Boston to look at the wreckage. Charles was wretched, wanting only to "hide in my hole and think of my own blind folly."

However, contrition and humility were not qualities which easily overcame an Adams, least of all Charles. As his brothers confronted him, Charles rediscovered his old bravado, especially after Brooks assailed him for his sins. He tried to fix the blame for the fiasco elsewhere, though later he vowed he would never forget the "very unpleasant two hours" spent at the family meeting. In that session, Brooks had been the most vocal in his shock and anger over the failed investments and huge indebtedness Charles had incurred in pursuing a new kind of fame. Worse than these losses was the probability that much of the precious ancestral land cherished by the early Adamses would have to be forfeited to pay debts.

The family eventually recovered, thanks mostly to Brooks, although Charles deserved some credit. He did not shrink from the humiliating duty of approaching his father's friends for the loans needed to prevent bankruptcy. Every member of the fourth generation thereafter carried financial and emotional scars from this experience. While the losses were never completely recovered, more was surrendered than merely money, for the family was never the same again. The distrust of Charles became so deep-seated that Brooks led Henry and Mary into a separate combination. There was also a tragic aftermath, for the catastrophe in 1893 hastened John's decline and death.

Guilt, remorse, and fright sent John into a mental and physical collapse. He hid at home in a pitiable state, being firm only about one issue. When Fanny volunteered her own considerable fortune to help the Adamses, John refused. Thus, it became easy for Charles to blame John for the plight of their trust. Had John "stood up equally well," Charles complained, "we should have gone through the ordeal in excellent shape—but as it was I had to carry him, lug him along by sheer force." John served as Charles' scapegoat. "He is no sort of good here," the guilty younger brother complained.

However, by late summer in 1893, Charles awakened to the seriousness of John's state. Acknowledging that his brother's "helplessness of condition alarms me more and more," Charles undertook to persuade Fanny and her sons to take John abroad, the traditional soothing antidote for troubled Americans of means. John refused to budge, so that in November Charles "took a hand" by summoning his nephew, George Caspar Adams, "and scared him nearly out of his mind about his father." For a time, Fanny stood out against the trip to Europe, telling Charles that John would "do better if he remains quietly at home." Fanny indeed was eventually proved sounder in judgment than her domineering brother-in-law. However, she finally consented to go with John, and the pair departed for England in January 1894, leaving Charles now convinced that John's wife was the main problem. She had moved too long "in the same nasty little circle." Said Charles to his wife: "Bah!—they make me tired."

John's doleful appearance upon embarking changed little during the next four months. Never had Charles seen "a more lugubrious looking individual" go aboard ship, he reported to Henry. "One would have thought it was a defaulter returned upon requisition rather than a gentleman going abroad for the first time in his life." Charles gave no mind to the distress John must have felt at being forced to exchange surroundings he loved for a continent he had always detested. In April Charles met John in England, finding him "a hopeless wreck." Soon afterwards, John returned to Quincy where his weakened appearance distressed everyone. In early August 1894, he suffered a slight stroke, and then at dawn on 14 August John died peacefully after a second stroke. It was the day his tightly-knit household intended taking him to his beloved seaside haunt at the Glades. Death had come at age sixty. An autopsy revealed that his condition was a rapidly progressing arteriosclerosis particularly evident in the brain tissue.

"It was a thunderbolt," Charles said when he heard the news in England. "He lying there dead!—I never knew life without John. I don't accustom myself to the idea." Charles left no record of any remorse for having pushed John to collaborate in using family resources for land speculation. However, he knew that doing so left John with guilt his aging frame could not sustain. Charles was obviously relieved when the doctors informed him, after he pressed to know, that John's degenerative condition had existed for over two years. Freed from the thought that he might have precipitated John's premature death, Charles announced triumphantly that his brother was a victim of the same disorder that had claimed their father. He then bade John farewell with the revealing summation: "Early in life he dominated me," said Charles, "later, I dominated him."

After John's death, Brooks led the rest of the fourth generation in tying Charles' hands so that new follies might be avoided. Their goal was to cut free of him. Acting through her son, also named Charles, Fanny Adams joined Brooks, Henry, and Mary in this action since she had inherited all of John's part of the Adams Trust. Fanny's concern was seasoned with resentment, for soon after her husband's death, her son George Caspar Adams also died. He was another victim of that dread Adams foe, alcoholism. As senior member of the family, Charles had been urged to help his thirty-seven-year-old nephew. He refused to intervene, talking of "whiskey working on a soil without moral sand" and "how we throw back, like dogs." Insisting that "George wasn't a bit of Adams," Charles claimed his nephew "threw back" to someone "of a different stock."

Fanny had placed her husband next to their two children in the Mount Wollaston cemetery where George now joined them. Thereafter, she lived in quiet elegance until May 1911 when, despite her apparent splendid health, she died of a coronary attack. Her daughter Abigail, known as Hitty, reported to Henry: "We buried Mamma on a beautiful summer afternoon among all the spring flowers that she loved so well. It was lovely over in the little cemetery and I was glad to think of her safely by my father's side at last—and among her long-lost children once again."

Galling though Fanny's viewpoint and opposition were to Charles after the family came apart in 1893, his sterner feelings were for Brooks, the leader of the rebellion against him. The elder brother had striking recollections of Brooks' behavior in the crisis. "Brooks," Charles was still groaning years later, "was a croaking raven—before, behind, beside me!" Not only had Brooks become an ardent pessimist, which greatly annoyed Charles, but he was an astute businessman who was as cautious as Charles was extravagant. With relentless pressure Brooks moved the family toward one goal which, as he reminded Charles in 1894, was "to wind up our joint concerns." No one, least of all Charles, was to "embark on anything new either in joint or several account, so long as any of our economic obligations are outstanding." One reason for Brooks' vehemence was that he shared their father's reverent and cautious view toward the Adams landed assets. The voice of Charles Francis spoke through Brooks when he bluntly expressed the family's indignation to Charles: "My objection is now, as it has been for years, to carrying very highly speculative stocks . . . I call that gambling, and I can't go for it. I believe it to be ruin sooner or later. When one pays cash one may gamble for a rise, but not on credit."

So Brooks tried to emancipate the family from Charles. Brooks told Henry that if their brother had been "willing to live reasonably," they might have settled the family's financial problems as early as 1897. "Charles has, however, squandered money at such a rate that he has exhausted his funds, and has drawn very deeply on the surplus, nor have we been able to check or control him in the least." Brooks was blunt to Henry, saying that as long as Charles retained any element of discretion over family money, both present and future security were imperiled. Meanwhile, the more Brooks preached the word "restraint" to Charles, the more furious the elder brother became.

Since neither Henry nor Mary felt in a position to challenge or supervise Charles, it was left to Brooks to keep track of him. Under the youngest brother's eye, Charles was allowed nominal command at Adams headquarters in Boston's financial district. It was not easy for anyone to accost Charles, but Brooks persisted. It was he who made Charles first see that his family no longer trusted him. In 1897, Brooks had to insist that he rather than Charles should hold the collateral which had been set aside to cover unprotected notes. The family plainly feared that Charles would throw the collateral after his other imprudent investments. As graciously as his nature permitted, Brooks pressed for "dividing the estate," while apologizing for seeming so "fault-finding and meddling."

By July 1902, a new family combine had been created which did not include Charles, who remained heavily in debt to it. Even Henry joined the new union, assured by Brooks that "we no longer have reason to fear Charles." The price was high; they had not only to unload the worthless western properties which Charles had fastened like a mill-stone, as Brooks said, around the family's neck, but also to sell the Court Street property in Boston which a youthful John Adams had bought and where the family had kept its business center. Although Charles pleaded against such a painful and drastic measure, his broth-ers and sister insisted. It was a bad moment for Charles. Rather than leading the Adamses to a glorious career, he had created a nightmare. He reluctantly agreed to the sale, acknowledging that "I am a back number"—"a wholly new order of things is impending."

The rest of the family watched Charles closely and not without sym-pathy, for he was now the eldest surviving member of the fourth gen-eration. Brooks felt his brother would come out of the ordeal satisfac-torily, but only if "he will face the music, cut down as the rest of us have, live on what he has and not on what he dreams of, and make no more debt." The problem, as Brooks put it succinctly, was that Charles "will never face realities while he can avoid discomfort." For his pains, Brooks earned the everlasting animosity of his brother. Charles' ful-minations against him usually took a personal turn: "Those gaunt, wearied eyes, that anxious furrowed face, that croaky voice and tense utterances. . . . The world is large enough for him and for me,—he may now go to the left, and, if he does, I will strike out decisively for the right." By 1903, Charles had fastened the blame for his own down-fall upon Brooks and his vituperation was extreme. "Such a compound of incompetence, egoism, pessimism and self appreciation," Charles growled.

When the tangle of debt and ownership had been smoothed, the Adamses were half a million dollars poorer. Unfortunately, however, the troubles for the fourth generation were far from over. The breach between Charles and his sister and brothers widened. It was discovered that Charles was operating high-handedly with a family treasure more precious than the money—the cherished letters, diaries, and other manuscripts which had been passed so lovingly from generation to gen-eration.

There were two issues. The safekeeping and organizing of these doc-uments, as well as granting outsiders permission to use them. Once again, the elder brother considered himself in command. According to the will of Charles Francis Adams, all these materials, along with the libraries accumulated by the second and third generations, were to remain in the safety of Quincy, in the stone library adjacent to the Old House. Like the family money, Charles treated these materials as if they were his own, and he allowed the papers to be read by persons whom his relatives considered "strangers," but who were mostly scholars.

What especially troubled Brooks and Henry was that these family documents, so liberally displayed, contained the unhappy side of Adams history. Mary, too, was indignant at his casualness. She, Brooks, and Henry began another campaign against Charles. They agreed that something must be done at once, lest (as one family member put it) Charles might deem it "his special duty to publish everything in sight." This danger roused Henry much more than financial peril had, and it was he who proposed that the papers be placed with the Massachusetts Historical Society and closed for fifty years until a later generation could assess the situation.

It fell to Brooks to bear this ultimatum to Charles who responded in a scene Brooks never forgot. "Instantly," he said, Charles "set off at speed upon my father, the Monroe Doctrine, and Worthington Ford. He gave me a diffuse sketch of what *he* had done and the Governor [CFA] had not, of how much superior his grandfather was to his father, of how much superior he was to either." After threatening to destroy all the manuscripts, claiming he had used the worthwhile parts, Charles demanded a full hearing before all the family. At this point in the scene between them, Brooks reported, "I wilted. I can't stand that sort of horrible infliction. I fled with my tail down."

This encounter took place in 1902. Eventually, a compromise was reached. The papers were removed to the Massachusetts Historical Society, but Charles remained in full command. There he supervised everything, destroying much, and employing an aging Lucy Baxter— whose eyes might be trusted—to help with the sorting. It was not a satisfactory solution. Reports reached other family members that Charles was still genially allowing public access to the documents. Clearly, a legal understanding was essential if the elder brother was to be restrained. With Henry's backing, Brooks undertook in the autumn of 1905 to get signatures on a document placing both the manuscripts and the Old House in a trust. He worked ardently, convinced he was putting out of reach letters "which ought never to have been kept and which no one now should ever see."

Charles Francis Adams in 1879. Portrait by Frederick P. Vinton. The fix of the eyes in this likeness was an alarming sign to his family of Charles Francis' loss of mental power. *Courtesy of the National Park Service, Adams National Historic Site.*

Abigail Brooks Adams in 1872. Charles Francis considered this portrait, by William Morris Hunt, to be a masterpiece. *Courtesy of the National Park Service, Adams National Historic Site.*

Left, Fanny Crowninshield Adams, wife of John Quincy Adams 2d, in an early photograph. *Right,* Mary Gardner Adams, daughter of Charles Francis and Abby. Photograph probably made near the time of Mary's marriage to Henry Parker Quincy in 1877. *Both, courtesy of the National Park Service, Adams National Historic Site.*

The second Charles Francis Adams in a portrait painted in 1876 by his good friend, Frank Davis Millet. *Courtesy of the National Park Service, Adams National Historic Site.*

Left, Brooks Adams and his wife, Evelyn "Daisy" Davis, pictured around 1890. *Courtesy of the National Park Service, Adams National Historic Site. Right,* Henry Adams writing in the summer house he and Clover cherished at Beverly Farms. The photograph was taken by Clover, probably in 1883, two years before her death. *Courtesy of the Massachusetts Historical Society.*

The monument for Clover which Henry Adams commissioned Augustus Saint-Gaudens to create and which stands at the graves of Henry and Clover in Washington's Rock Creek cemetery. It must be seen to be appreciated. *Courtesy of the National Park Service, Adams National Historic Site.*

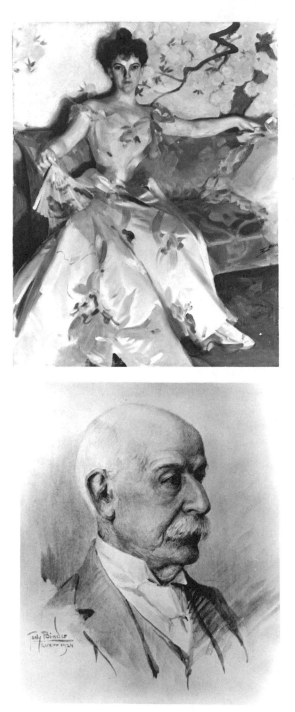

Elizabeth "Lizzie" Cameron in the famous painting by Anders Zorn in 1900. She was just past forty when this portrait was completed. *Courtesy of the National Museum of American Art, Smithsonian Institution.*

Brooks Adams in 1924. From a crayon drawing made in Luxor, Egypt. *Courtesy of Mrs. Wilhelmina S. Harris.*

In explaining to Charles what the family had in mind, Brooks took as his text "the dead cannot defend themselves." It was, said the younger brother, a solemn duty to guard their ancestors against those who clamored for admission to the papers, individuals whom he dismissed as interested only in "malicious gossip." "Our ancestors have suffered enough from indiscreet publication and breach of private confidence," Brooks asserted. He feared Charles' carelessness so much that he began talking of destroying all the indelicate material, documents which told of the sad lives of Charles and Thomas Boylston Adams from the second generation, and George and John Adams from the third.

This latest family disagreement stemmed from the growing belief among Henry, Brooks, and Mary that Charles did not have their affection and regard for earlier generations of Adamses. In fact, Brooks declared himself defender of "the family reputation" against all molesters, including their elder brother. Seeing that his family "all want to pitch into me," Charles said, "I simply won't fight." But he remained annoyed at the family's suspicion that he would open the private letter books and diaries to strangers. Meanwhile, Charles was not the barrier to an agreement on the manuscripts. John's widow held back, fearing that in sealing the documents for fifty years she might surrender something her children would cherish. Now supporting the plan, Charles fumed in impatience, grumbling to himself, "Oh! Hell!—more delay." He said he had no sympathy for anyone like Fanny, who was "simply absorbed in 'my children' and the family instinct."

Eventually, Brooks persuaded Fanny to sign the deed of trust, although the matter did not rest there. Even after the agreement was executed and the precious family manuscripts presumably in safe repose at the Massachusetts Historical Society, the controversy occasionally flared anew. As an old man, Charles could not resist poking around in the documents, letters, and diaries and talking of opening them to others. Sometimes the pact was so strained that the Historical Society had to convey the opposing opinions between the two family camps—Charles on one side, his brothers and sister on the other.

Such animosities reinforced the position Charles had taken among his relatives since childhood. By taste and impulse he had been an outsider, one who could never find a role comfortable for himself within the Adams world, and yet could never break free. He surely had tried—through wartime service, through business, through life-style, through friends. But he had never succeeded. After 1893 brought controversy over money and manuscripts, the tension between him and his family seemed never ending, so that Charles remained restless and unhappy. Once, after his seventieth birthday, he tried to explain himself to his wife by saying, "Of course I'm responsible—I always am, for everything that is set as it ought to be. . . . Its a great deal better to take all the responsibility, and have done! So I just assume all the blame for everything, and there isn't anything more to be said about it!" He did accept responsibility for what he considered to be the blunder of his life: "I should have cut clear from the family traditions."

When he lost his hopes for national leadership in business and finance, Charles had no recourse but to retreat into his study, there devoting the last twenty years of his life to those time-tested family consolations, historical writing and public commentary. He claimed to feel "safe from the world." "I seem to have got into harbor at last." From this refuge and like many an Adams before him, Charles began to study himself, conceding that nature did not intend him for a politician, an executive, a soldier, and, he added, "what I lament most of all" was his unfitness "for social life and enjoyment, and the easy and friendly mixing with men and women." It was the same complaint heard in every generation of his family.

Casting about for something to do, Charles decided that "it was for literary work that I was adapted." How unfortunate "that one discovers ones aptitudes and limitations only at 60; but my father certainly did not understand mine, when he was 57 and I only 28!" Amid the financial wreckage of the 1890s, the old need to excel returned to Charles. He was no more able to sit quietly in his study than any earlier Adams. Now he tried publishing and oratory. He talked loudly and in extremes on many subjects: education at Harvard, imperialism, race relations, Theodore Roosevelt. His stands hardly earned him an audience even among his own family. Once more he distressed them. He caused his sister and brothers genuine pain when he decided to make a biography of their father his main project.

For this task Charles prepared himself by reading Charles Francis' diary. Early in the process, during 1895, Charles confessed in his own journal, "I don't like my father!" He derided his father's domesticity, claiming that his mind was never exercised "by contact with men and life." Henry sensed what was happening and tried to soften his brother's approach, urging him to recognize the marvelous balance in Charles Francis' make-up and outlook. "Had there been a little more of him, or a little less of him, he would have been less perfect," said Henry. "As he stands, he stands alone." Not only did Charles not want to hear this, but he may not have seen Henry's point. While Henry begged, "let the Governor have his own say . . . show him as simply as he showed himself," Charles heeded only his own apparent determination to be as harsh and misunderstanding with his father as he felt the latter had been to him. This stemmed also from resentment that his father had achieved such mastery over self and life, an astonishing performance for an Adams and well beyond the younger Charles' capacity. As a result, he lost sight of his father entirely. "He was a Puritan," Charles sneered, "born, ingrained. That dreary taint permeated his life. Duty!—everywhere. Kindliness, geniality, sentiment, sympathy with man and nature nowhere. . . . I do not like him."

He vowed to write, however, with objectivity. "I will forget he is my father." But such moods vanished when Charles reminded himself of his appointed task: "I have to do for my father what he failed to do . . . save him from himself." In March 1897, the research for the biography was finished and Charles sat back to contemplate the parent about whom he must now write. His summary was chilling. Charles Francis appeared to this son as a person completely different from the man lovingly, even thrillingly recalled by Henry, Mary, and Brooks. For Charles, words such as hard, cold, selfish, hot-tempered, arrogant, dominated by family pride, and devoid of passion best described a father he claimed was ruined by "that New England meagreness and rigidity, that hard, cold narrowness of a biblical morality."

This hostility can be explained by remembering that Charles believed he had never been understood or appreciated by his father. Charles never forgave what happened when he enlisted as a soldier in the Civil War. It was the one move of his life in which Charles always took comfort. But his father had not been sympathetic. "Had my father no sense of manliness!" Charles would shout at his desk. "What milksops with pallid cheeks he wanted us to be!" To Charles' credit, however, he did pause at times and speculate over his own outlook. "I wonder whether I am equally purblind, and morbid, and put such an absurdly wrong measurement on events and development. I suppose so! What fools and imbeciles we mortals are!"

Charles contended that his own nature was much like the family member he most admired, his grandfather, John Quincy Adams, from whom he claimed to have inherited an inclination for passionate attack. Even here Charles was angered by his father who, he said, should have used more restraint in editing John Quincy's journal. Embarrassed that his father had published so much of his grandfather's hasty, tempestuous observations, Charles announced: "I cannot pardon it. It was a cruel, irremedial wrong to my grandfather."

However, when the biography of his father was completed, Charles had managed to control himself remarkably well, allowing only glimpses of his peevishness to appear. These are scarcely noticeable today, but in 1900 they were enough to remind his family of the grudge they knew rested beneath the book's pages. Brooks summed up their feeling: "Charles has got his life of our father out. I rather shrink from reading it." When he found the courage to speak to Charles about the book, he avoided the issue and settled for praising their parent in superlatives: "I have never yet met the man whose power of statement or methods of attack were so admirable. He never erred and he almost never failed." However, Brooks stopped short of a fight over the book, telling Charles, "enough of this. I take it you have your own view of my father pretty well digested by this time and have no particular interest in reading mine."

Relief that Charles had done no more damage to their father brought Mary to tell him she liked his work, and then immediately to beg him to discard his plans for a much longer study. She predicted a larger biography would be too personal and involve the remembrance of other individuals associated with the Adams family, and she cautiously suggested that her brother was "not sensitive on that score."

It was not Mary's praise, however, which Charles craved, but Henry's, the brother whose great talent and unique life Charles secretly envied. As soon as the book appeared in February 1900, Charles hastened to Washington to dine with Henry. It was to be "my ice-breaking evening," Charles said, in high hopes, but he was painfully disappointed. "[N]ot by word, look, or line" did Henry recognize that Charles had published the biography—and Charles himself was too uneasy to bring it up. A month later, there was another dinner with the same result: "pretty dull and very restrained—not a reference even to the 'Life,'" Charles mourned. To Brooks, however, Henry was candid. "Thank my miserable cowardice that I did not write it," he said, recalling that earlier the family had encouraged him to do so. He still believed that Charles Francis Adams should speak for himself, defying "definition" by others, especially kinsmen. Telling Brooks that no biographer should put an Adams under "the microscope or even the telescope," Henry said he also dreaded "the photograph and Sargent's analysis of character."

Meanwhile, Charles turned to another task which also caused uneasiness in the family. Having taken his father's measure, Charles now began an autobiography, which he completed many years before it was posthumously published. Charles always wrote well and had the happy talent of needing to revise very little of his first drafts. Consequently, the version of his autobiography which appeared in 1916 was very much as he had left it in 1900. "I prefer to be my own Boswell," Charles announced, acknowledging that his chief motive for an autobiography was to be certain his case was fairly presented to the world. Only Henry might have written his biography, Charles said, but "of late years" he feared Henry had "not been in touch with me or in any particular degree sympathetic." So Charles set out to explain himself, using the theme: "Taken as a whole, my life has not been the success it ought to have been." His explanation for this was that his father, with a "coldness of temper," had denied his son the freedom and encouragement needed for fulfillment. The *Autobiography* did acknowledge that there were additional family shortcomings to afflict him, and Charles seemed particularly to enjoy confessing "my utter lack of a nice, ingratiating tact in my dealings with other men and difficult situations." It had been an inescapable flaw as Charles saw it, "an inherited deficiency, a family trait," and thus "my great handicap and hindrance in life."

Charles went on to repudiate heatedly what he said was the widely held belief that success came automatically to a late-generation Adams. "In my case people have always been over-ready to talk of 'family influence' and all that sort of thing in an owlish way, so accounting for about everything I ever accomplished." This was wholly untrue, he insisted. In fact, it required some effort to find something positive about being a child of Charles Francis Adams: "I believe in the equality of men before the law; but social equality, whether for man or child, is altogether another thing. My father, at least, didn't force that on us."

After preparing these memoirs, Charles was soon drawn back to the apparently irresistible subject of his father. He began a much longer biography than had been published in 1900, struggling with this project until his death. In 1913 he employed professional help to speed the work in which he by then no longer moved with his accustomed zest. "My nerves just gave way," he moaned in 1914 and put down his pen, dissatisfied. "I am overworked!" A year later he was dead and the monumental biography of his father was left in manuscript with the Massachusetts Historical Society. The work is incomplete; the first third is a study of John Quincy Adams, and thereafter it is mostly numerous excerpts from Charles Francis' diary and letters, connected by random comments from his son.

This unfinished portrait was not greatly different from the briefer sketch Charles had prepared earlier. Some interesting new features, however, included Charles' insistence that his father had "a terror, almost morbid, of what is known as 'the presidential bee.'" He had called it a "demoralizing mania." Had the elder Adams been elected in 1872, the choice, Charles noted in passing, "would, humanly speaking, have proved an irreparable calamity"—an allusion to his father's rapid mental decline in the 1870s. In bewilderment as much as admiration, Charles reported how his father had coolly refused "to lift his hand" for the presidency. After acknowledging that there was "nothing like it in recent American history," Charles pushed politics aside to contemplate the burden old age put upon his father. Now at the end of the manuscript, an unaccustomed tenderness entered his tone, for by then he was also growing aware of the effects of advancing years upon himself. Lamenting his own weakening condition, Charles ended the battle against his father's ghost in something near to sympathy and reconciliation.

This unhappiness with family memory which kept disturbing Charles at his writing table also vexed him when he walked out of his study to look at the scenes around him after 1890, whether in Boston or Quincy. When his troubled presidency of the Union Pacific Railroad ended, consciousness of failure made Charles physically restless. Pen, paper, and books were not sufficient comfort. He wished to find a new milieu, where every turn did not bring accusations from the past. Once Charles had been happy to write essays about the Quincy area. His excellent collection, first privately published in 1883 as *Episodes in New England History*, had affectionately explored Quincy's story. In 1891, however, Charles' journal reveals that the town had lost "even for me all its individuality and interest." What remained was "sad and repellant. . . . The old, sleepy, quiet look is all gone, and an air of commonplace mediocrity has appeared as a poor substitute for it." Crassness, ugliness, inferior people—these features, so Charles thought, meant that "the time for us to go is near at hand."

Yet making any change troubled Charles, despite his customary talk of being impatient for progress. "Leaving forever the graves of my forebears, —is an awful consideration for me," he said. With an abruptness which suggested he was afraid to think overlong about it,

Charles moved from Quincy in 1893, setting up a new headquarters in South Lincoln, a woodsy area to the west of Boston near Concord. It was a "most painful" wrench, he said. "Quincy was bone of my bone—flesh of the Adams flesh. There I had lived vicariously or in person since 1640. . . . I felt as if I owned the town." Here, perhaps, is the point. Just as the political principles of the American republic seemed to disappoint earlier Adamses, in the late nineteenth century even Quincy appeared to turn fickle and ungrateful. The "stone cutters" who had rejected the authority in town meetings of Charles and his brother John were "good people in their way," Charles wrote, "but their way is not mine," so that "with the present Quincy, I have neither ties nor sympathy."

For $55,000, he sold his handsome house on President's Hill, grumbling at giving such a bargain but convinced that no one worthwhile would now wish to move to the town. In a mood half-haughty, half-sad, he departed from Quincy, completing one of the ironies in the story of his family. His father had helped change Quincy, for Charles Francis, as banker and landholder, had encouraged the coming of modern commerce, remaining to endure in sorrow the changes which followed. Charles took longer to see the cost. For a time he championed the economic transformation of America. However, when the sobering price of trading the simplified order for modern development was reckoned on every street in Quincy, Charles slipped sadly, angrily away, rebelling at a metamorphosis he chose not to analyze.

Something of Charles, however, never really left the community. Even today, the area which Adamses loved and helped for generations continues to benefit from Charles' gift of land for recreation, as well as from his leadership in making the lovely Blue Hills region, just to the west of Quincy, into a magnificent natural preserve. This achievement even his critical brother Brooks hailed as truly splendid. Charles also worked to save the two little farm houses at the foot of Penn's Hill where John and John Quincy had been born, eventually allowing the property to be cared for by the Quincy Historical Society.

Privately, he doubted the wisdom of this step, claiming that the people of Quincy were a singularly unresponsive group. "They never seemed to take any interest in the traditions of the town." Nevertheless, Charles littered the ancestral countryside with statues, tablets, and other commemorative devices to mark his family. He took some wry enjoyment from the story that Roscoe Conkling, erstwhile senator from New York, had, when visiting Boston, been taken mistakenly to Mt. Auburn cemetery when he asked to see the graves of the Adams presidents. He was shown the tomb of Alvin Adams, founder of the Adams Express Company. The startled Conkling is said to have exclaimed, "My God, such is fame!"

Even in his deeds as family memorialist, Charles rarely acted to the satisfaction of his brothers. In 1906, he had prepared tablets of inscription honoring his father and his ancestor, Colonel John Quincy, to be placed in the Quincy church. "For some reason, I do not know what, I did not even consult any other member of the family about them," he acknowledged. This act particularly outraged Brooks, who, as the only Adams then living in Quincy, should have been consulted and who found the plaque on Charles Francis Adams in poor taste. Charles paid no heed, believing that he gave tribute less to his father than to Colonel Quincy, whom he said he had held always "in great reverence . . . I have clung to my descent from him." Now, he had a memorial, "and I have given it to him."

As he aged, family sentiment seemed to pull Charles back to Quincy and occasionally drew from him some genuine kindness. There was his gentle care for two distant cousins, grandchildren of John and Abigail Adams. The aged pair, brother and sister, still lived in the ancient residence where their father Thomas Boylston Adams had moved in 1829 when John Quincy had returned to the Old House. They depended in their last years upon the financial help Charles arranged for them from his brothers and himself. Cousin Hull died in 1900, while Cousin Lizzie lived to 1903, when she died at ninety-five, after a life devoted to treasuring the memory of her parents and grandparents. In a touching scene, one which showed Charles at his best, he made the pilgrimage to sit beside this venerable cousin during her final moments.

"And so the very last tie was snapped which connected my generation with the generation of John Adams," Charles wrote. "I am not sentimentally given, but at my present age, and in the presence of this event, I do feel reminiscent and reflective." For a moment, Quincy scenes so stirred Charles that he was even moved to pay rare tribute to Brooks for living in the Old House and the ancestral countryside. "How Brooks endures it, I don't know," Charles said, adding that the family ought to be grateful that he did "retain this strong sentiment."

Meanwhile, moving away from Quincy brought Charles no nearer the contentment he sought. "Birnam Wood," his new residence in Lincoln, was splendid, especially its 320 acres of forest land and bridle paths. Charles made an advantageous purchase in 1893, except that he lacked funds to pay for the property and was forced to take such regrettable steps as selling his Back Bay residence in Boston. In Lincoln he found a complete contrast to the noise and tawdriness of Quincy, but the satisfactions did not long assuage his restlessness. He announced in 1896 that he needed new scenes in which to develop. "I must go away and have a change, or I shall stifle."

Now Charles grumbled because his wife and daughters opposed anything which might take them from Boston society. "And so we go on; and all the while, life is ebbing away." In these moods Charles resembled his grandfather, John Quincy Adams. He, too, had required a scene which promised action and even leadership. So it was no real surprise that Charles followed his forebear's example and chose Washington as the place to live in winter, away from Massachusetts snow. He did this in 1899 at age sixty-four, anticipating being part of the national scene. Henry, who already lived in Washington, was aghast. "[H]e can hardly help treading on my toes," he wrote Brooks. Since he did not wish "to seem cold" to Charles, he proposed to leave town at once. "At bottom we are very likely identical," he said, "but we differ seriously in matters of expression. It would be hard to keep the peace." Henry was then sixty-one.

On 2 December 1899, Henry flashed the bulletin to Brooks: "True enough, Charles is in the next street. *Que diable!*" The newcomer remained nearby, and in 1907 built a mansion at 1701 Massachusetts Avenue where he did his best each winter season to compete in Washington society. He also never lost his desire for Henry's admiration, so he summoned the younger brother to behold the new palace. Henry reported that "the effect on me was cruel." Such ostentatiousness, Henry said. "I have decided to suppress my 'Education' in consequence of this proof that education consists in going backwards. It is no use. In the house I see Charles Eliot, Harvard College, Theodore Roosevelt and myself rolled into one." Henry wrote to a friend, "when you see the inside of my brother Charles' new house, you will know what to think of him." Meanwhile, Charles was so strapped by his extravagance that he could not find the $12,000 needed to buy the fine furnishings he coveted.

Sixty years later, Charles' grandchildren could remember the festive scenes in the Washington mansion where their grandparents were always entertaining. After dinner Charles customarily announced to the guests that there would now be "general conversation," which soon proved to be a monologue. Nor did he restrict his lectures to the drawing and dining rooms. He took every opportunity to speak in public, to write pamphlets, to address editors, and to advise statesmen and financiers. No president of the United States went long without uninvited counsel from Charles Adams. He had copies of his utterances reprinted in great quantity, mailing them to everyone he thought should have his wisdom.

However, still more was needed to satisfy the craving for activity and attention in this grandson of John Quincy Adams. "I am bored," he cried regularly. What was required were "new interests,—an emancipation from the treadmill." He looked about Washington and sighed. "Oh Lord!—the emptiness of it all." This word recurred again and again—"the usual sense of emptiness." In 1912 he campaigned to be named ambassador to England. His stay could only be brief, for shortly after the preceding ambassador, Whitelaw Reid, died, President William Howard Taft left the White House to make way for the Democrat Woodrow Wilson. "I will frankly confess," Charles declared to Joseph H. Choate, whose political influence he was counting upon, "that it would be to the last degree gratifying to me to be the fourth in direct line of descent to have occupied that post." When Choate hesitated, Charles breezily told him that surely Taft and Wilson could work out an understanding about his role. Choate, however, believed otherwise. He told Charles: "I give it up!" There the matter ended, and Charles subsided.

Charles' many enterprises did not escape his brothers, whose letters frequently spoke with embarrassment and even disgust about him. Their voluble brother seemed to have an extreme opinion on everything. Often, Henry teased Brooks, advising him to be grateful that he had "escaped making a fool of yourself like your brother Charles." For the younger brothers, Charles' sin was "talking out loud." Brooks could be as indignant as Henry. "Really," he said, "Charles is pitiful; why can he not go away if he can't keep quiet." After all, "talking so foolishly is not what *you* can call good 'form.'" To the Adamses' old friend Lucy Baxter, Henry complained about the mortification Charles caused the family by his failure to learn "the wisdom of holding one's tongue."

The problem, the brothers agreed, arose from Charles' determination to prove himself to be, as Brooks put it half-seriously, "the greatest genius of the century." The latest offenses, along with Charles' earlier sins as family financier, archivist, and biographer, meant that even in old age, Charles was usually at odds with his brothers. In 1911 he was painfully conscious of this estrangement, which he sensed had become

a topic of public gossip. People believe "that there is some family difficulty," he reported to Henry; a "case of ill feeling." Consequently, he
felt "this must be put a stop to," and announced that he would appear
at Henry's at 8:00 PM the following evening for dinner. "We will thus
be set right before an inquiring world!"

✥ 16 ✥

Henry

Proposing to dine at Henry's house was not a wise step for Charles to have undertaken. He did not see how eager Henry still was to keep his Adams past at a distance. That meant Quincy and Boston as well, where none of Henry's relatives understood him. Henry confided to Lucy Baxter, the one person before whom he was most likely to drop his guard, that he knew his association with his brothers and sister was always mutually difficult. "I have never yet struck approval on my side," he said, and "as the years pass, I care less and less to challenge either approval or blame"; he was content "to let my brothers depart without further risking their disapproval."

For his family, as he later became for several generations of Americans, Henry was a sometimes exasperating but always fetching enigma. He would have wished for no greater epithet. Brooks came as close as anyone to the mystery of Henry Adams when he said that his brother perpetually required an anodyne, something to relieve pain, to soothe his distress. That may be why Henry was drawn to the unhappy story of his grandmother, Louisa Catherine. One of the treasures in Harvard University's Houghton Library is Henry's transcription of journal excerpts left by his grandmother. It is through her, in fact, that we may begin to understand Henry, for he had an outlook much like hers.

All her life, Louisa Catherine Adams had cherished a family setting, requiring for her own wellbeing both to give and receive help and comfort. The needs of others brought out her best qualities; she was solaced by feeling useful in a world she disliked. In this mode Henry, too, was apparently most comfortable. As a young man, he was a homebody, comforting to his parents, intervening on behalf of his brothers and sisters. Restlessness often touched him as it had his grandmother, but it never relieved him of his concern for persons dear to him. The similarities between the two are striking, and they were unique among the Adamses. As a male and living two generations later, it was possible, of course, for Henry to develop his qualities more liberally than his grandmother could. Henry appeared pessimistic, hypercritical and disdainful, as had Louisa. Yet beneath this veneer, he was gentle, loving, solicitous, and charming, all attributes of Louisa Catherine—and no other Adams. The English diplomat Cecil Spring Rice, who held a place in the family of friends and nieces which Henry created to substitute for the Adamses, wrote in 1893 that "goodness leaks and oozes out of Uncle Henry at every pore—all sorts of kindness and affection and self-sacrifice. It always does one good to see him because it rehabilitates the ten commandments." Only when Charles Francis Adams described his mother, Louisa Catherine, had such a tribute been paid to a family member.

Henry built up several "families": the one he, Clover, the John Hays, and Clarence King created in Washington; his association with Elizabeth Cameron and her daughter; and his gathering of Clover's nieces and other young women in his old age. From all of these he drew what he needed, and gave much more in return. Over and over, the letters and reminiscences about Henry from both men and women stress his sympathetic nature, his patience and kindliness, his good humor despite occasional bouts of depression, and his physical vigor. He sought to bring out the best in those for whom he cared, and his encouragement and generosity were legendary among them.

On the other hand, he was often restless, deeply unhappy, and generally at odds with the world. He sometimes sat in silence, and his writing projects often kept him isolated. He carefully disciplined the use of his time and, like his father, he fought to subdue a quick temper, seeking the calm exterior and gracious tones which Brooks and Charles lacked. More revealing, perhaps, was the irony, detachment, and forced gaiety behind which he hid his deeper feelings. This stratagem baffled and diverted scholars as well as his own relatives. His nephew, the second Henry Adams, chose to call it his uncle's way of "letting off steam." His brother Charles considered it "talking through his hat," his term for Henry's indirect style.

Why did Henry Adams speak through a hat? Part of the answer rests in his appreciation of his own and the world's shortcomings. This con-

sciousness, which nearly all family members shared, kept enticing Henry
to withdraw from a society he came to consider false and futile. Other
Adamses, Brooks and Charles, for example, usually managed to speak
out directly and to be active. Henry had the family need to advise and
admonish the world, but he did so circuitously. This was certainly true
of *The Education of Henry Adams* and *Mont-Saint-Michel and Chartres;* but
it also holds in his two novels, and particularly in his elaborate scientific
explanations of society's impending decay. In this trait, he differed from
his grandmother, who preferred to be blunt about human weaknesses.

Henry told one of his brothers, "I regard the universe as a preposterous fraud and human beings as fit only for feeding swine; but, when
this preliminary understanding is once fully conceded, I see nothing in
particular to prevent one from taking a kindly view of one's surroundings." For Henry, Boston society provided a notable example of how,
on the surface, men insisted upon public endeavor and the idea of
progress while, underneath, doubts abounded on the perfectionist spirit
America had taken from the eighteenth century. Behind the "apparent
dogmatism and self-esteem" of Bostonians, Henry claimed, was "the
same self-distrust and absolute depreciation of self that has marked the
whole puritan stock." It was a public hypocrisy which, especially after
Clover's death, made him keep his distance.

Henry wrote few letters to family members and expected none, saying to Lucy Baxter, "no one needs me." He came to Boston in 1899 as
a gesture to his sister Mary when her husband, Harry Quincy, died.
With some envy, he observed of the deceased that he "was almost the
only man in my family who enjoyed every hour of life, and was never
bored or disgusted." It was not the funeral which bothered Henry, but
"the struggle to face the family." Still, if his help was needed, as the
bereaved Mary felt it was, Henry served. In fact, he entered the controversy over sealing the family papers because Mary begged him to.
"She has been very good and mild with me. I am grateful," Henry told
Elizabeth Cameron. "She irritates me less than my brothers do." When
the manuscripts had been safely put away, Henry, now past sixty-five,
fled, hoping that "the family is satisfied with me now, for I have done
my best to please them all, and need rest."

The disdain and affection which mingled so bewilderingly in Henry's
view of his heritage was explained in a remarkable way by Mabel
Hooper, one of Clover's five nieces. Like all her sisters, Mabel claimed
to have special wisdom about Uncle Henry's ways, and in 1897 she put
her belief in a letter to Evelyn "Daisy" Adams, whom Brooks had mar-

ried in 1889, complying with his mother's last wish that he find a wife;
Evelyn was sister to Mrs. Henry Cabot Lodge. She hid the epistle after
marking it "A Curiosity!" Mabel described Henry as being very lonely
and yearning to have "his own family" as his companions. Granting
"how queer Adamses are with each other," Mabel insisted that "Uncle
would care enormously to have them, his family, come out and meet
him half way at least." Since it was unlikely that either side would make
the gesture, Mabel proposed that she and "Aunt Evelyn" create a "kind
of Adams confederating society (secret if necessary) —anything to make
them dependent on each other." As bait, Mabel confided that Henry
had recently said he would "give *anything*" if Brooks and Evelyn would
move to Washington and be near him, but "he's too shy to tell you so."
Mabel's interpretation evidently was too outlandish for Brooks' wife to
accept, for she shared the general view that her brother-in-law enjoyed
evading anything and anyone associated with Quincy and Boston. If
Mabel was correct, Henry's predicament was that he could not reveal
his heart even to his immediate relatives.

When Henry wrote to members of his family after 1890, he often
spoke melodramatically about melancholy, weariness, isolation, silence,
and welcome death. Having seen everything, Henry told Brooks, he
now had only one way out of boredom, which was to "strike into paradox." He recalled nostalgically his old life, before 1885, "when I loved
and hated and the world was real." Any regret Henry may have allowed himself for avoiding a life of recognizable emotions and direct
statement he kept to himself. However, his genuine feelings may have
crept into a letter to his niece Hitty, daughter of John and Fanny Adams,
when her mother died in 1911. Hitty must now accept leadership for a
new era, he wrote. "I hope you will not feel about it as I did when I
flatly refused to go on, and stuck to lying on my back, and telling every-
body to go to the devil." He had never again stood up, Henry said,
since he could depend on others to do what should have been his work.
Hitty, however, must not run away but must take her part, "smile graciously on us all, and be head of all the families. . . . Don't forget it!"
he begged her.

Shortly before this, Henry had told another person that since Clover's death, the careers of all his family and friends had broken apart
while he looked on, unable to help them. The appreciation of such
futility seemed to him enough to justify resignation in the face of doom.
"It is never one's fault" when dear ones go amuck, Henry said. "We
have been always the victims, never the causes. Disease, Insanity, Vice,

Stupidity have ruined our lives through those on whom we depended; we have been bankrupted by our partners, and commonly our partners have suffered first and worst. . . . We are just the raft of the wreck."

Sometimes his air of composure failed him and Henry was heard to wonder if he himself might have been the reason why "every household on which I depend should, one after another, with mathematical certainty, fall to ruin." By this time he meant not only his own tragic bond with Clover and with his Adams relatives, but his relationship with the Camerons and the Hoopers, two families which he adopted after 1885. In those thirty years until his death, Henry's strong impulses of love, solicitude, and responsibility took command of him even though he spoke and wrote all the while about his hopelessness.

Henry's friendship with Elizabeth Sherman Cameron remains today, as it was in their time, a cause for speculation. Their association was resented by Henry's brothers and particularly by his sister Mary, who, as late as 1912, said impatiently, "there has been disagreeable scandal enough about that affair." The Adamses, like most observers, never quite understood the relationship. The fact that it seemed indiscreet was enough to offend Henry's relatives. When, soon after Clover's and his parents' deaths, Henry allowed Mrs. Cameron to draw him even further from the Adams family circle, his brothers and sister were dismayed.

Elizabeth Sherman, called Lizzie, was born 10 November 1857, making her nearly twenty years younger than Henry. She was the youngest child of Charles Sherman, about whose financial plight, it is said, his famous brothers, Senator John Sherman and General William Tecumseh Sherman, became so concerned that they arranged in 1878 for Lizzie to become the second wife of Senator James Donald Cameron, political boss of Pennsylvania. In effect, Lizzie was purchased by her husband, over the protests of his own children, an arrangement which evidently tore her from the man she loved, a young New York attorney. Don and Elizabeth Cameron were an unhappy match from the start, she having begged until the very eve of her wedding to be released from the arrangement.

Lizzie must have been a stunning bride. It is easy to see why Don Cameron and many other prominent men were drawn to her and why Clover Adams had admired her fascinating beauty. The Swedish artist Anders Zorn, who had a gift for capturing the quality of the women who sat for him, finished a portrait of Mrs. Cameron in 1900 which reveals a woman still young enough to have beauty and style. The extended bare arms seem both to invite and to recoil. At the center are dark, abstracted eyes and a pursed mouth, which speak of hurt, cunning, or wisdom, depending on the beholder's viewpoint.

Little of Lizzie's splendor was lost even when she passed her sixtieth birthday, for there are still persons today who remember her talent of making every gesture captivating. All her life she remained clever, sharp-tongued, and very lovely. In other words, Lizzie was all Clover Adams had been and beautiful besides, which might account in fact for Henry's intimacy with her and dependence on her. Clover and Henry took a fancy to Elizabeth Cameron, young wife to an aging senator, when her husband brought her to Washington. Clover visited Lizzie with a gift of flowers, a last social gesture before her suicide. When a baby was born to the Camerons on 25 June 1886, Lizzie wanted to name her Marian after Clover. Since Henry was traveling and could not be consulted, the little girl was instead called Martha after one of the Senator's grandmothers.

Lizzie Cameron turned from a loveless marriage to devote herself intensely to her daughter, and to succeeding as a woman whose charm compelled the attention of almost every man in Washington. A notable exception was Henry James who found Lizzie hard, a person who sucked the lifeblood of his friend, Henry Adams. One of her brothers-in-law, Wayne MacVeagh, was also no admirer: "If only to her other gifts there had been added a little sincerity and consideration for others!" he complained. Lizzie apparently needed the attention of men of all ages and her flirtatious charm seemed to return many middle-aged married men to their youthful manner. John Hay was one who expressed his adoration of her.

Senator Cameron went his own way, and kept his wife at a distance. Their estrangement was saved from complete rupture by the demands of convention and by little Martha. Lizzie built her life upon the attention of others, as she delighted men and intimidated women. Basically, however, she came to depend upon Henry Adams and he upon her. They served each other in genuinely familial ways. During most of the years 1886–1918, Lizzie and Henry's relationship was a partnership of two individuals in need of help, each of whom was peculiarly suited to the other. Lizzie knew Henry's outlook thoroughly. Shortly before he died, she displayed this by remarking to him, "It is a curious faculty you Adams' have of inspiring terror! It must be because you are frightened yourselves, and communicate it."

At the outset of their long friendship, neither Henry nor Lizzie Cameron was certain of the other or of themselves. The four years after Clover's death had been ghastly for Henry, a time in which he suffered severely from depression. This condition was hardly relieved by his having to watch his parents die in pathetic circumstances and by the emotional difficulties which overcame Clover's brother, Ned Hooper, and her sister, Ellen Gurney.

In addition to these horrors, Henry had to complete his *History of the United States of America,* which ran to nine volumes. The writing was hardly begun when Clover died, although the project had been Henry's great joy when it was started in 1878. It had so absorbed him that he may not have given Clover the attention her condition required. Realizing this after 1885 and also perceiving that the history was unlikely to receive the acclaim he hoped for, Henry forced himself to write on. In 1890, he reached the end, never quite certain how he managed to do so. The *History* remains in the opinion of many scholars the finest single work by a historian of the United States.

This achievement was not immediately apparent to Henry. In fact the tepid initial response to his labor of a decade enlarged his despair. Clover, his parents, his good friends, his aspirations—all now seemed gone or in disarray. As an escape from his misery, Henry began helping Mrs. Cameron. He knew that in its way, her life was as difficult as his, though there seemed to be a vast incongruity between them. He was now past fifty, nearly bald, short of stature, and given to withdrawal and observation. She was tall, graceful, barely thirty, and a dazzling presence. These were the qualities, along with her pitiful captivity in a loveless marriage, which had drawn both Henry and Clover to her. In turn she had looked to the Adamses, both needing them and using them.

Of course, Henry was no less struck by Lizzie's physical appeal than were most men, and he did try briefly to make theirs a love match in the fullest sense. If he succeeded, however, the union was a brief one. In the time around 1890 when Henry felt ardently about Lizzie, the hurt from her marriage and her social ambitions apparently combined to bar her from any complete romance. However, other needs kept the pair dependent upon each other, so that Henry's friendship with Lizzie was as near as he ever approached another person, except for Clover and Lucy Baxter.

The hundreds of letters which passed between Henry and Mrs. Cameron show that his concern for her was tender and compassionate, even fatherly. On her side, Lizzie seemed often to use her relationship with

Henry to help her vanity, and even to assist her social and financial needs. He knew this, of course. What is beyond our perception, in observing the depth of Henry's willing, patient response, is whether by kindness to Lizzie he eased the guilt over his failure with Clover. Something as complex as Henry's affection for Elizabeth Cameron remains part of the self which he held back from everyone after Clover's death. How completely he had surrendered himself in his marriage, of course, lingers as the greatest mystery of all.

After finishing his *History* in 1890, Henry realized that his tie with Lizzie was disturbing them both more than it was comforting them. His solution was a trip to Hawaii, Samoa, Tahiti, and the Fiji Islands which kept them apart until a reunion in Paris in October 1891. During this interval, Henry sent wonderful letters to her, mingling candor and introspection with description. He claimed that she had sent him away, although it was probably he who took the step because neither of them could propose a way in which she might belong to him. "I think and think, and go on thinking a great deal, and for my life I can see no way out of it," he said, in trying to resolve the fact that they were more than friends but less than lovers. "I feel too much or too little when I ought to see as a matter of course what is the correct and proper conventionality."

While professing affection and a wish to see him, Lizzie had some words about their parting. "As for throwing you over—how could I? You are bound to me in no way. You went your way free as air and I have no claim on you, but the claim of the weak on the strong. It is for you to throw me over, not I you. The dependence is wholly one-sided as proved by your going away." Usually, however, she wrote to Henry in a light style, save for an occasional heart-tugging call. Once, after a particularly dismal encounter with her husband, Lizzie said she was almost at the point of summoning Henry back to her. "I could stand it no longer," she said. "If I had written, I should have said *come.*"

But, as Henry surely noted, she waited until the mood passed, and then put the burden of deciding their future upon him. Not for worlds, she insisted, "would I bid you return if you must return to hopelessness and unhappiness. I simply cannot bear to see you as I have seen you here sometimes." Then her coquettishness reappeared as she scolded herself for writing in a candor which risked getting him "unsettled." Had she done so, "I shall *never forgive* myself."

These exchanges went on during a year-long separation, at the end of which, in the summer of 1891, Henry impatiently crossed the thousands of miles from Australia to France where Lizzie awaited him. As he steamed toward the reunion, he sent ahead his thoughts on the possible outcome of their meeting. "If you say so, I will return to Washington to try the old experiment of living," he promised, seeking once more to make her arrange a solution. If her answer was no, then he pledged to resume his "wanderings," for in them, at least, he had found "my old self; morbidness gone; nervousness and nervous excitement subsided; depression vanished; sleep and appetite good."

However, the closer Henry drew to Paris, the less confident he became. He began talking of never returning to America. He hinted that Lizzie might come away with him to the Pacific Isles, a romantic notion he acquired from having observed Robert Louis Stevenson's life with a wife in Polynesia. He worried over the disparity in age between Lizzie and himself. "I feel more than ever the conviction that you cannot care to see anyone who is so intolerably dead as I am." He wondered if he would not bore her. As he wrote, he had been watching ship-board flirtations between older women and younger men. This, however, was treacherous ground and Henry quickly withdrew the point, settling for the safe concession that while he might not be "young, mischievous and tormenting," and had found no woman on shipboard with whom he cared to spend half an hour, still, Lizzie should know that "I am not so *ennuié* but that I could get into mischief if I could select my own companions."

After all of Henry's hope and anxiety, the Paris reunion in 1891 was a disappointment. Lizzie by letter was quite different from Lizzie at heart. Whether their meeting in France ever involved more than suppers, theater, shopping, and strolls is unlikely since Mrs. Cameron kept herself exasperatingly busy and surrounded by a protective galaxy of relatives, friends, and her child. Henry was always sharing her, so that the pair parted again unhappily in November. Henry remained for medical care and a visit to England while Mrs. Cameron embarked for America. She claimed to be "furious" at the results of their much-anticipated time together. Henry simply acknowledged that the "Paris experiment" had not met their expectations.

Whatever the demands or proposals Henry may or may not have made in Paris, it was clear that he had left Lizzie uncomfortable and unhappy. Perhaps her flirtatious language had no longer been sufficient and she had been compelled to confront Henry's insistence on a deeper relationship, one she could not enter, for emotional and practical reasons. Or quite possibly Henry may have been the culprit, if he had disappointed Lizzie by not in fact urging such a tie. From a spot he loved, Wenlock Abbey in Shropshire, England, Henry hinted that it was he who drew back. In a lengthy and candid letter, written 5 November 1891, he stated that a fully desirable association between them would never be possible. Acknowledging he had the duty to see that they avoided "mischief," Henry said he did not relish simply a friendly relationship, for "no matter how much I may efface myself or how little I may ask, I must always make more demand on you than you can satisfy, and you must always have the consciousness that whatever I may profess, I want more than I can have."

In this remarkable letter, Henry also reproved Lizzie for adding to their problem through her coquettishness. He wished he could just once "look clear down to the bottom of your mind and understand the whole of it." She must join him, he said, in acknowledging they were doomed as a couple and must keep an association "as innocent as the angels, yet as unhappy as the wicked." It meant that she must live knowing that although he accepted a platonic association, he would in his heart be unsatisfied. Was she prepared, he asked, thus to see him, knowing that "I, who would lie down and die rather than give you a day's pain, am going to pain you the more, the more I love"? The day after Henry dispatched this long declaration, he sent another letter which discussed Lizzie's complaint over their parting. Had she forgotten that their real farewell had been in the words shared in her rooms? he asked. Then, probably to underscore the importance he attached to his letter of the previous day, Henry announced he should not have written to her as he did. Ah, well, he concluded, "let fate have its way."

Once he had taken to safer ground, Henry showed that he could be as playfully vague as Lizzie had been. Reporting that his English friends were urging him to find a wife, he told Mrs. Cameron that he would not do so, for "in forty years of search I have never met but one woman who met me all round so as to be a real companion." Lizzie was left to speculate whether he meant Clover or herself. She responded by recommending his marriage, saying that a woman might help him "lose some of the bitterness in your veins." Henry replied with rising bravado, "I am past marriage, more's the pity! I would not marry now—no, not even you—if I could." However, he could not resist adding the kind of extreme comment which so vexed his Adams relatives—"life has for me no more interest or meaning; and never had any except in marriage."

Soon Henry was back in America, beginning a new life in which a social alliance with Mrs. Cameron was crucial. It enabled him to watch people from a safe distance, as she brought him stories and small talk garnered from her active, flirtatious role in Washington society. Lizzie still shared her husband's home, a few steps from Henry's residence on Lafayette Square. In this situation Henry described himself as a tame cat whose role was "to lie still and purr." He was serious, and once when Lizzie tried her teasing allurements upon him, Henry impatiently reproved her "for talking so invitingly about my coming to you. Why do you say such things?" Had he not learned to exist, knowing she was out of reach? "Why bring back the paroxysms! If you are contented, I ask no more." Nearly a year passed after their unhappy meeting in Paris before Henry and Lizzie became accustomed to the extraordinary alliance their friendship had become. He occasionally allowed himself such final taunts as "you can no longer punish me much," and "if I thought you cared, I should care more."

Meanwhile, Henry continued on cordial terms with Senator Cameron, frequently being his guest on an island off the South Carolina coast. He began using the pet names "Dobbitt" and "Dordy," adopted from Lizzie's little daughter Martha's words when she saw a postage stamp picture of George Washington which she seemed to think resembled Henry. From babyhood, Martha claimed Henry's close attention. He sent her affectionate messages which Lizzie might interpret as she chose. This tactic also allowed Henry to pretend he was a boy Martha's age and to write to Mrs. Cameron as a mother, which he often said was the best role for a woman.

After 1892 Lizzie suffered increasing unhappiness in her marriage, obliging her at times to forget pretense and be candid with him. This sort of letter is not found among her manuscripts, for she evidently chose to destroy these years later when she sorted through their correspondence, preparing it for posterity. She did not, however, discard other letters which mentioned these outbursts. Such moments of awful candor were doubly painful for Henry, since he felt he was watching another woman he cherished face breakdown. Recalling Clover's collapse and his own severe depression, he told Lizzie that at such times she must "sit still and let things pass."

Henry knew that Lizzie was now worrying over a strain of mental weakness in her family. She was also suspicious about her husband's affairs. Of this, Henry said, "what of it? What do you really care?" He told her: "You have your own and Martha's interests to manage and your power is the greater in proportion to the errors and blunders of other people." When she and Don continued tormenting each other, Henry announced, "I am at my wits end to devise some means of helping you." Lizzie and her husband became estranged in 1897, leaving her with mysterious physical ailments. Henry fussed over her like a doting father until she pledged to try a reconciliation with Don, something she never attained. Not even with Henry could she talk of the details of her marriage, she said. Neither, on the other hand, could she "say anything of all that I feel to you." Someday, Lizzie promised, she would "go on my knees to you and humbly kiss your hand," although, she said, not even that gesture would adequately display her feeling for him. He was her "very dearest."

Henry now easily kept the distance he preferred between them. In 1898, Mrs. Cameron arrived in Europe to hear that Henry had rejected her request that they travel together. "I need not say how disappointed I am," she complained. Soon afterwards, their association came close to a breakup as Lizzie, now older and more uneasy about her powers, began making demands upon Henry. She tried coaxing him into spending more time with her in Europe and even into sharing the Paris apartment they had always scrupulously used on separate schedules. When Henry drew back, she wanted an explanation. He resisted any candid reply, reminding her that they must avoid direct talk on this subject, which could injure and lead to misunderstandings. Surely she knew "it was better to say nothing," he said, "all the more because you and I know all there is to say." She should remember his obligation to convention. "Never can I escape my respectability," Henry said, "and just because it is the real truth." He could have added that their friendship had created ample gossip as it was.

Now sixty-five years old, Henry was fretful about his health and only humorously concerned about romance. When his good friend, Clover's cousin Sturgis Bigelow, became infatuated with a Parisian dancer, Henry remembered his own circumstance a decade earlier. He shook his head as "the bald-headed, white-bearded weary philanderer of fifty dreams that still someone might love him and fill his life for him." As Henry grew more distant from Mrs. Cameron, the predictable happened. Lizzie threw off her own restraints in a romance with Joseph Trumbull Stickney, twenty years her junior, handsome, well over six feet tall, and a poet.

Stickney was among the Americans of literary and artistic tastes who spent much of their time in Europe. To this circle Henry had become a beloved figure, so that the group was very uncomfortable when it came out that Lizzie had gone off to Florence for several weeks with her "Jo." Before departing she was heard to speak unkindly of how Henry was aging. Saying nothing, Henry watched the affair through clouds of cigar smoke, knowing that she would return after her fling.

And indeed, a year later, Lizzie wrote penitently to Henry of how her life had become a "hopeless muddle," lamenting "my wreck of an existence." Fretting over what Don might do to punish her, Lizzie pleaded, "I have no energy left. You must now brace me up and provide 'go' for us both." In 1904, Stickney died of a brain tumor in America, leaving a memory which Henry and Lizzie never openly mentioned in their correspondence. She settled into the role of an aging charmer who frequently begged him to be kind. "No one else has the power to hurt now," she told him. "And you know better than anyone that my nerve is gone."

At fifty, Lizzie had found no peace. Much of her time went to seeking the "cure" at Europe's watering places. When she was not thinking of herself, she fretted over her daughter. As a young woman, Martha had complicated physical and emotional disorders which left her lazy and peevish. Henry patiently helped both daughter and mother, usually from a discreet distance. He delicately made purchases, paid rent, and even provided an auto and chauffeur, to Lizzie's delight, after she had issued a broad hint. In 1909 Martha married Ronald Lindsay, an English diplomat who seemed almost more interested in the mother of the bride. Soon Martha entered a sort of withdrawal familiar enough to Henry in his experience with Clover and her relatives. Lizzie kept imploring him for advice, and Henry began to refer to the mother and daughter as his "invalids."

Henry was also accumulating other serious cares, these in Clover's family. Once he put his Pacific wanderings behind him, his main concern, besides Mrs. Cameron, was Ned Hooper and his five daughters. Because of Ned's emotional weakness, Henry attended to the five nieces of whom Clover had been very fond, making an often annoying burden for himself but which he used as a handy escape when Lizzie Cameron became too demanding and oppressive. However, the device forced him to exchange one set of problems for another. The five "Hoopcats," as Ned's daughters were sometimes called by those who resented the time Henry gave them, reminded Henry of the precarious mental balance which Clover had inherited.

Ned Hooper had always been dear to Henry. Despite his frail nerves, he was a well-intentioned if delicate parent. More tranquil, gentler than his sister had been, Ned lived quietly when he was not traveling with Henry or their male friends. He looked after his daughters, collected art with a talented eye, served as treasurer for the Harvard Corporation, and watched over Henry's money. By 1898 Ned's bouts with depression became so severe that he talked of leaving his Cambridge home and of resigning his Harvard office. In the spring of 1901, he became incapacitated. Henry could not bring himself to come to Boston to see him in this state. Instead, he tried to advise Ned's children from Washington. When Ned evidently attempted suicide by leaping from an upper-story window, Henry wrote to one of the children: "The complications are now beyond foresight. . . . When one gets into these snarls one learns what religion is. My only resource is to do what I'm told, and pray to the virgin."

Remaining in Washington and delaying his spring departure for Paris, Henry awaited news about Ned's prognosis. He feared that Ned would have to be confined in the asylum at Somerville, the institution which Clover had so dreaded, and indeed, by the end of April 1901 it was necessary to hospitalize him. "[S]ooner or later in Boston we all go, or ought to go, to the Asylum," said Henry. After talking of how his father "lost his mind at seventy and lived ten years," Henry spoke frankly with one niece: "You must remember that for twenty years I have been more or less expecting the present situation and fearing that it would come before you were grown." To Lizzie he confided during this crisis that knowing the Hoopers had made him live "with a queer sense of waiting for the skies to fall," especially after Clover's death "left my nervous ganglia all tangled up." Now, not only was Ned slipping away, but Henry's intimate friends, Clarence King and Augustus Saint-Gaudens, were also nearing the sad end of their own bouts with melancholia. "Positively it gets to be a joke, and one becomes Rabelaisian in the face of these spectres of Bedlam," Henry told Lizzie. "After a certain point one loses the sense of horror."

With Ned in Waverly Asylum—at least he was spared Somerville—Henry made a belated departure for Paris, pondering his enlarged responsibility for Ned's five daughters who bore such a fearful emotional legacy. Their father may have failed suicide but his two sisters had not. "Congenital imbecility," Henry called the family condition as he watched Sturgis Bigelow sink into depression and psychosomatic affliction. Ned did not linger long, dying in June 1901. He "had always been my best ally and most necessary friend," Henry wrote, sadly, praying "that this long and difficult chapter is closed forever." There was no more to say.

The three troubled children of Dr. Robert Hooper were all at rest. There were now Ned's five daughters for Henry to hope that he could help, using the wisdom of sad experience.

What Henry could not very well express to his nieces was the all-absorbing apprehension—would mental illness spare them? Another delicately balanced young person in the circle around Henry, George Cabot "Bay" Lodge, the poet son of the senator, put the concern succinctly: "The way that inheritance has gone thro that family is fearful." Ned's first daughter, Ellen, had been born in 1872, followed by Louisa—"Loolie"—in 1874; Mabel in 1875; Fanny in 1877; and Mary—"Molly"—in 1879. After 1901, these women became famous—notorious in some quarters—for clustering lovingly around their Uncle Henry. One of Henry's most devoted admirers, Cecil Spring Rice, summarized the situation for Mrs. Henry Cabot Lodge: "Uncle Henry is well occupied with the Hoopers and Mrs. Cameron—I like that sentence—it has such endless possibilities for witty remarks after it. How I tend to hate the Hoopcats."

Henry, of course, was fully aware of his situation. Would he keep this new family in safe dimensions or would it overpower him? What he feared, he conceded, was the nieces "trying to run me, for the instinct of the women of all sexes and epochs is to run somebody." He lived in hope that he could marry each one off and set them "to making babies of their own." Each niece would have several thousand dollars a year income from inherited money and from Henry's generosity. Although the Hooper girls did marry, they never gave up their devout care for Uncle Henry. John Hay called them a "body guard" which never relaxed its vigilance. "At whatever personal inconvenience they are determined to do their duty by their Uncle 'who is so lonely without us.'" Bay Lodge was less gentle in his appraisal of the "Hoopcats." "No house should be full of unmarried females except a harem and the Hoopcats are moulded by the Almighty in such a form that they are not eligible for such an institution—at least not for mine." Yet he conceded that Henry presided with gentle grace in "the Seraglio of H Street." Actually, Henry's solicitous nature did not prevent his sharing very privately his worries over the nieces, and his continuing fear that they would smother him, especially during the several nervous breakdowns among them before 1912. Somehow, these were overcome, and, finally, all the women were married. The situation had its amusing side, for Henry came to behave and sound like an indulgent and worried father.

The final episode involving Clover's nieces came in the last summer of Henry's life, 1917, when he went back to his house at Beverly Farms for the first time since 1885. At last he found himself able to be happy even with the shadow of Clover all about. However, to this beloved ground trooped the women he had inherited from Ned and Clover. "It is a holy spectacle to see all these Hooper nieces," he told Lizzie, noting that they seemed even then no different from their forebears of forty agonizing years earlier. "I feel as though I were the Archangel Michael marching them all out as I did their parents with no sensible differences in their forms of Hooperism and with no apparent prospect that there will ever be a change in . . . Hoopers till the world's end." Then, looking down the shore to Quincy harbor, Henry could not resist adding, "As for Adamses, I shudder to think what new and more hideous terrors are coming on the other side of Boston Bay."

Retreat though he might in person from the Adams world, Henry still meditated upon the legacy begun by old John and Abigail. It was during the years of Lizzie's affair with Stickney, Ned's collapse, and the wearisome Hoopcat duties that the remarkable *Education of Henry Adams* was composed. In it Henry tried to perfect his role as loiterer in the world while remaining safely distant from any commitments. Consequently, the book was as baffling to his family and friends as was the author's daily style of evasive talk. Also in this period, Henry wrote his other masterpiece, *Mont-Saint-Michel and Chartres*. This book was a great success when it was published in 1913 amid lavish praise from the American Institute of Architects, but Henry made certain that his *Education* did not go to the public until after his death. It appeared in 1918, and it was a startling success. In 1919, it received the Pulitzer Prize. In the following years, both the *Education* and the *Chartres* have continuously been acclaimed as two of the most important and triumphant books written by an American.

Each book, no matter how often reread, never fails to teach something more about Henry's rich personality and mind, as well as about his Adams nature. While the *Education* is the book which was closest to Henry as an Adams family member, the *Chartres* could not avoid being affected by this part of him. To all intents, it is a powerful analysis of medieval civilization, including architecture, art, the spiritual and the temporal. The author addresses mankind's largest issues by using as his subject the great Gothic places and the people of the twelfth and thirteenth centuries. Yet even from such a dizzying intellectual height and remote moment in history, Henry's manner comes through as the Adamses had taught it.

As the book opens, Henry assumes the pose he most enjoyed—the distant instructor. *Chartres,* he wrote, was conceived to teach his nieces. The tie between son and father was too close, whereas nieces afforded a relationship "capable of being anything or nothing at the will of either party." Henry could not resist the aside that this might increasingly also be said of an American marriage.

So Henry began his great book on medievalism with the blithe announcement—"The uncle talks." In style as well as in content, Henry displayed his preference and admiration for women. Starting with the quip that, unlike nephews, nieces read in their youth, Henry proceeded to a logical description of the power of Mary, Queen of Heaven, for whom the great cathedral around which the book is centered was but yet another tribute. Henry claimed the church was built in an age when the universe was drawn to this supreme woman, the Mother of God. In the background as he wrote was Henry's own devotion to his grandmother.

As a young man he had read carefully many of her private memoirs, all of which displayed her courage in the face of sorrow and disappointment. For Henry she represented "the ideal of feminine grace, charity, and love" which otherwise he found only at the heart of the twelfth century. It was from Louisa, with her French Roman Catholic education, her simple devotion to the faith of the Anglican Prayer Book, and her reluctance to accept modernity, that Henry claimed to derive his "half exotic" nature. He also knew how she had suffered and how "thoroughly weary" she had been "of being beaten about a stormy world." Thus, the memory of her life and of his own tragedy brought Henry at the close of the *Chartres* to assert that the Gothic cathedral he had described so marvelously was at its base simply the greatest cry of faith and anguish, of aspiration and self-distrust, which he could show his nieces.

"You can read out of it whatever else pleases your youth and confidence," Henry said in closing his book, and added, "to me, this is all." Louisa Catherine would probably have seen at once what her grandson meant. She, however, had also chosen a triumphant faith. Henry may have found the twelfth century a satisfying retreat from his disheartening memories, but he apparently never found in the historical pattern a final reassurance. He remained a stoic, feeling spiritually and emotionally inadequate like all the male sex he professed to scorn. It was Louisa Catherine who, among all the Adamses, triumphed in the quest for faith so many of them had undertaken.

By giving time, talent, and money to venerating the world as it had been eight hundred years before, Henry went as far and as boldly as possible from the Adams legacy. More subtle was his way of hiding in *The Education of Henry Adams.* Many readers saw it as only autobiographical. After all, his name was in the title, although Henry claimed privately, "I have deliberately and systematically effaced myself, even in my own history." In 1907 Henry sent copies of a private printing of the *Education* to relatives and friends asking for their comment. His brothers found the book bewildering.

Like the *Chartres,* the *Education* was intended mostly as social and philosophical commentary. Henry used his own story only as a means of instruction, but this was in itself a remarkable decision. Except for John Quincy Adams who, until the day of his death, remained in the public eye, the Adamses had preferred to keep themselves quite private. Henry himself had once stated to Brooks that silence was "the only sensible form of expression." Yet the *Education* reveals much about the Adams family story; Henry kept silent only about the years 1872–92, his life with Clover and the years of shock after her death. The *Education* begins with Henry's delightful recollections of early life and goes on to sections more didactic and philosophical. The final fifteen chapters, which begin with Henry back from the Pacific and Paris escapades, contain his explanation of the world's unfolding. In this book, Henry pitted the Virgin's power against the laws of energy. All of this speculation in the last pages seems to the reader far from the early section on Quincy—as, of course, Henry intended.

Not everyone could see his point, least of all his relatives. Charles called much of the *Education* "simply silly," while Brooks felt let down by the brother he nearly worshipped. He charged that Henry had written a "jocose" book. Henry tried not to be surprised, predicting that his brothers would continue to point out "that my art fails of its effect." Typically, Brooks and Charles disagreed over the *Education,* the older brother announcing his thrill at the early chapters on Quincy and Boston: "Lord! how you do bring it all back! How we did hate Boston! How we loved Quincy!" Brooks, on the other hand, praised the concept of civilization, as well he might since it reflected his own bleak outlook. However, he was more intent on having Henry "eliminate the apparent effort to write fragments of biography," and to concentrate instead on telling a story which was "a huge and awful tragedy," the end of man's struggle with nature.

One of Henry's dutiful nieces, Loolie, undertook to explain the *Education* to a friend. "Most people took the book more personally" than her uncle intended, she said. "He says that few take in what he's driving at." One of the rare, direct statements about his *Education* which Henry finally did make was offered to Mary Cadwallader Jones, his and Brooks' cherished friend. In the midst of the first, disappointing response to the book he told Mrs. Jones that she should "bear in mind that I don't mean any harm." The motive behind the first part, he said, is "to acquit my conscience about my father," by which he apparently meant that he wished to correct impressions left by Charles' biography. The next section is a tribute to John Hay, he said, leaving only the last three chapters for *"Ego."* Along with these three he urged Mrs. Jones to read the final three chapters of the *Chartres,* these six, he maintained, being "my little say in life. . . . With that—and St. Gaudens' figure—adoo!"

Henry knew before he died that these messages he left to the world—six chapters and a cemetery monument—would be often misconstrued. The twentieth-century mood was uncomfortable with a philosophy of peaceable resignation before inevitable decay in the social and material world. Henry was human enough to wish for a wider acknowledgment of what he was saying. His books and the Saint-Gaudens sculpture were meant to speak out much as his great-grandfather, his grandfather, and his father had done. Each in his own way had worked and written for the American democracy, putting aside his pessimism to do so in a direct, conventional fashion. Henry, because of his skepticism, the inadequacies he saw in himself, and his personal tragedies, preferred a whimsical career, claiming that even the simplest gestures carried risk enough.

Henry was either the wisest or the weakest of the Adamses. Or perhaps he was both, since each of these qualities may imply the other. Indignant with his forebears, he claimed they had left him a heritage of duplicity since they had accommodated to the world even when they could not believe in the ultimate success of democracy, capitalism, and nationalism. At heart, the *Education* was an indictment of his family for having behaved as if it had hope, despite its private pessimism. His great forebears had refused to take what to Henry seemed the logical step, that of withdrawing to the sidelines to declare their pessimism. Henry also resented the fact that the family had made each generation of Adamses discover for itself the odds against republicanism and capitalism.

What Henry never disclosed to his friends and family (and possibly never quite acknowledged even to himself) was whether the patience, generosity, and kindness he bestowed upon others from the shadows of his retreat were in reality a manifestation of his belief that dire though the human condition might be, there still was such a thing as moral obligation. Louisa Catherine Adams had discovered this through her Christianity. Faith was her answer to the family's tension between futility and ambition. Her grandson, a skeptic, was unable to be direct about any helpful gesture he made. Consequently, while Louisa never really retreated, Henry had to run away from the family's policy of doing something, as well as away from Boston and from normal social exchange, finally trading the tawdry present for the comfort of the stark harmonies and lofty spires of the twelfth and thirteenth centuries.

～ 17 ～

Brooks

Henry's wish to keep away from the Adams world required him to try to avoid his brother Brooks. To some friends this seemed strange, since of all the family Brooks appeared to be a bridge into Henry's retreat. Brooks adored his older brother and Henry, in turn, admired the pessimistic revelations which Brooks made about history, economics, and the future. There was difficulty for Henry, however, because the younger brother took ideas in earnest and was eager to put them into practice. Henry invariably became uncomfortable when Brooks arrived to argue that the world should be taken seriously. Acknowledging that many of the concepts in the *Education* and the *Chartres* originated with Brooks, Henry had been also struck by his brother's statement about the Gothic French churches. As he sat in them, Brooks recalled, "I confess I disgraced myself. . . . I really and truly did believe the miracle, and as I sat and blubbered in the nave, and knelt at the elevation, I did receive the body of God." The Gothic presence was the "greatest emotional stimulant" in the world, and Brooks felt at one with the medieval figures.

Another side of Brooks equally impressed Henry, who exulted in Brooks' book *The Law of Civilization and Decay*. Said Henry: it "reduces all history to a scientific formula. . . . My admiration for it is much too great to be told." Meanwhile, Brooks was praising Henry in similar tones. Each learned much from the other, as their enormous correspondence shows. It is hard to determine who did most for the other. Henry's deft imagination allowed him to see and use for himself many of Brooks' points, often before the slower, younger man fully appreciated his own insights. This happened most notably in the case of Brooks' thoughts about the inevitable decline of civilization's material, spiritual, and æsthetic energies. Henry built these ideas into those final chapters in both the *Chartres* and the *Education* which he prized and recommended to Mrs. Jones. He even used this fatalism from Brooks in his two essays on historical theory of 1909 and 1910, "The Rule of Phase Applied to History" and *A Letter to American Teachers of History.*

As Brooks developed his pessimistic philosophy in 1895 with publication of *The Law of Civilization and Decay,* which he called "An Essay in History," Henry stood by with the encouragement Brooks sorely needed. This latest volume was actually an enlarged approach to the ideas which had been considered in Brooks' early book *The Emancipation of Massachusetts.* These two volumes remain the most imaginative of his many books and essays, which included *The Gold Standard, America's Economic Supremacy, The New Empire,* and *The Theory of Social Revolutions.* In *The Emancipation of Massachusetts,* Brooks posits that fear is essential to social energy. Afterwards, he discovered that in more complex stages of civilization, greed is the larger force. He emphasizes in *The Law of Civilization and Decay* that societies move inevitably through emotional, martial, and economic organic stages. Ultimately, he says, the capitalistic energies so drained the organism that civilization must decay and perish. The primitive remnants, according to Brooks, would therefore rest in a dormant biological state until the infusion of what he termed barbarian blood.

Brooks, said Henry, was a person "struck with the agony and bloody sweat of genius. . . . I stand in awe of him." But there were disadvantages to Brooks, at least for Henry, who needed calm and quiet. Frivolous talk and activity appealed to Henry when he was not alone at his writing table. But when Brooks visited Henry, he brought a raw vigor, a crude style, and a craving for encouragement and reassurance. His life would have been worthless without Henry, Brooks liked to say. "[F]rom the old days in England when I was a boy, you have been my good genius." He looked to Henry for understanding and support, requirements he searched for all his difficult life.

Henry tried as patience allowed to calm and bolster his brother, chiding him lightly in 1903 with the remark that "you still conceive of me as twenty years old and yourself as ten. Perhaps its so!" In a famous comparison, Henry delightedly reported that he had just found a picture of two donkeys resembling himself and Brooks, the latter "looking sideways as though preparing to kick; while I look straight at you, patient and resigned." To defend himself, Henry was usually pushed to do more, lest, as he put it, he be reduced to "wood-pulp" by Brooks' ravings. At such moments he tried to beat his brother's game—when Brooks "screams that America has lost every vestige of mind or will," Henry said, "I have to show him that she never had any."

As a result of their contrasting personalities, Brooks never came to know how much he had stimulated his older brother. So uncomfortable was Henry in Brooks' fiery company that he often made excuses to avoid seeing him, calling him "naturally a little of a genius, with bad manners and sensitive nerves." This description fit Brooks admirably, for Henry despaired over Brooks' constant, frantic excitement. "How is it possible for a really large mind like Brooks' to get into pink fits about a provincial teapot like this election—I cannot comprehend," Henry marveled in 1896.

Each of the brothers of the fourth Adams generation remaining in 1895—Charles, Henry, and Brooks—was disconcerting to other people. The three were famed for unusual temperaments, and of them, Brooks was the most difficult, as Charles never tired of declaring. None of this escaped Brooks, who looked back to what he called his "grand break-down in 1882" as the onset of his persisting unhappiness. Before the 1880s had brought Heloise's illness and death, he remembered living agreeably in Boston. This was all changed by unhappiness, and Brooks belatedly joined his brothers in finding Boston repellent. In that city and elsewhere after 1882, Brooks began sensing hostility in others. "I grow jealous of strangers—they don't see the pictures I do, and if I ever try to show them, they only stare," he said.

By 1890, Brooks claimed that even in Quincy he was treated as "a crank, and when one is always treated as a crank there is danger of becoming one." To Henry he conceded, "I thank God I have no one to come after me to inherit my nature and my shortcomings." As Brooks knew, many persons questioned the sanity of a man who wrote fore-casting disaster. Consequently, he was mostly ignored until the mone-tary crisis in the 1890s suggested that he might have been right. This drew him a little curious attention from the public. Brooks rushed to his brother-in-law and friend, Henry Cabot Lodge, crying exultantly: "I am not mad, I am after all like other men. I am not the victim of a delusion."

Such behavior kept Brooks estranged from Charles, who described his younger brother as "buzzing about like a big blue-bottle fly, and bent on doing just what ought not to be done." In contrast to Henry's high opinion of Brooks, Charles' evaluation was: "Poor Brooks. An insatiable ambition, intense egoism, no judgment, and only fair abilities—what a nuisance he is!" Caught thus beyond two brothers whom he made uncomfortable for different reasons, Brooks managed to find a sympathetic ear in Mary Cadwallader Jones, who kept his many letters, to the great benefit of his biographers. Brooks was more methodical than Henry in trying to efface his past, but he seems not to have urged Mrs. Jones to destroy his letters. They were unknown until recently when the Massachusetts Historical Society acquired them, all dating after 1890. These manuscripts clarify Brooks' complex personality as an Adams, for Mrs. Jones was the only friend for over a half century in whom he apparently was able to confide and with whom he could be himself.

Mary Cadwallader Jones, whose sister-in-law was the novelist Edith Wharton, is remembered as a tall, mannish woman with high cheekbones, a startling contrast to Mrs. Cameron. While she evidently took few pains with her personal or conversational style, Mrs. Jones had qualities which drew Brooks to her in a long friendship which began in the 1870s. Descending from the venerable Rawle family of Philadelphia, Mrs. Jones made a disastrous marriage at twenty—which produced one daughter and thereafter a long separation. In time, she became something of a popular writer, which Brooks envied greatly. He was also drawn to her because her personal life had been much troubled during the same time he was experiencing his own woes.

In corresponding with her, Brooks made as if to help her, although actually he was more often seeking advice and comfort for himself. At least once a year, Mrs. Jones paid a lengthy visit to Brooks and his wife. To her Brooks revealed his plight as a latter-day Adams. His problem, he said, was that he wanted to do something for America and the world but that he was a relic. He could reason "well enough" about the current age, "and see it all in theory," yet "when it comes to action I am all out." Rather wistfully, he assured Mrs. Jones that Henry could have had "almost anything he wanted, and he might have held any position." Yet what did his beloved elder brother do? "He shut his mouth, refused every advance, and tracks around Europe in the train of anyone he can pick up." As for himself, Brooks considered that he was still stuck with foolish, unrealistic dreaming that originated as far back as 1873 when he had first met Mrs. Jones.

Why was it, he wondered, that a person like himself could never enjoy life fully? "People who were adapted to the age" never had this sentiment—"they feel that life fits them," he went on enviously, and "they fit life, and they have no hesitation about it." Brooks said this in 1901 when he was fifty-three and suddenly struck by the likelihood that "life is behind me." "Ah! what a young fool I was with my dreams," said he, remembering the time when he had been caught up briefly in reform politics. Now, he was trapped in the crisis of mid-life, and he felt he was lucky to have one person—Mary herself—who could comfort him by hearing him out. She had the talent, Brooks assured her, to deal with wounded men "who have not quite gained the prize and who seek someone to ease their hurt a little."

Henry Adams and Mrs. Jones were the only persons to whom Brooks could address the question: "Should I have done better in another age." She patiently sent encouragement, but Brooks was not so easily comforted. He looked at such successful friends as Cabot Lodge, Wendell Holmes, and Teddy Roosevelt, and found himself by comparison "naturally a good, commonplace, plodding man, ruined by a spark of genius, which throws me out of step with my age." Much of Brooks' harping on this theme probably was intended to appeal to the sympathetic Mrs. Jones. The woefully insecure Brooks needed her comfort because from his wife Brooks received only the trivial praise which convention brought a man of the time from his mate. Actually, Evelyn Davis Adams was a woman of remarkable talent whose story became a pathetic feature of the last days of the Adams family.

Sometime before the death of Brooks' mother in 1889, the mysterious Heloise had died, and an aged Abby begged Brooks to find someone to marry; she sorrowed so to think of him alone once his parents were gone. So after his mother's funeral, Brooks obediently undertook to locate a bride. He went for advice to the woman whom he and his brothers admired extravagantly, Anna Cabot Mills Lodge, the wife of Henry Cabot Lodge. To "Nannie," Brooks confided that he wished to find a wife just like her. As Brooks later recalled, Nannie immediately suggested her sister Evelyn Davis, known as Daisy.

Nannie and Daisy were daughters of Admiral Charles Davis, who had been a classmate of Charles Francis Adams at Harvard. Brooks wasted no time in proposing, although he warned Daisy of the difficult person she would be marrying. Since she had no way of knowing that Brooks was not exaggerating, Daisy promptly accepted, and they were married within a few weeks. The wedding took place at the Lodges' summer residence on the north shore at Nahant, in September 1889. The bride was thirty-six, five years younger than the groom.

Most of the Adamses were present for the wedding, all behaving "as kind and pleasant as possible," Lizzie reported to Henry, who could not bring himself to attend. "I ought to have been there," he conceded, but he preferred the distraction of going to Canada with Lucy Baxter and the Hooper nieces on an excursion combining pleasure and historical research. "I should have enjoyed seeing Brooks submit to the conventionalities which he had hitherto made a business of swearing at." Penitent, Henry sent them a La Farge watercolor for a gift, noting with amusement that Charles had presented the newlyweds with a piece of Chinese porcelain "fit for an emperor." He also persuaded the pair to use his house at Beverly Farms as a honeymoon cottage.

A year later, Mrs. Cameron reported to Henry, then in the far Pacific, how Brooks and Daisy were faring. When she paid them a visit, "Brooks growled at me for an hour principally about you and was too tiresome for words," Lizzie said. As for Daisy, "she really seems to like him. And she shared his opinions till I hated her and him and their very atmosphere." Horrified, Lizzie reported that the colorful Brooks referred to Daisy affectionately as "idiot from hell." Actually, Brooks caused trouble even in his preference for his wife's real name. Daisy had rarely been called by her proper name, Evelyn, until the occasion of her wedding breakfast which brought together many Adamses and Davises. At this moment Brooks arose to announce that, henceforth, his wife must be called Evelyn. The demand was too much for one of Daisy's brothers, who blurted out to the groom, "I cannot give up the name I have been accustomed to using at the bidding of a comparative stranger." So Brooks was mostly alone in calling her Evelyn during their nearly forty years together.

Despite his often outrageous style and language, Brooks worked earnestly to please Daisy. "It is to me, of course, the most vital thing in life, that my wife should be happy," he told Cabot Lodge. "Whether I can succeed in making her so I can only know after I have tried." Being a man of moderate wealth, Brooks had given up any serious practice of law several years before his marriage, so that he and Daisy concentrated on his passion for writing, travel, and preserving the Old House and family relics. They lived in Boston during the cold weather, after which they hastened to the Old House to watch the ancient garden come to life. A trip to Europe was usually the next event, followed by a late summer return to Quincy and then to Boston, where, for most of their marriage, they lived at 33 Chestnut Street on Beacon Hill, a short stroll from the family's old haunt at 57 Mount Vernon Street.

Brooks seems to have studied himself and the process of wedlock closely during these years, sharing his views with Mrs. Jones. In 1913 he told her "that nothing but a fixed determination to make marriage a success can make it a success under the conditions which now prevail." Once upon a time, a family unit "cohered almost of necessity," while now, Brooks observed, it was often separated. His reaction to this was, characteristically, an arresting one—women should "love with whom they please, as these shop girls do," he said. Such an arrangement was better "than to enter loosely into a relation where existence demands a unity."

Predictably, existence with Brooks was an ordeal for Daisy, who her husband was calling a "poor little girl" in 1894. That she was having a "very hard time," Brooks attributed to "my nerves," while Daisy was "simple, honest, true and brave." Her letters had the unassuming tone of one eager to please, and gave much space to praising her husband. She also tried to help him, as in 1895 when the troubled Brooks had an emotional breakdown watching America's politicians and business men debate gold versus silver coinage. The issue was at the heart of Brooks' views about decaying civilization. At this difficult hour Daisy asked Charles Adams for help in seeing that Brooks was named librarian for the Boston Public Library—"he has been very anxious to find some occupation," she reported.

Charles replied at once and with characteristic bluntness. The library post for Brooks was preposterous, he said. "I should just about as soon think of taking charge of the steamship Lucania, or the Observatory at Cambridge. I know in advance that I couldn't do it. I do not exactly see how Brooks is any more qualified than I." He assured Daisy that he was "painfully" aware of Brooks' mood, but there was nothing he knew to suggest. Charles could only join Daisy in being concerned about what he called "Brooks' morbid, excited condition, and his inherited tendency to worry."

Meanwhile, Brooks lamented to Henry that Daisy "isn't finding marriage all a bed of roses. To nurse a cross and nervous invalid through a month of gout at Quincy, a collapse in Italy, and dysentery and bilious fever in India and Egypt, is paying rather dear for the whistle." Daisy herself often wrote to Henry, becoming, in her own way, one of those who took comfort from him. She once conceded to him that her husband was "selfish and ungrateful" as he dragged both of them "through the slough of despond every day and hour. . . . I wonder if divorce would cheer him," she asked, probably not very seriously; "it might give him a new lease on life." Then she was very much in earnest for a moment: "My life is indeed hard, but not as he suggests. You will sympathize, I am sure, with your poor sister."

Daisy undoubtedly tried to understand and help Brooks, but the strain proved too much for her. Her first breakdown came in the summer of 1899 when she suffered a physical and emotional collapse in Europe. This episode shook Brooks profoundly. He abandoned elaborate plans for visiting Russia and Asia, giving himself over to attending Daisy, which may or may not have helped her. "I don't think I shall ever look back on this summer without a shiver," he told Henry. Europe even lost its charm, and Brooks pined for Boston and Quincy. "I think I would accept home on almost any terms—even with Charles round the corner," he said. When Brooks finally could bring Daisy back to America in 1900, Charles was startled to see her looking "very old and worn."

Brooks now announced a new purpose in his life. He would devote himself to Daisy's health, which, for him, meant travel and protecting Daisy from persons whom he considered likely to upset her. Eventually, this included Daisy's cherished sister, Nannie Lodge, who in 1907 Brooks described as "daft." Mrs. Lodge paid them a visit that spring and promptly took Daisy's care away from Brooks, who was indignant. "I have kept Evelyn along pretty well till she came," he complained to Henry. "If she would only go back to Washington it would be a great relief." He grumbled that his sister-in-law's "perversity" kept her intruding. How grateful he was, said Brooks, that Daisy was such a contrast—she "does her work, holds her tongue, and obliterates herself to save others. I wish to God other women were like her."

From time to time, the strain of self-effacement and whatever else may have been required of her in her relationship with Brooks was more than Daisy could bear. Brooks meanwhile continued to send reports on his wife's plight to Mrs. Jones. Daisy "won't leave me," he wrote; "I'm sure I don't know why, for she says I nearly drive her to frenzy daily." And so the years passed, Brooks rejoicing in the intervals when Daisy's health and outlook improved, while little is known of what Daisy really thought. In 1915, Brooks told Mrs. Jones that there was one happy feature of existence after all: "Evelyn, thank God, keeps well." During the war years she gave receptions for war workers and was generally active, dutifully still showing an interest in Brooks' ideas and publications. But even this had to end, for soon after the war, Daisy's mind at last gave way irretrievably to strain and hardened arteries, a condition which Brooks tried vainly to overcome with more travel and attention.

In hope of keeping Daisy mentally alert, Brooks used as medicine his most cherished project, a determination to preserve in its historic state the old family mansion in Quincy. Daisy tagged along obediently as her husband became the self-appointed defender of the family name. The most important stratagem for him was saving the Old House with its stirring reminders of past happiness and lament. Brooks' determination stemmed in part from his belief that the Adamses' days as a notable family were over. Consequently, evidence of the great forebears should be preserved, lest men like John and John Quincy be forgotten and never properly appreciated.

Brooks' campaign was a lonely one, for he proved to be the only son of Charles Francis Adams who cherished the Old House. His elder brother John, to whom the place should properly have gone, had been elated to give way. Brooks had much work ahead of him, for he found the mansion "an abandoned ruin. . . . Dirty and dilapidated beyond words." An indifferent Charles had called the place "that old shanty." By 1898, however, a remarkable change was already evident, prompting John's widow Fanny to report to Henry that the Old House and the garden had become "quite as they used to look in your Mother's day."

Not only was there a physical rehabilitation of the premises, but Brooks also retrieved as many cherished family items from relatives as he could. The harvest of furniture, china, paintings, and other objects was placed in the Old House toward the day when, Brooks hoped, it would become a museum. He got some support in 1906 when Henry announced that "under no circumstances whatever do I consider myself at liberty to leave family property out of the family when it came to me from the family." Since he received very little encouragement from other family members, Brooks was greatly pleased by Henry's attitude. He urged Henry to come and see the marvelous change in the Old House, especially the paneled room where the restored mahogany walls were "magnificent," "like velvet." Henry, of course, stayed away.

Among Brooks' reasons for saving the mansion was its extraordinary representation of the first century of the American Republic. There was nothing to equal it, Brooks claimed: "It is complete, unique, and charming." He kept pestering his brothers and sister to contribute significant items from the family's past. It must be done now, he said; "I'm afraid the next generation knows little and cares less for the things which we have cared for." In 1910 Brooks considered himself "the trustee" of a collection of "the most interesting relics in the world," enhanced particularly by being "the record of four generations." To embellish the Old House, Brooks took to practicing many of the daily customs his parents had observed. He even sought contributions from Henry and Charles to help pay the sizable debt of the Quincy church, where Brooks now sat in the pew once occupied by his father and grandfather.

There were other duties, as family interpreter and preserver, which Brooks found difficult and painful. He attempted in 1911 to rescue his father from Charles' slender 1900 biography. Writing an essay for the Massachusetts Historical Society in 1911, Brooks concentrated upon his father's diplomatic triumph in England; his labor there was a "masterpiece." He likened his father to George Washington. This judgment must have astounded Charles. Notably absent from Brooks' monograph was the sharp and grimly fatalistic tone of his books on history and economics.

In 1904 Brooks had begun to read John Quincy Adams' papers, preparing to write a biography. He viewed the task as a near-overpowering trust; he pledged that the manuscript would be seen first by members of the family "in order to be sure that I have committed no error of judgment." Grateful at the time to have found a cause for his overflowing zeal, Brooks wrote Charles: "I am clear that if I have any duty further in this world, it is to try to the best of my power to save the family reputation for the next generation." The task was terribly difficult for him, since he shrank from exposing much of the family's past to the public, yet was determined to correct misunderstandings.

Brooks' decision to write about John Quincy was hardly pleasing to Charles. Not knowing of Brooks' plan, in 1905 he had told the staff of the Massachusetts Historical Society library that he was beginning a biography of John Quincy Adams himself. When he learned that Brooks had beaten him to the task, Charles responded glumly, "a plan shattered! poor J.Q.A." He retreated to the seaside at the Glades, where he cooled himself by taking "headers" off the rocks, all the while muttering "Too bad! —Too bad!" Ironically, had Charles written about his grandfather, the result probably would have been a eulogy since he identified with John Quincy. "For him I feel a genuine admiration," he said in 1901. "His courage was unconquerable; his ability great; his moral force admirable; his will abounding." Insisting that his ancestor had been correct on all the great issues of his day, so that a fair presentation of his case was imperative, Charles believed that only he could tackle the challenge. He never forgave the intrusion by Brooks.

If Brooks was aware that he had deprived Charles of his subject, he gave no sign. Instead, he worked earnestly until mid-November 1908 when he notified Henry that the long manuscript was nearly finished. He wanted Henry to read it at once, for he dreaded the possibility that he had committed "breaches of taste." This manuscript rests today in the Massachusetts Historical Society, 574 legal-size typed pages filled with many long quotations from John Quincy's papers. These excerpts are linked by Brooks' faltering attempts at interpretation. The work remains as he left it, incomplete. Brooks' nephew, the second Henry Adams, appended a note to the manuscript in 1938 saying that the material was "not to be taken out or shown. Brooks Adams did not feel satisfied with this." There is good reason why he did not. He had made the same mistake Charles had, of identifying too closely with his grandfather. The upshot was a curious eulogy: "A nature so complex presented an enigma to contemporaries, and as few could grasp all the premises from which he reasoned, so few could follow his deductions." Brooks mailed the manuscript to Henry and awaited his assessment apprehensively.

On 18 February 1909, Henry replied in one of the most fascinating documents to survive of the Adams family's private papers. It is also one of the most bewildering, and it certainly had that effect upon Brooks. To understand Henry's reaction, one must remember that Henry adored the memory of Louisa Catherine, the wife whom John Quincy had treated so selfishly. By no means could Henry see his grandfather as heroic. While Henry seemed on the one hand to encourage his brother's aim and to praise his manuscript, he overwhelmed him with a stunning criticism of John Quincy Adams, which contrasted sharply with Brooks' exercise in filiopietism. And although Henry urged his brother to continue working, he left Brooks' manuscript in a shambles. Brooks may have sensed what was coming when Henry sent a bulletin, in the midst of reading the biography: "Your picture of our wonderful grandfather is a psychologic nightmare to his degenerate and decadent grandson, . . . the psychological or pathological curiosity of the study takes possession of me." Of one thing Henry claimed to be certain. Their forebear's career "tells me never-never-never to be didactic." He acknowledged that he found John Quincy Adams as obnoxious as Brooks considered their brother Charles to be.

Henry advised Brooks not to be too adulatory in approaching John Quincy Adams nor abusive in treating Jefferson. This advice must have dumbfounded Brooks, whose slur upon Jefferson, said Henry, "shows a vicious temper altogether indecent in you." Jefferson, argued Henry, was far better than the venal Americans deserved—"For God Almighty's sake, leave Jefferson alone!" Rather, Henry urged, Brooks should try to see his grandfather more clearly. To help him, Henry sharply criticized John Quincy's career and character before 1830, destroying the exalted figure Brooks had clumsily created. Even John Quincy's poetic flights did not escape, Henry considering them pitiful sentiments which "turn me green with shame." John Quincy's performance as a professor of rhetoric was so inept in Henry's view that it would embarrass his grandchildren.

What Henry criticized most about his grandfather's career, especially before 1830, was how he had allowed ambition to drive him and self-deception to keep him filled with self-righteousness. "For my own part, I feel still more the old man's want of judgment," said Henry. "He had a nasty temper. Not that his temper was bad, but that it fooled him." In listing the deceit and errors he found in John Quincy's record, Henry was particularly critical about his support of Jackson during Monroe's administration and his backing of the Missouri Compromise, all moves, to Henry's view, born of ambition. He portrayed John Quincy as obsessed with the presidency. "JQA deliberately acted as the tool of the slave oligarchy (especially about Florida)," Henry observed, adding that his grandfather "never rebelled until the slave oligarchy contemptuously cut his throat." He meant here the election of 1828.

Although Henry's chief concern as a critic was to challenge Brooks' naive view of their grandfather, he also offered editorial suggestions, notations which today give glimpses of Henry's active regard for language and style. "Tut-tut-tut!" he exclaimed at one point. "Keep your temper! Such language even when strictly true, shows ill-breeding. Try to be a man of letters!" Although Henry tried to soften his criticisms by using indirection and instructive jest, he was still candid. At the outset he assailed the cruel practice of biography, saying that he shrank from recalling "the history of grandparents, whom I pity with the keenest sympathy and wish had never been born." And yet, though Henry felt that readers ought not to have details of family life, he still insisted that unless the inner workings of John Quincy Adams were known, his anguished story would never be understood and certainly not appreciated.

Henry admonished Brooks to remember how their family had always fretted about the character of John Quincy Adams—especially had it "worried his father and mother and wife." The cause of the concern was "that he was abominably selfish or absorbed in self, and incapable of feeling his duty to others." So spoke the Henry Adams who was at the time helping others as penance for his loss of Clover. "His neglect of his father for the sake of his damned Weights and Measures" was outrageous, Henry contended, "but his dragging his wife to Europe in 1809, and separating her from her children was demonic." This crime was made more grave, Henry contended, because their grandfather was running like a coward from defeat and from self-revelation. Thereafter, John Quincy Adams "loathed and hated America," Henry insisted; the exile returned in 1817 still filled with disgust. Yet this same man, said Henry, forced his children "to be trained to profess passionate patriotism which very strongly resembled cant." Their famous grandfather, he said, had fallen into "fraud and hideous moral depravity."

At last, however, Henry conceded that by 1830 John Quincy saw his life had been "a sentimental folly—a bitter absurdity," so he tried another career, this time less clearly driven by ambition. He himself, Henry said, had resisted recognizing the full truth about John Quincy's character and particularly his shameful fleeing in 1809 "because I have always felt in myself the sentimental weakness and have always avoided responsibility in consequence." In fact, said Henry, "the likelihood is great that I inherited some of the old man's nature because I loathe it so heartily."

Somehow, Brooks managed a reply to this shocking response from his revered Henry. Of course he would not publish the biography, he announced, in the face of these differing concepts of John Quincy. Although he felt he had been disinterested as a biographer, Brooks admitted that "I am showing him at his best." He began his rejoinder: "No one ever understood him, no one ever will understand us—but he was right and we are right." This was because John Quincy had a grasp of history which was, said Brooks, "Prophetic," "genius," "science"—all terms this grandson wished to earn for his own zeal to awaken and remodel civilization. "I cannot tell his life for him—I cannot even tell my own." Now Brooks' natural insecurity overtook him. "I have a deadly fear of failure," he wrote Henry on 6 March, and went off to ponder at leisure his brother's exposé of their grandfather.

A week later, Brooks wrote Henry again, showing what their exchange was affording him in introspection. "I have wrecked my life because I have been too hot. I have never known how to wait and bide my time until I could close and stab my enemy." For this reason, he said, John Quincy had been right to retreat to St. Petersburg. Brooks refused to believe, with Henry, that their grandfather's career was a patchwork of hypocrisy and blunders. Instead, he suggested that Henry realize how men were born to battle and kill. "I glory in the Old Man's fight from first to last, and I am willing to stand by him against all comers, be they who they may." Still, Brooks declined to publish such a controversial book. It was a hard decision and one which kept rankling him, for he followed his declaration of suppression by a mournful reminder to Henry that the Adamses had "made all we ever have made by the hardest kind of hard fighting, and when we renounce that we have little left to stand on." As an expert at renunciation, Henry could hardly have been impressed by this statement.

While Brooks continued to struggle with his decision to shelve the biography, his admiration for John Quincy Adams increased. He was eager to let the world know about a heroic ancestor who fought even when he knew he was defeated, for this was the mold, Brooks wanted to claim, that was his own shaping. Only this rapturous personal identification can reasonably explain why Brooks was unable to bring to the biography those powers of analysis and exposition so evident in his other work. In choosing to write about John Quincy Adams, Brooks wanted to make an emotional statement concerning himself and his family, and this left him unable to write the well-balanced work Henry called for. Brooks still wished that somehow the truth about the early Adamses could be seen. "It was the character of John Adams and of John Quincy Adams after him to be controversial," Brooks told Henry. After all, "our ancestors must stand or fall by what they were," he said. "I think they were pretty formidable men." Henry might not wish to join their vanguard, Brooks implied, but as for himself, he was proud to march behind the early generations.

Henry's reply to all this was scarcely helpful, for he decided to stop being candid and reverted to his accustomed jocular view. Certainly, he said, Brooks might well be correct about John Quincy Adams; but then he tossed off this concession as probably worthless since he had gone through life "in such an ocean of historical hogwash" that he would admit to anything being right, or wrong, or both. Henry said Brooks should do as he wished with the manuscript; he wanted no more part of it. He never wanted to "smell the old reek again. It is foul and I loathe it."

With Henry's abrupt, nearly rude withdrawal, Brooks had to find encouragement elsewhere. After looking over the manuscript during the spring of 1909, he mailed it to Mary Cadwallader Jones for her opinion. Never at a loss for zeal, Mrs. Jones sent the manuscript to the Macmillan publishing house for its assessment, omitting to ask Brooks' permission. What followed was almost comical. When Brooks found what she had done, he was frightened and angry, though he tried to soften his mood as he scolded her. Did she not realize that "the views I take are only tentative," he asked? Already he had changed his mind about many points. Even worse, however, the manuscript was "a private and confidential document" which no one should see until there was family agreement upon it. Furthermore, he would not consent to publication "unless I am convinced that it will raise my grandfather's reputation."

Brooks, knowing that the document was being passed about among strangers while his brother Charles and sister Mary had not even seen it, fumed. There would be trouble, he predicted, urging Mrs. Jones to retrieve the manuscript at once. "I cannot help shrinking from exposing my grandfather, not to say myself, unclad as it were." Brooks need not have added that he was "a little nervous about this book." Trying to make up for her blunder by fulsome praise of the manuscript, Mrs. Jones reclaimed it from Macmillan and returned it to Brooks, who put it aside for nine years, during which it was never long out of his thoughts.

When Henry died in 1918, Brooks believed that something must be done to evaluate both his grandfather and his brother. He failed to see that he was contriving the most bizarre combination in the family story, so fixed was he upon rescuing Henry from the misleading facade he had built to block the public's view of his greatness. "Poor Henry," mused Brooks, as he looked at "stacks" of reviews about the *Education,* being widely read in 1919. Henry had been betrayed, Brooks felt, by his penchant for sardonic humor. To correct this, Brooks prepared a book of three of Henry's unpublished essays on the nature of history. These were works with which Brooks agreed and which he praised extravagantly.

Brooks wrote a long introduction drawn from his John Quincy Adams manuscript, believing that in this way he might finally explain all things Adams. This ultimate statement, "The Heritage of Henry Adams," along with Henry's three essays, was published in 1919 as *The Degradation of the Democratic Dogma.* It still lacked the candid, balanced view that Brooks had lost in his approach to John Quincy. Aware of this, he tried to offset it by resorting to the fatalistic philosophy which both he and Henry had preached.

"I do not write on Henry at all," Brooks insisted to Mrs. Jones. "I write on him as a phenomenon, as my grandfather was and as I am." This was the great lesson he saw in the private world of the Adams family. Brooks contended, as Henry had earlier discovered, that his forebears knew the futility of believing in progress and in the perfection of democracy. The Adamses lost faith in the Republic and in God, Brooks insisted, although his essay fails to demonstrate this. Instead, he tied the story of his family to the development of a deterministic line of reasoning which Brooks himself had been advocating in his books for many years. At last he was free to rally great names of dead relatives to his side. "Of this growth [in fatalistic philosophy] my grandfather is the main figure—not Henry or me," he assured Mrs. Jones. "We are appendages only."

Now that Henry was dead, Brooks claimed that his brother had been wrong to make him suppress the biography of John Quincy, forgetting that Henry had relented at the last moment. Brooks wanted to blame someone else for his own lack of courage. He now also could indict Henry for running away from family and life, and for finding refuge in the frivolous and the obscure. "Henry himself was his own worst enemy as far as being taken seriously was concerned, by writing his 'Education,'" Brooks grumbled. "I can't help that," he complained, calling the *Education* "a jocose book of memoirs" which would have to be forgotten if Henry was to be "taken seriously as a thinker." Brooks considered himself "selfishly, stupidly and unreasonably irritated for my brother's reputation—and yet why should I care? Henry chose his own path, deliberately. He must walk it." With this part of Brooks' statement, Henry would probably have agreed emphatically. But he would also undoubtedly have been outraged by Brooks' determination to haul him from behind the facade which he had so patiently built.

Nevertheless, Brooks could not resist the temptation now that Henry and the other Adamses of note were in their tombs. Now he could explain them, although the brashness of the step occasionally troubled him. Significantly, he published *The Degradation* under Henry's name, late in 1919. "I have taken a large responsibility towards others—particularly my grandfather," Brooks conceded to Mrs. Jones. "The old man is dead and might have thought himself safe—particularly from his own children—but I have raked him out and exhibited him, and very like, made him ridiculous."

Having exhumed his brother and his grandfather, Brooks began to appreciate what he had done. His loneliness with a bleak philosophy

about man and the universe compelled him to seek assurance and justification by calling upon his defenseless relatives. Little had changed
for Brooks as an old man. He was as pathetically in need of his family,
real and imagined, as he had been as a young lad, born late in the lives
of his parents. "I have used my ancestor as a club," Brooks apologized
to Mrs. Jones, to whom he turned for help in improving his writing
style which seemed to fall apart when he addressed family subjects. His
essay in *The Degradation* owed a good deal to her.

Brooks had not failed entirely. He had seen in John Quincy Adams'
career and outlook something of the monumental achievement—and
the sorrow—of the Adams family. This was the rising belief that, as
Brooks himself put it, humanity was moving inexorably downward to
chaos and extinction. At the beginning of the Adams story, this outlook
had been a nagging doubt which shaped the family gospel. It had
clouded the hopes of John and Abigail as they looked ahead for their
family. Now, at the end of the saga, Brooks claimed the place of proclaiming this doubt which had so long troubled his predecessors. The
tough pessimism which the Adamses displayed at such cost in the face
of fashionable optimism was elevated to a universal vision by Brooks.
He saw society ruled by laws of decaying energy similar to those in the
natural world.

Brooks' view of the family's outlook failed to take into consideration
how the Adamses had possessed, until his own time, the ultimate comfort of a Christian design—including the possibility of redemption. This
Brooks never attained. Instead, his theory that civilization suffered from
diminishing energy allowed him to blame the decline of his family line
on forces beyond mortal control. Bleak though such fatalism was, it
may have been easier to bear than the prospect which John, John
Quincy, Charles Francis, and their wives had seen. For them, there had
been the fear that God might withdraw His personal support from the
family and from the Republic. This was lost to the latter-day Adamses,
so that the bonds of style and outlook which once had unified the family, and occasionally overwhelmed a member, fell slack after 1889. The
Adamses became as exhausted as Brooks foresaw the universe would
eventually be.

⁂ 18 ⁂

Endings

The three surviving brothers in the fourth generation were fascinated by the prospect of their own mental and physical decay. Even keeping his distance, Henry was no more free of this concern, particularly after his seventieth birthday. Like Charles and Brooks, he dreaded the possibility that he would share their father's fate—mental deterioration. The first sign was, of course, faltering memory, and every striking lapse brought Charles Francis to mind. Still, the catastrophe which came in April 1912 rather surprised the family. While dining alone at home in Washington on the twenty-fourth, Henry suffered a stroke, falling helplessly from his chair. As he later described what happened: "I dropped out of conscious existence for six weeks." With that event began the last chapter of the Adams family story in which all members shared. There is some irony in the drama since it was the fugitive of the family who inadvertently unified his divergent kinsmen.

During the spring of 1912, Henry had been tense with worry over his own problems and those of his various dependents—the Hooper nieces and Mrs. Cameron and Martha, particularly. He complained at the time that he had never felt so vexed by the troubles of others. However, when he fell gravely ill, it was not Henry's adoptive circle who provided the help and comfort his invalid state required. Instead, the Adams family took back its own, and Henry found he had not been so far away, after all.

Members of his family hastened to Washington and, with a few friends, kept vigil in his upstairs bedroom in the house on H Street. Through May and into June, Henry's condition offered little hope of significant recovery of mental capacity, although his physical powers began returning. Twice, as he gained strength, something in his darkened mind impelled Henry to try, as Clover's brother Ned Hooper had, to end his life by leaping from a window. Both times he was restrained. Usually with him were Minnie, Charles' wife, and his close friend Nannie Lodge. They listened helplessly as he cried out at intervals, "I want to go to Quincy."

The moment he learned of Henry's illness, Charles characteristically took charge of the situation. Recently, his boundless energy and desire to command had been hard put to find a satisfying outlet. Henry's crisis clearly was a summons to the titular family head. Charles may even have been slightly gratified that circumstances had delivered into his care this brother he had so long admired and pursued but never brought to earth. Away from Washington at the time the crisis occurred, Charles rushed back to take control away from servants, women, and even intimidating physicians. At first glance the situation appeared sad but simple. The doctors informed Charles that if Henry survived, he was doomed to suffer "from arterial decay, like John in /93." Thus, "a place" must be devised where he might be cared for, and Charles began plans for a pathetic invalidism, of the sort which Henry dreaded. As late as 20 May, the painful prospect remained the same, for Charles still found Henry "a babbling wreck."

However, by this time it also became distressingly evident to Charles that the task of mercy he had undertaken would prove more awkward than he had anticipated, for Henry's illness had produced much excited interest. Charles was pestered by advice from three women, his wife Minnie, his sister Mary, and Lizzie Cameron, who was in Paris. As if this were not enough to exhaust the seventy-seven-year-old Charles' patience, he also had to contend with Brooks, who looked upon himself as Henry's only family confidant. Even so, he might have endured all of this for Henry's sake. What took him most aback was the news that Mrs. Cameron was rushing home to take charge of Henry herself.

Upon learning of Henry's stroke Mrs. Cameron wrote to Charles: "I need not say what it means to me to think of him there alone." Since Henry had often shared with her his deepest fears of just this predicament, Lizzie announced that only she knew what Henry now thought and wished—or would, if he were able to be deliberate. She would handle everything in Paris, Lizzie assured Charles, including sorting Henry's private papers: "I shall destroy nothing." After closing his apartment, she announced she would start for America and come to Henry's side. In this and subsequent bulletins, Lizzie displayed supreme confidence in her place as the person closest to and most knowledgeable about Henry Adams. She even sought to reassure a thunderstruck Charles by telling him she approved of the interim steps he was taking in Henry's behalf. While Lizzie politely asked Charles for suggestions as to how she might best help, her goal was unmistakable. She would bring Henry to Paris. "[H]e can move right into my apartment which is relatively cool and comfortable. I have had my servants for several years and he knows and likes them and will feel at home here. He also likes the cook!"

For a time, Charles hoped that by being gentlemanly but discouraging, he could ward off Mrs. Cameron. There was nothing more now to do, he assured her several times: "It only remains for us to await developments." When Lizzie did not take this hint, Charles was more blunt, telling her that Henry would often "babble about. . . . his condition is very depressing," and that she should stay where she was. "You would be simply another supernumerary, so to speak, in and about his house." Clearly, said Charles, Henry was doomed to be a "helpless invalid under the care of his immediate relatives." It was "a very sad case, and must be dealt with accordingly."

By now, however, Charles had added reason to become direct with Lizzie. Learning of Mrs. Cameron's determination to rescue Henry from the Adamses, Mary was outraged, and she poured out her disgust upon poor Charles. To make matters worse, Mary had conferred with Brooks, and both were determined to have Henry moved to Massachusetts as quickly as possible. While Charles agreed to this, and much earlier had begun planning to convert a cottage on his estate as a hospital, his pride resented the stern lecturing his sister gave him. The worst moment for him was when she implied that Charles too might be susceptible to the famous wiles of Lizzie Cameron. Said Mary about Henry's plight: "Above all things we must have the control of everything about him. I object extremely to have any woman round him outside the family." Announcing that she personally had cabled Lizzie telling her to stay away, Mary put the matter bluntly. "I tell you fairly I won't have her, rather than that I would go there myself and keep her out."

Mary Adams Quincy then put into words what had annoyed the family for many years. "There has been disagreeable scandal enough about that affair and we certainly cannot permit people to say that in his last illness she came from Europe to look after him." Let the Adamses, said Mary, "take care of our own brother," adding for Charles' benefit, "I

expect you as the head of the family to support me. —If you can't
manage the women refer them to me. I'll settle it easily." Mary's view
was clear except for one feature: what would Henry prefer? It was
Charles who brought this up in behalf of the patient. In the face of
Mary's assault, Charles found himself urging that things be handled
Henry's way, perhaps the first time the elder brother had defended the
younger's preferences. He invited Mary to accept the fact that their
brother was "a regular Bunthorne," by which Charles meant the figure
in Gilbert and Sullivan's *Patience* who was always surrounded by adoring females. Mary was assured that Henry's lady friends were well-
behaved and that even Mrs. Cameron was "all right!" According to
Charles, the care of Henry would be readily and prudently disposed of
if Mary simply relaxed.

However, Charles' advice was not accepted, notably because Brooks
had begun warning that Mrs. Cameron must not be permitted to close
Henry's Paris affairs nor ship his books and furniture to America.
Brooks himself planned to be in Paris later in the summer and he
wanted to supervise these matters. Then Charles' wife Minnie decided
to thwart the campaign against Lizzie which Mary and Brooks had
mounted. Minnie saw that Henry had begun an astonishing improvement, and should therefore be consulted about his future care. She
appreciated Henry's longstanding design of replacing his Adams connections with an adopted family circle. Now was the time, Minnie advised Charles, to put Brooks in the background. "Keep him out! . . .
for heaven's sake! Don't bring Brooks in again! . . . I don't think Brooks
considers Henry at all in this matter."

Minnie sent this message on 23 May, and, after more impatient
thought, later that day mailed a second letter which urged Charles,
who was away in Massachusetts, to be conscious of "how absurd they
all are." She was referring to the "tantrums and turmoils" Mary and
Brooks had begun. They forgot "poor Henry's comfort and happiness"
in their determination "that Mrs. Cameron and others shall not get at
him." The public was watching, and Charles' prime responsibility was
"Henry's comfort and welfare," even if it meant allowing him the surroundings of his own home and the presence of his own friends.
"Please," Minnie begged, "don't consult Brooks or Mary about these
things." The letter was too late, however, for an exasperated Charles
had taken a step he quickly regretted. Out of patience, he turned Henry
over to Brooks and Mary, telling Mrs. Cameron that had Brooks been
in Washington when Henry fell ill, the latter "would have looked to

him in this matter." So now, said Charles, "it would be better" to hand Brooks the responsibility. Lizzie could expect to hear from him.

Elaborately, even sarcastically, Charles resigned command of this family crisis in favor of his youngest brother. "I made a mistake," he told Brooks. "This whole matter should be in your hands." With a flourish, he announced: "I will abdicate." Meantime, he was so annoyed at Minnie's intrusion that he sent further messages to her through his secretary, who was obliged to inform Mrs. Adams that her husband believed "There are altogether too many horses in the harness, and some of them seem to labor under the impression that they are the entire show." Matters might yet go well, Charles announced distantly, if "only people can be persuaded that their fingers are not essential ingredients in the pie." Minnie knew how to shift to better ground. Immediately she soothed Charles by assuring him that he had acted wisely in turning affairs over to Brooks, in view of the latter's "worrying and fussing." However, she added, "I am sorry for Mrs. Cameron." True, Minnie acknowledged, Lizzie was foolish in writing to everyone about her determination to nurse Henry. But for the Adamses to display such hostility toward her was sure to create "trouble and much talk." Brooks should realize, Minnie said, that, despite his suspicions, Henry "was very fond of Mrs. Cameron—that she has been accustomed to taking care of things for him in Paris." Henry would be outraged to know that Mrs. Cameron "was being thrust aside and treated like a stranger." In fact, mused Minnie, half-playfully, perhaps it might be worth letting Mary and Brooks know that Lizzie wished to move Henry into her own apartment—"then there would be fun!" Finally, remembering the scolding she had taken through Charles' secretary, Minnie had the last word with her husband. "It seems to me that it is I that should pitch into you instead of your pitching into me. —But the habit has become ingrained with you. —If things don't go your way, why, of course, it is my fault!"

Charles' stratagem with Brooks had been unwittingly brilliant. Astonished that Charles should ever capitulate on anything, the younger brother found himself very uncomfortable with the responsibility for decisions about Henry. While Brooks was always full of ideas, he never enjoyed having to carry them out. As he often observed, his nerves had not recovered from the great family financial crisis of 1893, when he had had no choice but to act. Now, the prospect of handling the artful Mrs. Cameron was enough to change Brooks' outlook at once. He tried, feebly, to be aggressive, but his letters to her breathed a spirit of courtliness and gratitude for her contribution. He was too late, it turned out, as far as the furniture in Henry's apartment was concerned, for Lizzie had shipped it to America. At this news, Brooks gave up and prepared to flee to Europe, quite willing to leave Henry once more under Charles' command. "Any arrangements touching Henry which are satisfactory to the doctors and you and Minnie are satisfactory to me. I have no suggestion of any kind to make," Brooks wrote.

With his authority regained, Charles found it was now possible to do more for Henry, who, in early June, showed a sudden and remarkable improvement. His mind rapidly grew clearer. On 17 June, amid elaborate preparations and using a private railroad car arranged by Senator Lodge, Henry was taken from Washington to South Lincoln, Massachusetts, where he was installed under medical supervision in the cottage on Charles' estate. To the family's delight, Henry could be found immediately after arriving, as Charles reported, "sitting up, clad, and refreshing himself with some ice cream," while talking of his relief at being away from Washington. For Charles, who had regularly pronounced Henry's case to be hopeless, this was nothing less than "phenomenal."

During this time, Charles began a series of letters to Mrs. Cameron in a softer tone. He seemed impressed by the efficient manner in which she had managed Henry's complicated Paris business, a marked contrast to the shambles in America created by family contentiousness. Also, he was touched by Lizzie's piteous wish to see Henry. No doubt he regretted having earlier stated to her that "you could contribute in no way either to his physical well-being or to his mental contentment." But by this time Mrs. Cameron had heard of Henry's improvement from other sources. It was all the news she needed. She set out at once for Massachusetts, planning to stay at an inn in Concord near Charles' residence. When she was informed while still afar that Concord had no rooms for rent, she was dismayed but unmoved in her determination. "I long to get there," she told Nannie Lodge. "I feel that I must not miss any more of him than can be helped."

Appalled at this development, Charles tried as tactfully as he could to dissuade Lizzie by stressing that Henry had ample company and that he was very happy, while reminding her that there were no convenient lodgings at hand. Charles also discouraged Brooks, who was on the eve of his departure, from visiting Henry. The elder brother held the view, quite correctly, that Brooks' excitable presence usually upset Henry. Indignant at this, Brooks retaliated by asking Nannie Lodge, whom

Henry adored, to remove Henry from the imprisonment arranged by Charles. Could not Henry go to stay with the Lodges at Nahant, which was near Beverly Farms? Brooks contended that Henry would be happier there. Thus the plot thickened, as Mrs. Cameron sent word to Mrs. Lodge, announcing that she was still coming to Henry's bedside, no matter what Charles might think. "I shall pay no attention to what he says," Lizzie wrote. On 1 July, this news reached Charles. "Mrs. Cameron comes!—the siege is to begin," he reported to his diary, adding, "fool woman!"

There was no choice now but for Charles to enlist Henry's help. After hearing about the situation, the patient, who was continuing to recover rapidly, dictated a letter to Lizzie which politely urged her to stay away, although Henry added that he was unwilling to advise her about making a brief visit. He told her he wanted now to see only his family; no one wished to display himself in weakness, so "I much prefer to hide myself." She must wait until he was at least strong enough for them to take walks. He closed: "I prefer not to exhibit myself to you or to any of my friends as more of an imbecile than I am made by nature. I really hope you will not see me until I am fairly visible." This was a remarkably pointed statement for Henry, who ordinarily spoke by indirection.

With this reinforcement, Charles himself could now be blunt. He told Lizzie that he detected "a lack of appreciation on your part of the condition of affairs here." Reviewing the history of strokes in the family, particularly in the cases of John Quincy, Louisa Catherine, and his brother John, Charles besought Lizzie to realize that Henry had "heard the first bell." Life would never be the same for Henry, he said; if Mrs. Cameron would only appreciate that prospect, she might help Henry realize it. The first step would be now for Lizzie to stay away. Charles assured her that Henry was content; his state "cannot be improved upon."

Unfortunately for Charles, even as he wrote, Lizzie was proclaiming joyfully to her friends: "I can read to him, write his letters, or walk or drive him out." To the Adamses, she carefully put her intention in terms of relieving Minnie and Elsie, Charles' daughter, who were in the circle around Henry. Said Charles, in defeat, "I have done my best" to head off the invasion "and to secure a reprieve for the victim." He told Mrs. Lodge on 5 July, "The enemy would be at our gates, and the siege begins, on or about the 20th."

By now, even Charles could see some humor in the situation. Henry was improving at an astonishing rate. "I think by the 20th the victim will be almost ripe for the altar," he laughed, and wondered whether Lizzie would "carry Henry off in triumph . . . to enjoy a somewhat superannuated honeymoon. But all this time where is Don?—Where?— oh, where?" (Lizzie, of course, had long been estranged from her husband.) Charles was even genial in receiving advice from Wayne Mac-Veagh, Lizzie's brother-in-law: "Don't allow Lizzie to bully you. . . . She has been the very incarnation of *selfishness* and will impose on the good nature of any body and go to any extent to accomplish her own selfish purposes." Charles replied that he was not Henry's keeper, and that the patient was almost fully recovered. Whatever understanding Henry and Lizzie might have, said Charles, in a rare teasing mood, the pair's plans "do not include an elopement, inasmuch as Henry's locomotive capacity is at present seriously impaired." Charles became more earnest when he told MacVeagh that the less talk there was about this reunion, the better. "I should deeply regret anything conducive to scandal or newspaper notoriety."

Meanwhile, Nannie Lodge chose to enter the drama, moved by the suggestion Brooks had made earlier. She proposed that Henry move to the Lodge estate beside the ocean at Nahant, a scheme which turned Charles' frivolous mood to sarcasm. His hospitality and capacity to care for Henry had once again been challenged, and he promptly replied to Nannie with a series of sardonic observations—the likelihood that Lizzie Cameron would take full command of the Lodge "castle-by-the-sea"; the importance thereafter of inviting Don Cameron, whom Charles knew the Lodges detested; and the desire of Minnie and himself to remain "mere lookers-on." Charles' hurt was evident in his assertion: "My own somewhat contemptible Lincoln arrangements, having been voted wholly inadequate, are relegated to the forgotten past. The game is now wholly in your hands."

This was mid-July 1912; finally matters improved for Charles' side, and the drama of Henry Adams' illness closed gracefully. Nannie Lodge apologized handsomely to Charles, retreating at once from what she saw was an unfortunate notion about moving Henry. She even conceded, cautiously, her own misgivings about Mrs. Cameron, whom she called much changed "from the old days when she had and needed any amount of courage and self-control." Further reinforcement came from Henry, who was surely aware of the discomfort Lizzie was creating. He continued to discourage her visit, warning her that after a correct and brief stay with Charles and Minnie, she must move to Boston.

He knew that from such a distance further visits to the cottage in Lincoln would be difficult. Lizzie now realized that she was not to be met with open arms. "I gather from Dordy's letters that he does not care particularly to see me nor to have me stay!" she told Nannie. "But I shall decide nothing until I have seen him."

Charles and Minnie received Lizzie as a weekend guest on 24 July, and the occasion proved to be remarkably pleasant and gratifying for all but Lizzie, whose feelings were mixed. She discovered she had miscalculated what she had understood to be Henry's antipathy toward all things Adams. In fact, she found him very happy; and while he was pleased to see her, he was clearly without any desire to detain her. Soon, after driving a few times from Boston to the patient's side, Lizzie became bored—as Henry had foreseen. She departed for the social excitement of Bar Harbor, Maine; from there she wrote to Henry, a bit too pointedly, of the good time she was having.

Victorious, Charles tried to be magnanimous; he had enjoyed seeing Mrs. Cameron, "a good deal of a woman of the world, full of experience and observation," he said. Henry, meanwhile, told Nannie Lodge that he was "singularly happy." Here, in what he called "the society of my youth," he could elude the depressing realities of 1912. He was especially grateful to escape the streetcars which ran in front of his Washington house; "when I was ill last May the trams almost did for me and I wept for the one horse car of our youth."

By August, Henry was taking motor rides and long walks. In the area around Charles' residence, he said, "the silence is thick. I love it. I walk for hours in the woods, devoured by gnats and mosquitoes, just gulping the silence." He and Charles renewed something of their old warmth, lost for half a century, talking endlessly in the dusk each evening. Finally, in November 1912, Henry returned to Lafayette Square in Washington, pronounced by Charles as "better both physically and intellectually, than he has been for some years past." Henry was pleased that his illness had precipitated a reconciliation with Charles. The elder brother meanwhile often looked wistfully at the now vacant cottage where he had seen to Henry's care.

Certainly Henry was profoundly grateful to Charles for the aid rendered during 1912. "You have enabled me to close up some old chapters that needed it." Whether Henry included in this statement his former relations with Lizzie Cameron is uncertain, but she clearly took a place in the background of Henry's life after she moved on to Bar Harbor. She continued as a kind of manager for his Paris residence,

but he had only two more seasons to spend in Europe. Although Henry
returned to his former existence with a laconic motto—"Play the
game!"—one change was noticeable in the rest of his life. He summoned a new ménage around him which did not include Mrs. Cameron—a younger generation whom he could help while they aided him,
a group led by Elsie Adams, Charles' unmarried daughter, and Aileen
Tone, a friend of the Hoopers, who charmed Henry with her ability to
sing the medieval French songs which brought the twelfth century alive.
Thus, Henry resumed his retreat to the soothing shadows of ancient
cathedrals and the figure of the Virgin. With increasing but disguised
impatience, he bore the frequent visits and letters from the Hooper
nieces. At such times he found it prudent to use what Charles called
"all his old affections." On the whole, Henry's life after his illness was
so pleasant that for the first time since 1885 he conceded he felt happy
even in December, a month he had always dreaded because of memories of Clover's suicide.

Somewhat resentfully, Charles watched this development, telling Mary
that Henry was again "as fantastic as of yore, indulging in boundless
paradox and in efforts at conversational and other affectations and eccentricities, of which my more sober, and possibly a bit critical judgment, fails wholly to approve." Meanwhile, as he reported it, Henry's
feminine courtiers "sit around looking at him with admiring eyes, and
say:—'Oh, how delightful he is!'" Brooks was even more extravagant,
insisting that when Henry embarked for Europe in the spring of 1913,
"Every foot on the ship was filled with ladies, of every age, race, color,
and former condition of servitude. No room remained for passengers."
And Henry himself explained to Lizzie that such surroundings helped
him retain a calm mind, by which he could ignore the present.

Lizzie sought occasionally to reestablish herself in charge of Henry,
but each time he delicately resisted, although he still looked for ways
to aid her financially. Eventually, she withdrew with considerable grace,
speaking only infrequently to him of "your attendant females," and of
how she must "drift around" and then "turn up" when "perhaps you
will need me more." There was one last benefit she did provide, however, for Henry fled war-stricken Europe in August 1914 to the peaceful English country estate, Stepleton, in Dorset near the southern coast,
owned by Lizzie's son-in-law. Lizzie, however, was absent.

Completely cut off from her aged husband, she spent the war years
in Europe, doing volunteer work and hovering near her ailing daughter. Henry never saw her again, although he sent her money and letters

from time to time. So ended a complicated relationship which had afforded Henry a familial comfort when he most needed it. It was a bond to which he always gave much more than he received, but it hardly deserved at any point Charles' scorn: "The Cameron-Adams arrangement is flagrant! As the expression goes, it 'smells to Heaven' under my very nose."

How differently Henry felt was clear in a remarkable letter he wrote Lizzie late in 1915, a last testament to the unique family he had created around Lafayette Square. After telling Lizzie he had not been replying to her messages because he had nothing to say beyond the last chapter in the *Education,* he announced that he had reversed his stand about the letters they had exchanged for thirty years. He no longer wished them destroyed, as he had instructed her in 1900. Instead, he said he wanted the letters to immortalize his earlier days with her. He talked with special fondness of a precious quartet from those days: Nannie Lodge and John Hay, who Henry said had been hopelessly in love with each other, and Lizzie and himself. Then he added a fifth member. "I do want you and Nanny to stand by the side of John Hay and Clover and me forever—at Rock Creek, if you like—but only to round out the picture." All that he now had left, Henry said, was "to feel you there, with Clover and me, and Nanny and Hay, til St. Gaudens' figure is forgotten or runs away." It was an extraordinary letter, requiring great emotional effort to write. Henry confessed it took "a week's hard work."

Lizzie was overwhelmed by Henry's gesture. "I think of my reckless, wasted life with you as the only redeeming thing running through it," she replied; he had been her source for "the sustaining power to keep going, always keeping me from withering up. . . . Whatever I have or am is due to you, to that never failing, never ending, never impatient nor exhausted friendship. I wish I were more credit to you." Now, she said, the balance of her life would be spent in England where she wished to die, while their letters would be safe there with her until the proper time. Lizzie did live for many years at her son-in-law's residence where she died a very old woman. She remained a striking figure, nourished by recalling her daughter Martha, who died during the war, and by recounting her glories as the queen of Lafayette Square.

Even as Henry wrote his last important letter to Lizzie, the story of the Adams family was closing. Despite the fact that Henry had beaten him to the warning bell, Charles was first to die. To his last moments, he remained as active and aggressive as always, particularly with regard to family members. While he had scant success with Henry and Brooks, his relationship with Mary was different. Mary's husband, Henry Parker Quincy, had died of kidney failure in 1899 at the age of sixty. Mary needed to lean upon someone, for she had been indulged by her parents and then by her husband. Left alone with two young daughters, Mary accepted Charles' instructions and attentions with considerable grace. For once, the elder brother found a family member who agreed with him occasionally, since Mary shared his belief that an Adams should observe the proprieties of life, especially in Boston. Brooks, who was less concerned about such matters, said of his sister, "Mary is the only one of us now who has the true Boston instinct."

Mary became rather more lively after 1900, widowhood releasing her from the seclusion necessary in the later part of her marriage. When her niece Abigail Adams Homans, "Hitty," paid her a visit in 1910— Mary was sixty-five at the time—the younger woman reported: "Aunt Mary was in fine form—dressed in the brightest summer clothes, masses of jewels, and well painted as to her face. I was enchanted with her appearance." Charles called on his sister whenever he visited Boston in the winter, knowing that theater outings were a special pleasure for her.

However, in 1912, Mary's elegant existence failed to keep her from worrying, a trait she may have acquired from her mother. Charles learned that his sister had developed a dependence on drugs, the "morphine habit!" he grimly reported to his diary. He tried to be tactful in helping her, speaking only of the "very vague but disquieting impressions" he had been receiving. Surely she knew, he said, that if troubles of a domestic, financial, "or of a nervous character" were afflicting her, "I am the proper person for you to call upon." He was, he said, "unreservedly at your service," and willing to do anything "in the power or province of an older brother." Charles even unbent a bit, telling her that he, too, had been where she now found herself. "I have had my period of anxiety and many wakeful nights. I have known what it was to have no one to turn to."

Mary replied that she was "very much touched and pleased" by Charles' letter, conceding that her own health and that of other family members alarmed her. However, she insisted the problem was only temporary. "You may be quite sure if I need help I will come to you at once." Charles could not wait, fretfully reminding Mary that, above all, she must not go back to worrying. He suggested she turn her many fiscal problems over to him: "let me be the surgeon!—It means sleep!"

Although Mary relied upon her nephew, Charley Adams, for her

financial management, she looked to Charles in other ways, urging him
to call her often for a visit at luncheon or dinner. Mary outlived Charles
by thirteen years, and Brooks by one, so that it was the younger brother
who was left to look after her. The fourth generation of Adamses always expected women to require help facing life, a condescension Sister Lou and Clover had experienced, no doubt adding to their distress.
More like her mother, Abby, Mary evidently learned to accept female
dependency.

Charles' concern for his sister's health was modest compared to the
attention he gave to his own physical state. There was an almost boyish
quality to his proud reports of horseback riding, woodchopping, walking, taking "headers" into the chill Atlantic water at the Glades, and
even in "wheeling," as he called bicycling. He enjoyed being pleased
with himself. "Good for me!" he exclaimed as he neared seventy. Once
more there were flattering triumphs over his father, who Charles said
would "have considered it in some way undignified to don leggings and
go at a tree hatless and stripped to the undershirt." Old John Quincy
Adams was always beside him in sympathy as Charles chopped wood
for ninety minutes a day—"pretty well for a gentleman in his sixty-
ninth year." He recorded each triumph of exercise by trumpeting in
his journal, "None of my forebears could have begun to equal it."

Eventually, though, even Charles failed. By 1912, aged seventy-seven,
he was writing: "Am sorely troubled about myself; signs of weakness,
old age, and rheumatism." Now his passion was riding horseback no
matter what the weather. He took particular delight in doing so in
Washington where he grandly put to shame his less robust—or more
prudent—contemporaries and neighbors. But Charles knew his end was
approaching. Early in 1915, he brought himself to talk estate matters
frankly with his twin sons, John and Henry, neither of whom had quite
met Charles' high expectations. Now there was no alternative but to
turn to them, and Charles did so with relief. "My practical abdication—
Laus Deo!"

He began to try to see Henry more often. Henry found he could
even tease Charles again in ways recalling their boyhood spirits. When
Charles returned from England where he had received an honorary
degree at Oxford University in 1913, the story was widely circulated
that the Boston customs officials had detained the scarlet robe he had
worn in the ceremonies until cloth experts could determine if it were
of excessive value. Hearing this tale, Henry joyfully sent a telegram to
Charles: "Greatly regret painful controversy at custom house of Ox-

ford gown. Hope you are in no trouble with officials. If so call on me."

It gave Charles great satisfaction when he passed the age at which his father had died. Soon afterwards, however, the death in 1915 of his son-in-law, Grafton Abbott, took Charles to Boston, in a cruel Massachusetts winter, and there he was weakened by chills. He told Justice Oliver Wendell Holmes that, as he stood beside Abbott's grave, he endured a "harsh and gritty West wind which made one feel a strong desire to be put away—nicely packed in cottonwool." Another trip back to Boston a few days later made Charles complain to his diary, "I am tired, chilled, heart-sick, and homesick!" However, once again in Washington, the old compulsions won and he dragged himself out daily to ride Red Oak, his horse, for an exact two hours. "It was chilling cold, and I, out of gear and dead tired," he whispered to his diary.

One of Charles' last journal entries, on 15 March 1915, described how he had managed to stay on Red Oak for the prescribed time, "but was very tired and out of heart." He closed the day with these lines: "Dined with Henry—he and I alone and very dull. Glad to get to bed at 10:30." Henry later recalled that, as Charles left him for what proved the last time, he turned on the stairs landing, looked up at Henry and said, "Brooks was right. We have lived to see the end of a republican form of government. It is, after all, merely an intermediate stage between monarchy and anarchy, between the Czar and the Bolsheviks." Thus, reunited with Henry and, even more astonishing, agreeing with Brooks, Charles departed.

Five days later, he was dead, carried off gently while he slept by a coronary seizure. Ill with influenza, for three days he had lain alone in his big Washington house until he finally summoned Minnie from a visit in New York. She sat beside his bed and was unaware when he quietly died. Charles' final diary entry contained two words which may say more than all the pages of his Memorabilia—"heartsick," "homesick." He departed life more embittered and thwarted than anyone knew.

There was circulated after Charles' death a venomous "Memoir" of him, written evidently by a distant cousin who bore him a grudge. A copy of the memoir had been sent to Charles in 1913, warning him it would be published upon his death, a threat Charles took goodhumoredly, objecting only that the document called him "knockkneed." The point of the attack was Charles' famed contrariness, which the slanderer claimed was worsened by a harsh, arrogant nature, acquired through the "rough, farmer blood" which coursed in Adams veins. The memoir was a petty, worthless assault, although it may actually have come close to explaining the part Charles played in the private world of his family. From childhood on, the critic claimed, Charles invariably took views wildly opposite to others. Of this attribute, Abby was quoted as saying, " 'Well, that is like Charles, but you know, nobody pays any attention to Charles.' "

On 23 March 1915, Charles was brought back to Quincy for burial. A large crowd packed into the old Adams church for his funeral, after which he was buried on a cemetery knoll up from where John and Fanny rested. On his tombstone was the acknowledgment he had coveted: "he left to his descendants an honorable name worthy of those which had before him shone in the annals of the state." Minnie survived him by twenty years, continuing to enjoy a gracious life at the Birnam Wood estate with four maids in waiting, until she, too, went to the Mount Wollaston cemetery after her death on 23 March 1935.

Eight months after Charles died, the Massachusetts Historical Society held a memorial service in Boston's First Church for this Adams who had been its president for so long. The more restrained occasions which customarily honored departed members were deemed inadequate. Henry Cabot Lodge presented a eulogy, after which the audience sang the not inappropriate hymn "Give ear, ye children, to my law." Henry was absent, but he later thanked Lodge for the memorial address, adding, "As you know, I loved Charles." His brother, Henry said, had been a person of action, one "with strong love of power, while I, for that reason, was almost compelled to become a man of contemplation, a critic and a writer." While the two had moved apart, they were always ready to help each other, Henry said, remembering how Charles in 1912 had been "the one to see me through."

Henry survived Charles by three years; both he and Brooks expressed envy of their brother's easy death. The winters Henry spent on Lafayette Square in the house the famed architect H. H. Richardson had designed for him. In the summers, however, he crept closer to familiar Massachusetts scenes, taking his ménage to Dublin, New Hampshire, in 1915, and Tyringham, Massachusetts, in 1916. Then, as if to be reconciled with the past, in his last summer Henry lived in the house he and Clover had built at Beverly Farms, just northeast of Boston. It was his first stay there since 1885, and it went well, after which Henry returned to Washington in the late autumn, bidding goodbye to Brooks and to others. "None of them looked as though they much expected to see me again," he noted. In February 1918, he celebrated his eightieth birthday, surrounded by flowers and women. He took bets with all his physicians that they could not keep him alive for more than three months. In his last note to Lizzie he said, "we try to be cheerful and whistle our twelfth century melodies." He managed still to take a daily walk as well as a drive, and to see guests either at luncheon or dinner.

The doctors lost their wager. Henry Adams died peacefully soon after dawn on Wednesday morning, 27 March 1918. It was Holy Week, and Aileen Tone, an ardent Catholic, made certain that Henry in death was surrounded by the medieval atmosphere he had craved intellectually and emotionally. The funeral was held in his library, with lilacs on his coffin, and with what was left of the Adams generations and a few friends present. In contrast to the torrential rain which had fallen when Clover was carried to Rock Creek cemetery, the day was soft and lovely. Afterward, as one observer wrote to Lizzie, "the inscrutable face of the great statue looked quietly over his grave."

On Easter Sunday 1918, two days after Henry's interment, Brooks Adams' wife was writing notes and watching her husband uneasily as he mourned the loss of his brother. "It is very hard for Brooks," Daisy told Isabella Gardner. "Henry was his dearest friend as well as brother, and it leaves him very lonely." Indeed, Brooks had been dreading the death of Henry, often begging his older brother not to go first. From the family's days in London during the 1860s, he had depended upon Henry and admired him extravagantly. "You always were the best of us four brothers," he told him in 1915 after Charles' death. "You are so still now that we are reduced to two." Brooks found himself bereft and "an old man," trying to take comfort in what Henry had written, particularly the *Chartres.* Henry had often heard Brooks say, "thank God that you have redeemed our generation." As for the others, Brooks insisted, "The rest of us leave nothing."

Brooks settled back to a few more years of life, spent mostly in the Old House with its reminders of the past. There was still time for travel in Europe and Central America—Brooks and Daisy now had the great advantage of a talented companion and social secretary. Wilhelmina Sellers was a young woman from Alabama who had come to Boston to study music, but gave it up for life with the last of this fourth generation. She had a difficult existence. Increasingly, Brooks was tormented by the sight of Daisy's loss of mental powers, bringing to mind a prediction of Henry's to Lucy Baxter in 1894 "that Brooks rather wears on Evelyn, and will end by killing her with his constant neurosis." By

1920, Brooks' own emotional state grew more volatile. He tried to overcome this by taking Daisy on long trips to familiar scenes, therapy which annoyed her relatives, who did not know that Brooks believed his father had decayed because he lacked stimulating sights and thoughts. For family observers, it was often difficult to determine whether Brooks or Daisy had the most "seriously shattered mind."

In time, Brooks saw that his efforts to revive his wife were futile, so he left her in the hands of doctors and nurses and doggedly kept traveling in search of stimulating sights and thoughts. Away from Daisy, who reminded him of his father at the end, Brooks recovered some stability and calm. It gave him time to ponder a fact which fascinated him: "It does seem hard luck that, in a single family, both my brother Henry and I should have had wives whose minds gave way." This observation he shared in 1921 with Mrs. Jones. His wife's fate, he said, was "unutterably sad." His own situation offered him little consolation. In these last grim years, Brooks made an art of reproaching himself for being a failure in his generation; how much better the earlier Adamses were! He frequently asked, "Do you think I deserve to stand with the rest of my family?" And just as often, he repeated an aphorism which had been dear to his brothers: "Families must run out." Not that this comforted him as he watched his books and essays enter oblivion. "Everything I ever did has missed," he insisted. "As I look back on my life and my boyhood it seems to me as if all were a failure." The only pleasing recollection for him was that "one or two women have been happier because I have lived." Preoccupying himself with the Adams story and its echoes of duty completed, stewardship carried forward, and valiant deeds pursued, Brooks said, pitifully, that he had done little to justify his existence. "I leave nothing behind when I die. I go out like a candle."—"Do you think I deserve to stand with the rest of my family?"

The greater his despair, the more devotion Brooks gave in his last days to family traditions and scenes, and especially to the conservation of the ancient mansion in Quincy. Even though Brooks particularly admired John Adams, he made no effort to preserve life in the house as it was when his great-grandfather was there wishing to be a simple farmer. For instance, he chose to see that the parlor maid wore three different uniforms daily, white for breakfast, grey for luncheon, and black taffeta for dinner. Brooks himself appeared for the evening meal in full dress, including tails. At such times sentiment would overcome him as he recalled that by living in the Old House he was complying with his mother's dying wish. "[A]lmost her last words to me were to make a promise that I would live in this house." If there were anyone at dinner to listen, Brooks would call up ghosts from the family's past. They came sometimes to cheer him, at other times to accuse him.

Aside from family stories, the aged Brooks had few other satisfactions. When his unstable temperament permitted, he enjoyed educating his secretary, Miss Sellers, through talk, reading, and especially travel. Occasionally, when he had calmer moments, he was taken on his favorite drive through the large nature preserve in the Blue Hills, just west of the Old House, which had been a favorite family haunt since John Adams' day. Brooks liked to remember that Charles was responsible for this conservation. He often insisted on stopping to count the numerous cars passing through the reservation, and then asking with a chuckle how many fewer persons were likely to have read the *Education* that day.

Brooks also took time for religious concerns. He shared with Henry a fancy for Roman Catholicism, whose services reminded him of the days when he haunted Europe's venerable cathedrals. In one of his last letters, Brooks wrote, "I am great friends with cardinals and the like, and I yield to no one in my admiration of their churches and their chants, as for instance the gregorian." However, he seemed no closer to the goal he described to Oliver Wendell Holmes: "I wish I could have faith in Christ. It is sad to have no inner light as we near the end and I have none."

After years of lamenting the fate of living too long, Brooks neared his end by fighting for life to its close. He had painful surgery in 1922 for a prostate disorder which turned into the cancer that terminated his life. "I am not afraid of death, I care nothing for that. But the boredom and pain of dying appall me," he wrote Mrs. Jones. His mood was not improved by having to watch Daisy in a disorder which not only destroyed her mind, but left her with a terrible form of muscular rigidity. She was cared for in a sanitarium until her death on 14 December 1926. Brooks was comforted that she had died first, for, he had feared, should it have been he, Daisy's relatives might have tried to use her dower rights to break into Brooks' share of the Adams estate. He was determined that the family money must pass to another generation of Adamses, and in more generous quantity than when he had received it.

By late summer of 1926, Brooks was gravely ill from cancer and incapacitated by a broken hip he had suffered since he insisted upon using an unsteady chair of his father's. The accident occurred at the Old House, which made Brooks all the more determined to die there. His servants were put to work by Miss Sellers to make the mansion ready for the first winter's stay in a century. In November, however, the weather became so cold that the huge fires in hearths which once warmed John and Abigail proved insufficient to please Brooks' servants, who demanded the comfort of Boston.

Taking some of his favorite family portraits to keep him company, Brooks moved to his Chestnut Street home on Beacon Hill where he slipped peacefully out of life on 13 February 1927, surrounded by the ease and medical attention commanded by the wealth he had accumulated through years of prudent investment. In his last moments, he continually called for two persons from his youth. These were not Abby, nor Charles Francis, nor any of his brothers. Instead, Brooks murmured the name of his love from long ago, Heloise, and of his nurse, Rebecca, who had comforted him when his life as the youngest Adams child was so difficult.

Brooks had often talked of wishing to be buried next to Heloise, but to the relief of his nephews and nieces, he left no instructions about his funeral or other final matters. He had refused to discuss such arrangements. His relatives assumed that Uncle Brooks wished to have a funeral in the old Quincy church, beneath which his revered grandparents and great-grandparents rested. Even in death, however, Brooks could not proceed without stumbling. He had often teased his friends who believed that the only proper place for the last rites was Trinity Episcopal Church at Copley Square in Boston, a parish where it was the custom to play John Henry Newman's hymn "Lead, Kindly Light!" at funerals. Brooks liked to ask his Episcopalian acquaintances if they really believed a song which seemed to talk of a light proceeding into a dark hole was a particularly Christian way of sending a person into death.

Since the organist at the Quincy church was deemed unfit to play for the funeral of an Adams, the musician at Trinity Church was fetched; the family had carefully selected the music. With the busts of John and John Quincy Adams looking down upon the funeral, all went well until the organist faced a few unexpected moments when more music was required. What choice was there but to peal forth reflexively the melody:

Lead, kindly Light!
Amid th' encircling gloom,
Lead Thou me on . . .

Brooks' coffin was then taken to the Mount Wollaston cemetery, several
miles from where Heloise was buried, and placed properly beside Daisy
and only a few feet from his parents' tomb. Nearby were the graves of
his brothers John and Charles. Henry lay far away in Washington.

Devising a fitting statement to be carved on Brooks' tombstone was
not easy. Most of the world, as well as his survivors, considered him
eccentric, wild, even mad. However, a niece, Molly Adams Abbott, said
she felt up to the task. Molly was the last Adams to have been born in
the Old House. Her work probably would have pleased Brooks:

Essayist and Historian.
He set forth the Truth as he saw it,
undeterred by precedent,
undismayed by criticism.

Appropriate though this farewell was, it failed to memorialize the
achievement which brought Brooks Adams what little satisfaction his
life contained—his devotion to his family's heritage. After the emotional storms of the 1880s, there had been scant repose for him except
as he sat in the Old House, struggling with his effort to rescue Henry
and John Quincy, and to make the public aware of Adams achievements. Brooks believed much could be learned about the family through
seeing the venerable mansion and its furniture, china, and paintings.
There it was easier to interpret the family's philosophy, which also occupied Brooks. Sitting in Quincy, he would repeat how the Adamses
for four generations had realized that history taught mankind's limita-
tion, not perfection. It pleased Brooks that as far back as the 1880s, he
was writing about the impulses over which humanity had no control,
particularly "fear and greed." Brooks knew he shared this belief with
the family founder, John Adams.

These two, John and Brooks, the first and the last of the Adamses to
live in the Old House, saw the beginning and the end of a bleak family
outlook. Utterly different in personality, the two men were alike in their
deep love for their family as well as in their capacity to read clearly the
distressing lessons of history and of human nature. Progress was a delusive dream. Man, democracy, society—all were doomed. In his last
moments, Brooks lived quietly with this realization, one he could per-

suade few others to share. His spirit, like John's a century before, drew contentment from the consoling atmosphere of the Old House. Restless, angry, thwarted though his life may have seemed, Brooks had one final satisfaction. The last family duties were done, and he had seen to their doing. It was the same source of comfort to which his father, his grandfather, and his great-grandfather had eventually turned. This time, however, it was the end of a great family's story, for the next generation had no one like Brooks, Henry, Charles, and their famed forebears.

Postscript

What occurred after Brooks' death probably would have pleased him immensely. While John, Charles, and Mary left descendants who displayed few of the qualities which had made the Adamses one of America's greatest families, these later generations shared with Brooks a concern which ultimately brought the realization of his hope that the Old House would survive. Members of the fifth generation created the Adams Memorial Society in 1927 to ensure the care of the Old House as they sought to make the place a museum and shrine. It proved to be an expensive tribute; the annual cost by 1944 was $4000 and the proceeds from admissions only $600. In 1946 the Adamses presented the Old House to the people of America, a gift made on the careful premise that the house and its contents were to remain as the first four Adams generations kept them: "the whole idea of giving this to the Government is to preserve the house as it is. If it does not then this gift is a failure."

The family warned that the Old House must be considered "not a representation of one man or any one generation. It is a representation of the lives of four generations and what each of them has done should remain unchanged." Consequently, the National Park Service was informed that the Old House must be continued "just as Brooks Adams the last of the family to live in it had left it." There was even talk about the Old House being a place for "fostering civic virtue and patriotism."

Early in 1947, the American government, through the Department of the Interior and the National Park Service, took responsibility for the Old House and its environs, restoring and preserving them with meticulous care. The mansion is annually opened to the public from April to November, once the family's traditional season of residence. This preservation succeeded largely because Brooks Adams' secretary, Wilhelmina Sellers, was persuaded by the National Park Service in 1948 to direct the management of the historic site. She had lived in the Old House throughout Brooks' last years, after which she had married Colonel Frank Harris, raised three sons, and been widowed. In 1950 Mrs. Harris became Superintendent, a post she still holds with marvelous benefit to the traditions of the Adams family and to the American nation.

Brooks Adams had felt that his discovery of Wilhelmina Sellers was a happy event, and he carried her painstakingly through the customs and traditions of the Old House as he himself relived the past. The experience left the young woman forever aware of what the Adams family cherished, enabling the National Park Service with her guidance to maintain the finest Historic Site in America. The Old House is graced today with not only those things which were in it over four generations but also with an understanding of what they meant. In addition, the National Park Service received from the city of Quincy in 1978 the two farmhouses at Penn's Hill where John and John Quincy had been born and where so much of the early life of the family had been centered. In 1980, the federal government agreed to accept the old Adams church with its presidential crypt as part of the Historic Site. These scenes and buildings enable visitors to see how this nation's most famous family lived from colonial days until 1926.

However, still unanswered was the question whether scholars could be allowed to use the family manuscripts, a treasure-trove of incalculable historical value. Charles' son, the second Henry Adams, had guarded these at the Massachusetts Historical Society, mindful of his Uncle Brooks' fear lest the documents be seen by the irreverent or unsympathetic. This faithful guardian died in 1951. The next year, two sixth-generation trustees for the manuscripts, Thomas Boylston Adams and John Quincy Adams, invited a group of academic and public figures to meet in the library at the Old House. There, these family spokesmen stated that the entire body of Adams family papers would be opened without restriction to historians. The fifty-year seal placed at the start of the twentieth century by the fourth generation had expired.

Under a deed of gift dated 4 July 1956, the Adams Manuscript Trust and the Massachusetts Historical Society began using microfilm and letterpress publication to make the family papers available to the world.

Thanks to those editors of high distinction, Lyman Butterfield, Marc Friedlaender, and Robert J. Taylor, and to those friends of scholarship at the Massachusetts Historical Society, Stephen T. Riley and Louis L. Tucker, the story of the Adams family can now be appreciated as Charles Francis Adams saw it—"one of great triumphs in the world but of deep groans within, one of extraordinary brilliancy and deep corroding mortification."

Sources

This essay has two objectives—to describe the manuscript collections which support the material in this book and to list some of the printed editions of documents and the biographies which will help readers learn more about the Adamses as a family. I shall not mention here the literature which presents the Adamses as statesmen, diplomats, historians, businessmen, and authors.

The first fourteen chapters of this book are based upon manuscripts in the 608 microfilm reels of Adams Papers which include documents from 1639 to 1889. My use of quotations from the filmed Adams Papers is by permission of the Massachusetts Historical Society, to which I am grateful. These reels of microfilm are divided into four parts. Part I contains the diaries of John Adams, John Quincy Adams, and Charles Francis Adams; I used these 88 reels with great profit. Part II, 90 reels, has letter books of the three Adams statesmen and includes both public and private material. I read only those portions pertaining to personal affairs. Part III, designated simply as "Miscellany," has 162 reels of autobiographical material, financial records, legal papers, literary efforts, and political writings. Here I caught many glimpses of all the family members through such sources as John Quincy's account books and poems; Louisa Catherine's diary, autobiography, and poems; the scant diary of Thomas Boylston Adams; the fragments from the career of George Washington Adams; a variety of account books, literary notes, and transcripts prepared by Charles Francis; and even from such documents as funeral records, scrapbooks, and the manuscript legacies from obscure family figures. However, it is Part IV which helped me most, with its 265 reels of letters and other items received by family members until 1889. Here opinions are exchanged among Adamses themselves so that these reels are the best example of the family's private relationships. An excellent overview of these manuscripts is found in Lyman H. Butterfield's essay, "The Papers of the Adams Family: Some Account of Their History," in *Proceedings of the Massachusetts Historical Society*, LXXI (1953–57), 328–56.

A splendid printed edition of the Adams Papers is being published by the Belknap Press of Harvard University. The initial editor-in-chief was L. H. Butterfield. He was succeeded by Robert J. Taylor, while the editorial enterprise continues to be sponsored by the Massachusetts Historical Society. This project will require many years to complete; it involves three series of publications, the first of which will include the diaries of John Adams, John Quincy Adams, and Charles Francis Adams. The second series will contain the Adams family correspondence, although not every document can be printed due to the sheer quantity of the material. General correspondence and other papers of the Adams statesmen will make up the third series. In addition, a special delight are the two volumes of portraits published under the auspices of Adams Papers: Andrew Oliver's *Portraits of John and Abigail Adams* (Cambridge, Mass., 1967), and *Portraits of John Quincy Adams and His Wife* (Cambridge, Mass., 1970).

No editorial project has more general significance for American history than the publication of the Adams Papers. The brilliant editorial introductions and annotations in these volumes alone are a valuable source for understanding the Adams family and their times. A detailed discussion of the philosophy and plan for this venture in editing and publication can be found in L. H. Butterfield and others (eds.), *Diary and Autobiography of John Adams* (Cambridge, Mass., 1961), I, pp. xii–lxxiv.

There are several other collections of source materials, both manuscript and printed, which are important in reconstructing the family's story before 1889. Stewart Mitchell (ed.), *New Letters of Abigail Adams, 1788–1801* (Boston, 1947), contains numerous revealing messages which Abigail sent to her sister, Mary Cranch. Since most of Nabby's papers were lost in a fire in the mid-nineteenth century, some items no longer found elsewhere are in Caroline A. de Windt, *Journal and Correspondence of Miss Adams, Daughter of John Adams* (New York and London, 1841–42), in two volumes. There are helpful Cranch family papers at the Massachusetts Historical Society and the Library of Congress. From both of these collections I learned about the youthful years of the second generation. See also "The Journal of Elizabeth Cranch," *Essex Institute Historical Collections,* LXXX (January 1944), 1–36. A few of the fragments that remain concerning Thomas are in Charles Grenfill Washburn (ed.), "Letters of Thomas Boylston Adams to William Smith Shaw, 1799–1823," *Proceedings of the American Antiquarian Society,* New Series, XXVII (April 1917), 83–126, and in Victor Hugo Paltsits, "Berlin and the Prussian Court in 1798: Journal of Thomas Boylston Adams," *Bulletin of the New York Public Library,* XIX (November 1915), 803–43.

The Massachusetts Historical Society also possesses the Peter Chardon Brooks manuscripts concerning Abby Brooks, wife of Charles Francis. Here, too, are the Lunt Family Papers showing the close relationship of the Adamses during John Quincy and Charles Francis' time to the family of William P. Lunt, pastor of the Unitarian church in Quincy. Especially valuable for my purposes were the letters exchanged between Charles Francis and his son John while the former was in England, 1861–68, with a few dating from the time the father returned to Europe for the Geneva Arbitration in 1871–72. This extensive collection was acquired too late for inclusion in the film edition of the Adams Papers. It is retained in the Adams Papers editorial office and designated as the Homans Acquisition. That office also has two boxes of material drawn from the Adams family business files in Boston, which cover all four generations. This group of documents, known as the Adams Office Papers, helped me with questions on estates and finances. One more document, unfilmed and in the care of the Adams Papers, is a diary kept by Abby Adams from 20 June 1868 to 4 April 1871, the period of Sister Lou's death. For that episode I was also aided by the Alward-Babcock Accession held by the Massachusetts Historical Society.

The Adams Papers incorporate manuscripts only from the lifetime of the first three generations, so that 1889 marks their limit. Therefore, to follow the fourth generation beyond that date, I used the unfilmed sources in the Massachusetts Historical Society, a collection generally designated "Adams Papers—Fourth Generation." Here emerge the personal sides of Henry, Brooks, and Charles. Far less material remains from John, although there are some glimpses of John and his wife, Fanny Crowninshield, in the Crowninshield-Magnus Papers, also held by the Massachusetts Historical Society. Charles Adams left a vast assortment of manuscripts, mostly in twenty-four boxes called "Loose Papers," these covering the years 1890–1918, and in four more boxes of miscellaneous documents. However, these are by no means all of Charles' presence in the Adams Papers—Fourth Generation, for there is also his pocket diary in twenty-nine tiny volumes, extending from 1861 to 1915. His "Memorabilia," a journal begun in 1888 and ended in 1912 after running to over 3000 pages, was of enormous usefulness to me. It too is held by the Massachusetts Historical Society and includes his "Autobiography" which, with few changes, was published soon after his death: *Charles Francis Adams, 1835–1915: An Autobiography* (Boston and New York, 1916).

Persons interested in Henry Adams can now find his papers on microfilm, available through the Massachusetts Historical Society, prepared with the aid of the National Endowment for the Humanities. This set of thirty-six reels includes all of the correspondence originating with or directed to Henry which is in the Massachusetts Historical Society, the Houghton Library at Harvard University, and the John Carter Brown Library at Brown University. I thank the Massachusetts Historical Society for permission to quote from the microfilm edition of the Henry Adams Papers, as I do for its consent to quote from the Adams Papers—Fourth Generation. The Harvard University Press plans to publish many of Henry's letters. These are being prepared by J. C. Levenson, Ernest Samuels, Charles Vandersee, and Viola Winner in the Henry Adams editorial project at the University of Virginia. The first three volumes of this series are in press, taking Henry's story to February 1892. Six volumes in all are expected.

At the South Caroliniana Library on the campus of the University of South Carolina are letters of Henry and John Adams to Lucy Baxter. These are in the Sarah Strong Baxter Hampton Papers, and I am grateful to Mr. E. L. Inabinett, Librarian of the South Caroliniana Library, for permission to use and quote from these. Other manuscripts I relied upon for material on Henry Adams were the Henry Cabot Lodge Papers at the Massachusetts Historical Society, the George Cabot Lodge Papers at the Massachusetts Historical Society, the Isabella Stewart Gardner Papers at the Gardner Museum in Boston, and the John Hay letters to Elizabeth Cameron in the Massachusetts Historical Society. The Lodge papers, both of father and son, while particularly revealing about Henry, were also helpful for Brooks and Charles. I thank the Massachusetts Historical Society for permission to quote from these manuscripts.

For Clover Adams, the Marian Hooper Adams letters to Dr. Robert William Hooper are the major surviving clues along with a few letters which Clover wrote to Anne Palmer. All these are in the Massachusetts Historical Society, to whom I am grateful for permission to quote from this material. Most of the Marian Hooper Adams letters to her father were published in Ward Thoron (ed.), *The Letters of Mrs. Henry Adams* (Boston, 1936). A comparison of this edition with the original documents, however, shows that a number of passages were silently omitted in the printed version.

Many of the useful Brooks Adams papers are now readily available in the microfilm edition of the Henry Adams letters. However, there are other important manuscripts left by Brooks which the Massachusetts Historical Society possesses and whom I thank for permission to use and quote from them. Indispensable are the many letters Brooks wrote to Mary Cadwallader Jones between 1891 and 1924, some so candid as to match in significance the messages Brooks sent to Henry. Nor can Brooks be comprehended without reading his manuscript biography of John Quincy Adams which the Massachusetts Historical Society owns and which it graciously allowed me to study. Mary Gardner Adams Quincy can be traced in the Adams Papers—Fourth Generation. Additional manuscripts are in the possession of Dorothy Quincy Beckwith Nelson of Weston, Massachusetts, Mary's great-granddaughter. Her permission to use these papers I much appreciate. I owe a special debt to L. H. Butterfield, who allowed me to study notes and an assortment of photographs which he had intended to use in writing a book about the Old House. That the volume was never completed because of Mr. Butterfield's retirement and death is our great loss. Finally, I thank the Harvard University Archives for allowing me to examine Corporation and College records and papers, as well as Reports to Overseers, for glimpses of the Adams generations as they studied at and later served Harvard University.

Among other printed selections of Adams letters the most significant are the many volumes edited by Charles Francis Adams. These include: *Letters of John Adams, Addressed to His Wife* (Boston, 1841), 2 volumes; *The Works of John Adams* (Boston, 1850–56), 10 volumes; *Familiar Letters of John Adams and His Wife Abigail Adams, during the Revolution* (New York, 1876); *Letters of Mrs. Adams, the Wife of John Adams* (Boston, 1848) 4th ed.; and *Memoirs of John Quincy Adams* (Philadelphia, 1874–77), 12 volumes. The best of the printed sources for the general reader interested in the family's early days is L. H. Butterfield, Marc Friedlaender, and Mary-Jo Kline (eds.), *The Book of Abigail and John* (Cambridge, Mass., 1975). Another view of family life is in a selection of letters exchanged

between Charles Francis Adams and his sons Henry and Charles: Worthington C. Ford (ed.), *A Cycle of Adams Letters 1861–1865* (Boston and
New York, 1920), 2 volumes. Portions of Henry's correspondence are
in Worthington C. Ford (ed.), *Letters of Henry Adams 1858–1891* (Boston, 1930); Worthington C. Ford (ed.), *Letters of Henry Adams 1892–
1918* (Boston, 1938); and Harold Dean Cater (ed.), *Henry Adams and
His Friends* (Boston, 1947).

From the biography and special studies about the Adamses, the following titles can be helpful. For genealogical purposes, Andrew N.
Adams, *Genealogical History of Henry Adams of Braintree, Massachusetts*
(Rutland, Vt., 1898). James Truslow Adams, *The Adams Family* (New
York, 1930), is simply a group of biographical sketches, now badly out
of date. Jack Shepherd, *The Adams Chronicles* (Boston, 1975), is valuable
for its illustrations. The best way to learn about John Adams is to read
Page Smith, *John Adams* (Garden City, N.Y., 1962), 2 volumes, and Peter Shaw, *The Character of John Adams* (Chapel Hill, 1976). There are
two recent biographies of Mrs. John Adams: Charles W. Akers, *Abigail
Adams* (Boston, 1980), and Lynne Withey, *Dearest Friend* (New York,
1981). Remarkably satisfying and accurate is Irving Stone's biographical novel of John and Abigail, *Those Who Love* (Garden City, N.Y., 1965).
The one study of Nabby, Katherine Metcalf Roof, *Colonel William Smith
and Lady* (Boston, 1929), is marred by adoring sentimentality. An interesting glimpse of Nabby's story is in George T. Tanselle, *Royall Tyler*
(Cambridge, Mass., 1967).

There are no biographies of Charles or Thomas Boylston Adams,
but for their famous brother there are Samuel Flagg Bemis' two splendid studies: *John Quincy Adams and the Foundations of American Foreign
Policy* (New York, 1949), and *John Quincy Adams and the Union* (New
York, 1956). A single-volume biography which some readers may find
helpful is Marie B. Hecht, *John Quincy Adams: A Personal History of an
Independent Man* (New York, 1972). For Louisa Catherine Adams there
is L. H. Butterfield, "Tending a Dragon-Killer: Notes for the Biographer of Mrs. John Quincy Adams," *Proceedings of the American Philosophical Society,* 118 (April 1974), 165–78. A study of the Adams family from
an interesting perspective is David F. Musto, "The Youth of John Quincy
Adams," *Proceedings of the American Philosophical Society,* 113 (August
1969), 269–82. Much less revealing is Robert A. East, *John Quincy Adams:
The Crucial Years 1785–1794* (New York, 1962).

For Charles Francis Adams there is a biography devoted to his public
career: Martin Duberman, *Charles Francis Adams 1807–1886* (Boston,

1961). Charles is sympathetically treated in Edward C. Kirkland, *Charles Francis Adams, Jr., 1835–1915: The Patrician at Bay* (Cambridge, Mass., 1965). Understanding Henry Adams as a family member must begin with the splendid three-volume biography by Ernest Samuels: *The Young Henry Adams* (Cambridge, Mass., 1948); *Henry Adams: The Middle Years* (Cambridge, Mass., 1958); and *Henry Adams: The Major Phase* (Cambridge, Mass., 1964). Also valuable is the fine treatment in J. C. Levenson, *The Mind and Art of Henry Adams* (Boston, 1957). More recent studies include William Dusinberre, *Henry Adams: The Myth of Failure* (Charlottesville, 1980); Earl N. Harbert, *The Force So Much Closer Home: Henry Adams and the Adams Family* (New York, 1977); Otto Friedrich, *Clover* (New York, 1979); and Eugenia Kaledin, *The Education of Mrs. Henry Adams* (Philadelphia, 1981).

The best treatments of Brooks Adams are not in book form. Readers should consult Wilhelmina S. Harris, "The Brooks Adams I Knew," *Yale Review*, LIX (October, 1969), 50–70, and Marc Friedlaender, "Brooks Adams *en Famille*," *Proceedings of the Massachusetts Historical Society*, LXXX (1968), 77–93. Two older biographies suffer because they were written from limited material: Thornton Anderson, *Brooks Adams: Constructive Conservative* (Ithaca, 1951), and Arthur F. Beringause, *Brooks Adams* (New York, 1955). Similarly handicapped is Timothy P. Donovan, *Henry Adams and Brooks Adams: The Education of Two American Historians* (Norman, Okla., 1961).

Descendants of fourth-generation Adamses have recorded memories of their forebears. Of these, the most significant is Abigail Adams Homans, *Education by Uncles* (Boston, 1966). The *Proceedings of the Massachusetts Historical Society* contain three essays of recollection: Elliott Perkins; John A. Abbott, M.D.; and Thomas B. Adams, "Three Views of Charles Francis Adams II," LXXII (1957–60), 212–37; John Adams, "Memories of an Old Man," LXXII (1957–60), 294–99; and John Adams, "Notes to a Grandson," LXXIX (1967), 89–96.

Wilhelmina S. Harris is the author of a booklet describing the traditions and contents of the Old House. It is being published by the National Park Service and will be available through the Adams National Historic Site in Quincy, Massachusetts.

Index

Since subentries in this index must refer often to many prominent Adamses, using the following forms for their names will be helpful. Such a plan was first devised by the Adamses themselves and used in their correspondence. It was refined by the editors of the Adams Papers for the benefit of their readers. Only slight changes in the Adams Papers format have been needed here.

First Generation

JA	John Adams (1735–1826)
AA	Abigail Smith (1744–1818), *m.* JA 1764

Second Generation

AA2	Abigail Adams (1765–1813), daughter of JA and AA, *m.* WSS 1786
WSS	William Stephens Smith (1755–1816), brother of Mrs. CA
JQA	John Quincy Adams (1767–1848), son of JA and AA
LCA	Louisa Catherine Johnson (1775–1852), *m.* JQA 1797
CA	Charles Adams (1770–1800), son of JA and AA
Mrs. CA	Sarah Smith (1769–1828), sister of WSS, *m.* CA 1795
TBA	Thomas Boylston Adams (1772–1832), son of JA and AA
Mrs. TBA	Ann Harrod (1774–1846), *m.* TBA 1805

Third Generation

GWA	George Washington Adams (1801–29), son of JQA and LCA
JA2	John Adams (1803–34), son of JQA and LCA
Mrs. JA2	Mary Catherine Hellen (1807–70), *m.* JA2 1828
CFA	Charles Francis Adams (1807–86), son of JQA and LCA
ABA	Abigail Brown Brooks (1808–89), *m.* CFA 1829

Fourth Generation

LCA2   Louisa Catherine Adams (1831–70), daughter of CFA and
                ABA, *m.* Charles Kuhn 1854
JQA2   John Quincy Adams (1833–94), son of CFA and ABA
Mrs. JQA2   Fanny Cadwalader Crowninshield (1840–1911), *m.* JQA2
                1861
CFA2   Charles Francis Adams (1835–1915), son of CFA and
                ABA
Mrs. CFA2   Mary Hone Ogden (1843–1935), *m.* CFA2 1865
HA   Henry Adams (1838–1918), son of CFA and ABA
MHA   Marian Hooper (1842–85), *m.* HA 1872
MA   Mary Adams (1845–1928), daughter of CFA and ABA,
                m. Henry Parker Quincy 1877
BA   Brooks Adams (1848–1927), son of CFA and ABA
Mrs. BA   Evelyn Davis (1853–1926), *m.* BA 1889